H.S. Kasana · K.D. Kumar

Introductory Operations Research

T0238829

Harvir Singh Kasana · Krishna Dev Kumar

Introductory
Operations Research

Theory and Applications

With 58 Figures

 Springer

Prof. Dr. Harvir Singh Kasana
Thapar Institute Engineering and Technology
Department of Mathematics
147 001 Patiala
India

Dr. Krishna Dev Kumar
Korea Advanced Institute of Science and Technology
Department of Aerospace Engineering
Kusong-dong 373-1, Yusong-gu
305-701 Taejon
Republic of Korea
kumar@fdcl.kaist.ac.kr

ISBN 978-3-642-07270-3

Springer is a part of Springer Science+Business Media

springeronline.com

© Springer-Verlag Berlin Heidelberg 2010
Printed in Germany

Cover-Design: deblik, Berlin
Printed on acid-free paper   62/3020 Rw 5 4 3 2 1 0

Preface

This book is an outcome of the lectures delivered by the authors for engineering and management students at the Birla Institute of Technology and Science, Pilani, India. However, the text started when the author shifted to the Thapar Institute of Engineering and Technology, Patiala and the coauthor shifted to the Indian Institute of Technology, Kanpur. During the teaching of this course, the authors realized a need of a good text on "Optimization Research" and its applications which may give comprehensive idea of various concepts and can be used as a companion for problem solving techniques.

The primary purpose of this text is to bring this new mathematical formalism into the education system, not merely for its own sake, but as a basic framework for characterizing the full scope of the concept of modern approach. The authors have tried all contents of this book utilizing four hours a week in one semester as a core course. The level of this book assumes that the reader is well acquainted with elementary calculus and linear algebra. Being a textbook, we have taken enough care so that the reader may attempt different type of problems.

Any one who aspires to some managerial assignment or who is a part of decision making body will find an understanding of optimization techniques very useful. The book is applied in orientation with a concentration on engineering and management problems. Each concept has been discussed with sufficient mathematical background supported by examples. A set of problems has been added in the end of every chapter.

Because of the imposed restriction of writing a relatively brief textbook on an extensive subject area, a number of choices had to be made relating to the inclusion and exclusion of certain topics. No obvious way of resolving this problem exists. The basic thinking is centralized about the theme how the reader may continue to read advanced level textbooks by self-study to develop research oriented thoughts. Hence, the fundamental techniques have been emphasized while highly specialized topics should be relegated to secondary one.

During the course of writing this book we have received remarkable encouragement from our esteemed colleagues Prof. S. R. Yadava, and Dr. S. P. Yadava at BITS, Pilani, India.

H. S. Kasana
K. D. Kumar

Contents

Chapter 1

Formulation

We start with the introduction of linear programming and illustrate
the preliminary concepts which form the basic foundation of optimization. The concentration will remain focus on the formulation of linear programming problems. In the end, some nonlinear programming
problems have also been formulated.

1.1 The Scope of Optimization

Optimization means the mathematical process through which best possible results are obtained under the given set of conditions. Initially,
the optimization methods were restricted to the use of calculus based
techniques. Cauchy made the first attempt by applying steepest descent method for minimizing a function over the domain of definition.
A contribution but very little was made by Newton, Leibniz and Lagrange in this direction. Also, as early as 1939, L. V. Kantorovich
pointed out the practical significance of a restricted class of linear programming models for production planning and proposed an algorithm
for their solution. Unfortunately, Kantorovich's work remain neglected
in the USSR, and unknown elsewhere until after programming had
been well established by G. B. Dantzig and others.

During the second world war the subject, *'Optimization Techniques'* in the name of *'Operations Research'* gained a momentum. The
development of famous simplex method for solving the linear programming problems was first conceived by Dantzig in 1947 while he was
working as mathematical adviser to the United States. This method

gave a real boost to the subject. In 1975, the topic came to public attention when the Royal Swedish Academy of Sciences awarded the noble prize in economic science to L. V. Kantorovich and T. C. Koopmans. Based on simplex algorithms various linear systems have been studied in detail. In 1979, Khachian proved the ellipsoid method of Shor which ultimately exhibited polynomial-time performance. The time performance of simplex method is exponential. The theoretically superior, ellipsoid method could not be popular in practical use, even though its time performance is better than the simplex method. In 1984, a real breakthrough came from N. Karmarkar's "projective scaling algorithm" for linear programming. The new algorithm not only outperforms the simplex method in theory but also shows its enormous potential for solving large scale practical problems. Karmarkar's algorithm is radically different from simplex method-it approaches an optimal solution from the interior of the feasible region. However, this research was limited to linear optimization problems.

The pioneer work by Kuhn and Tucker in 1951 on the necessary and sufficient conditions for the optimal solution laid the foundation for researchers to work on nonlinear systems. In 1957, the emergence of dynamic programming by Bellman brought a revolution in the subject and consequently, linear and nonlinear systems have been studied simultaneously. Although no universal techniques have been established for the nonlinear systems, the researches by Fiacco and McCornik proved to be significant. Geometric programming was developed by Duffin, Zener and Petersion in 1960. Later on, Dantzig and Charnes developed Stochastic programming.

The process of optimizing more than one objective led to the development of multi-objective programming. During the meantime problems on network analysis essentially useful for management control came into existence. Well known games theory has been successfully applied for different programming problems. Multi-objective problems with specified goals in the name of Goal programming has been the topic of recent interest. At the moment fuzzy logic is being extensively used for studying various linear and nonlinear systems.

The subject has been fully exploited to solve various engineering, scientific, economics and management problems. We mention a few as

1. Design of aircrafts and aerospace structures for tolerating environment resistance.

2. Setting up of pipelines and reservoirs for flow of different items

at required points.

3. Decision making for maximizing industrial outputs.

4. Selection of machining conditions in different industrial processes to minimize the production cost.

5. Optimal production planning, controlling and scheduling of projects.

6. Optimal designing of chemical processing equipment and plants.

7. Shortest route problems under varying conditions.

8. Planning the best strategies to obtain maximum profit.

9. Design of pumps, electric machines, computers etc., for minimizing the cost.

10. Transportation of materials from places of manufacture to places of requirement so that the cost of transportation is minimized.

11. How the jobs should be assigned to workers so as to have optimal efficiency of the system.

12. Allocation of resources and services among several activities to maximize the profit.

13. Inventory problem deals with the demands at specific time, here we have to decide how much and what to order.

14. Queuing problems deal with customers at service stations. The direct increase in service stations increases the service cost but waiting time in queue is reduced. However, waiting time also involves cost. In such type of problems we seek optimal number of services so that cost of service and waiting time is minimized.

Least to say, in every walk of life, optimization techniques are being extensively applied in day to day practice. Operations Research Society, USA defined OR as

> *"Operations research is the systematic applications of quantitative methods, techniques, tools to the analysis of problems involved in the operation of systems"*

1.2 Introduction

The mathematical formulation of our thoughts to optimize profit, loss, production etc., under given set of conditions is called mathematical programming.

The mathematical programming problem (MPP) is written as

$$\text{opt} \quad = f(X)$$
$$\text{subject to} \quad g_i(X) \geq, \; =, \; \leq \; 0, \quad i = 1, 2, \ldots, m \qquad (1.1)$$
$$X \geq 0, \qquad\qquad\qquad\qquad (1.2)$$

where $X = (x_1, x_2, \ldots, x_n)^T$ is the column vector in n-dimensional real linear space $\mathbb{R}^n$.

Thus, $X^T = (x_1, x_2, \ldots, x_n)$ is the row vector. In the text, column vectors and row vectors will be represented by a column matrix and row matrix, respectively.

Now, we define

(i) The function $f(X)$ to be optimized is termed as objective function;

(ii) The relations in (1.1) are constraints;

(iii) The conditions in (1.2) are nonnegative restrictions;

(iv) Variables $x_1, x_2, \ldots, x_n$ are decision variables.

(v) The terminology opt (optimize) stands for minimize or maximize.

The symbol $\geq, =, \leq$ means that one and only one of these is involved in each constraint.

The mathematical programming problem (MPP) is further classified into two classes, viz.,

1. Linear programming problem. If the objective function $f(X)$ and all the constraints $g_i(X)$ are linear in a mathematical programming problem, we call the problem a linear programming problem (LPP).

2. Nonlinear programming problem. If the objective function $f(X)$ or at least one of the constraints $g_i(X)$ or both are nonlinear functions in a mathematical programming problem, then the problem is termed as a nonlinear programming problem (NLPP).

Remarks. An integer programming problem is a particular case of the LPP or NLPP in which some or all the decision variables $x_1, x_2, \ldots, x_n$ are integers. A quadratic programming problem is also a particular NLPP in which the objective function $f(X)$ is a quadratic but all the constraints $g_i(X)$ are linear functions.

Let us discuss the linear programming problems in detail. Any LPP has the general form:

$$\text{opt} \quad z = c_1 x_1 + c_2 x_2 + \cdots + c_n x_n$$
$$\text{s.t.} \quad a_{i1} x_1 + a_{i2} x_2 + \cdots + a_{in} x_n \geq, =, \leq b_i, \quad i = 1, 2, \ldots, m$$
$$x_1, x_2, \ldots, x_n \geq 0,$$

where c_k, $k = 1, 2, \ldots, n$ and b_i, $i = 1, 2, \ldots, m$ are real numbers (may be negative).

From now onward "s.t." stands for "subject to" in the whole text.

Standard form of linear programme. The standard form of a LPP is written as

$$\text{opt} \quad z = c_1 x_1 + c_2 x_2 + \cdots + c_n x_n$$
$$\text{s.t.} \quad a_{i1} x_1 + a_{i2} x_2 + \cdots + a_{in} x_n = b_i, \quad i = 1, 2, \ldots, m$$
$$x_1, x_2, \ldots, x_n \geq 0, \; b_1, b_2, \ldots, b_m \geq 0$$

or

$$\text{opt} \quad z = c_1 x_1 + c_2 x_2 + \cdots + c_n x_n$$
$$\text{s.t.} \quad a_{11} x_1 + a_{12} x_2 + \cdots + a_{1n} x_n = b_1$$
$$a_{21} x_1 + a_{22} x_2 + \cdots + a_{2n} x_n = b_2$$
$$\vdots$$
$$a_{m1} x_1 + a_{m2} x_2 + \cdots + a_{mn} x_n = b_m$$
$$x_1, x_2, \ldots, x_n \geq 0, \; b_1, b_2, \ldots, b_m \geq 0$$

or, in the matrix form:

$$\text{opt} \quad z = C^T X$$
$$\text{s.t.} \quad AX = b$$
$$X \geq 0, \; b \geq 0,$$

where

$$C = (c_1, c_2, \ldots, c_n)^T \text{ (cost vector)},$$
$$X = (x_1, x_2, \ldots, x_n)^T,$$
$$A = (a_{ij})_{m \times n}, \text{ the coefficient matrix of order } m \text{ by } n \text{ and}$$
$$b = (b_1, b_2, \ldots, b_m)^T.$$

Converting to standard form. The standard form of a linear programme deals with nonnegative decision variables and linear equality constraints. Here we explain the means how to convert the linear programme into the standard form in case any or both of these conditions are not available in the LPP.

Linear inequalities. A linear inequality can easily be converted into an equation by introducing slack and surplus variables. If the ith constraint has the form

$$a_{i1}x_1 + a_{i2}x_2 + \cdots + a_{in}x_n \leq b_i,$$

we can add a nonnegative variable $s_i \geq 0$ to have

$$a_{i1}x_1 + a_{i2}x_2 + \cdots + a_{in}x_n + s_i = b_i.$$

Here, the variable s_i is called the *slack variable*.

Similarly, if ith constraint has the form

$$a_{i1}x_1 + a_{i2}x_2 + \cdots + a_{in}x_n \geq b_i,$$

a nonnegative variable $s_i \geq 0$ is subtracted to have

$$a_{i1}x_1 + a_{i2}x_2 + \cdots + a_{in}x_n - s_i = b_i.$$

This time s_i is termed as the *surplus variable*.

Note that $b_i \geq 0$ in the above inequalities. If $b_i \leq 0$, then multiply by -1 before introducing the slack or surplus variables.

Restricted and unrestricted variables. If a variable x is restricted, i.e., for $x \geq p$, this implies $x - p \geq 0$. Taking, $x' = x - p$ implies $x' \geq 0$. So, we replace x by $x' + p$, and in a similar way, for the case $x \leq p$, replace x by $-x' + p$ to have $x' \geq 0$.

However, if a variable x is unrestricted in sign, i.e., $x \in \mathbb{R}$ (may be positive or negative), we write $x = x^+ - x^-$, where x^+ and x^- are defined by

$$x^+ = \begin{cases} x, & x \geq 0, \\ 0, & x \leq 0, \end{cases} \quad \text{and} \quad x^- = \begin{cases} 0, & x \geq 0, \\ -x, & x \leq 0. \end{cases}$$

Obviously, for each real number x we can find nonnegative real number u and v such that $|x| = u + v$ and $x = u - v$. Here, u and v play the role of x^+ and x^-, respectively.

Example 1. Write the following linear programme into the standard form:

$$\begin{aligned} \text{opt} \quad & z = x_1 + 2x_2 - x_3 \\ \text{s.t.} \quad & -x_1 + 2x_2 + 3x_3 \geq -4 \\ & 2x_1 + 3x_2 - 4x_3 \geq 5 \\ & x_1 + x_2 + x_3 = 2 \\ & x_1 \geq 0, \ x_2 \geq 1 \text{ and } x_3 \text{ is unrestricted in sign.} \end{aligned}$$

Here x_2 and x_3 are restricted and unrestricted variables, respectively. Replacing x_2 by $x_2' + 1$ and x_3 by $x_3^+ - x_3^-$, the above LPP is written in the standard form as

$$\begin{aligned} \text{opt} \quad & z = x_1 + 2x_2' - x_3^+ + x_3^- + 2 \\ \text{s.t.} \quad & x_1 - 2x_2' - 3x_3^+ + 3x_3^- + s_1 = 6 \\ & 2x_1 + 3x_2' - 4x_3^+ + 4x_3^- - s_2 = 2 \\ & x_1 + x_2' + x_3^+ - x_3^- = 1 \\ & x_1, x_2', x_3^+, x_3^-, s_1, s_2 \geq 0. \end{aligned}$$

Note that x_1, x_2', x_3^+ and x_3^- are now the decision variables, s_1 slack variable and s_2 surplus variable when the LPP has been written in the standard form.

From now onward it will be understood that the slack or surplus variable s_i means it is associated with the ith constraint.

The above discussion reveals that, in general k unrestricted variable produce $2k$ nonnegative variables to write the problem in the standard form. This will substantially increase the size of the problem. However, under certain conditions we develop a better technique in which k unrestricted variable can be replaced by $k + 1$ nonnegative variables to express the LPP into the standard form.

Proposition. In a LPP, let k variables out of n variables be unrestricted in sign and are bounded. Then the problem can be converted into the standard form by using $k + 1$ nonnegative variables in place of these k unrestricted variables.

Proof. Let $x_1, x_2, \ldots, x_n$ be n variables of a LPP. Given that k of these variables are unrestricted in sign. Without loss of generality we may assume $x_1, x_2, \ldots, x_k$ are unrestricted in sign.

Define $y = |\min\{x_1, x_2, \ldots, x_k\}|$. Then we observe that

$$y_1 = x_1 + y \geq 0$$
$$y_2 = x_2 + y \geq 0$$
$$= \vdots$$
$$y_k = x_k + y \geq 0$$
$$y \geq 0$$

This implies, $x_1 = y_1 - y$, $x_2 = y_2 - y, \cdots, x_k = y_k - y$, and the constraints

$$a_{i1}x_1 + a_{i2}x_2 + \cdots + a_{in}x_n \leq, =, \geq b_i$$

are converted into

$$a_{i1}y_1 + \cdots + a_{ik}y_k - (a_{i1} + \cdots + a_{ik})y$$
$$+ a_{i,k+1}x_{k+1} + \cdots + a_{in}x_n \pm s_i = b_i$$
$$y_1, y_2, \ldots, y_k, y, x_{k+1}, \ldots, x_n \geq 0.$$

Here, $i = 1, 2, \ldots, m$ and s_i is a slack or surplus variable and in case of equality constraint $s_i = 0$.

Example 2. Illustrate the above proposition by taking a particular LPP:

$$\begin{aligned} \max \quad & z = x_1 + x_2 + x_3 \\ \text{s.t.} \quad & x_1 - x_2 + x_3 \leq 5 \\ & 2x_1 - x_2 + 2x_3 \geq 7 \\ & x_1 - x_2 - 3x_3 \leq 9 \\ & x_3 \geq 0 \text{ and } x_1, x_2 \text{ are unrestricted in sign.} \end{aligned}$$

Let $y = |\min\{x_1, x_2\}|$. Then $y_1 = x_1 + y \geq 0$, $y_2 = x_2 + y \geq 0$, and we have

$$\begin{aligned}
\max \quad & z = y_1 + y_2 + x_3 - 2y \\
\text{s.t.} \quad & y_1 - y_2 + x_3 + s_1 = 5 \\
& 2y_1 - y_2 + 2x_3 - y - s_2 = 7 \\
& y_1 - y_2 - 3x_3 + s_3 = 9 \\
& y_1, y_2, y, x_3, s_1, s_2, s_3 \geq 0.
\end{aligned}$$

Example 3. Linearize the following objective function:

$$\max \ z = \min\{|2x_1 + 5x_2|, |7x_1 - 3x_2|\}.$$

Let $y = \min\{|2x_1 + 5x_2|, |7x_1 - 3x_2|\}$. Hence,

$$y \leq |2x_1 + 5x_2| \Rightarrow u_1 + v_1 \geq y \text{ for some variables } u_1, v_1 \geq 0.$$

and

$$y \leq |7x_1 - 3x_2| \Rightarrow u_2 + v_2 \geq y \text{ for some variables } u_2, v_2 \geq 0.$$

Combining the above inequalities, the given objective function can be written in the form of LPP as

$$\begin{aligned}
\max \quad & z = y \\
\text{s.t.} \quad & u_1 + v_1 - y \geq 0 \\
& u_2 + v_2 - y \geq 0 \\
& u_1, v_1, u_2, v_2, y \geq 0.
\end{aligned}$$

Note that $2x_1 + 5x_2$ and $7x_1 - 3x_2$ may be nonnegative or nonpositive, since we are silent about the nature of x_1 and x_2.

1.3 Formulation of Models

Learning to formulate the mathematical programming problem using the given data is the first step for optimizing any system. If we fail at this stage, then it bears no fruitful results. The modeling of the problem includes

(i) Decision variables that we seek to determine.

(ii) Construction of the objective function to be optimized.

(iii) Constraints that satisfy various conditions.

(iv) Nonnegative restrictions and their nature.

The proper definition of the decision variables is the most sensitive part toward the development of a model. Once the decision variables are defined the construction of the objective function and the constraints from the given data is not laborious. For incorporating $\geq, =, \leq$ in the constraints one has to be careful about the phrases: at least or minimum, exactly satisfied, at most or maximum or no longer than, etc.

In this section we have formulated various problems which are in common use.

Linear models. Here we formulate some well known problems as linear programming problems.

Diet problem. A medical practitioner recommends the constituents of a balanced diet for a patient which satisfies the daily minimum requirements of Proteins P units, Fats F units, and Carbohydrates C units at a minimum cost. Choice from five different types of foods can be made. The yield per unit of these foods are given by

Food type	Protein	Fats	Carbohydrates	Cost/unit
1	p_1	f_1	c_1	d_1
2	p_2	f_2	c_2	d_2
3	p_3	f_3	c_3	d_3
4	p_4	f_4	c_4	d_4
5	p_5	f_5	c_5	d_5

How the patient should select the items so that he has to pay minimum.

Suppose x_i = the number of units of the ith food which the patient selects. The objective function is

$$\min \ z = d_1 x_1 + d_2 x_2 + \cdots + d_5 x_5,$$

and the constraints are

$$p_1 x_1 + p_2 x_2 + \cdots + p_5 x_5 \geq P$$
$$f_1 x_1 + f_2 x_2 + \cdots + f_5 x_5 \geq F$$
$$c_1 x_1 + c_2 x_2 + \cdots + c_5 x_5 \geq C$$
$$x_i \geq 0, \ i = 1, 2, \ldots, 5.$$

Product mix problem. A manufacturing process requires three different inputs viz., A, B and C. A sandal soap of the first type requires 30 gm of A, 20 gm of B and 6 gm of C, while this data for the second type of soap is 25, 5 and 15, respectively. The maximum availability of A, B and C are 6000, 3000 and 3000 gm, respectively. The selling price of the sandal soap of the first and second type are $ 14 and 15, respectively. The profit is proportional to the amount of soaps manufactured. How many soaps of the first and second kinds should be manufactured to maximize the profit. Assume that the market has unlimited demand.

Let us put the data in the tabular form:

Type	Inputs/unit			Selling price/unit
	A	B	C	
I	30	20	6	14
II	25	5	15	15
Max availability	6000	3000	3000	

Let x_1 and x_2 be the number of the first and second types of soaps to be manufactured. The profit from selling is given by $z = 14x_1 + 15x_2$. This is subjected to the availability constraints given by $30x_1 + 25x_2 \leq 6000$, $20x_1 + 5x_2 \leq 3000$, $6x_1 + 15x_2 \leq 3000$. The decision variables are $x_1, x_2 \geq 0$, and in addition, these must be integers.

Thus, the required LPP is

$$\max \quad z = 14x_1 + 15x_2$$
$$\text{s.t.} \quad 30x_1 + 25x_2 \leq 6000$$
$$20x_1 + 5x_2 \leq 3000$$
$$6x_1 + 15x_2 \leq 3000$$
$$x_1, x_2 \geq 0 \text{ and are integers.}$$

Bus scheduling problem. IP Depo runs buses during the time period 5 AM to 1 AM. Each bus can operate for 8 hours successively, and then it is directed to workshop for maintenance and fuel. The minimum number of buses required fluctuate with the time intervals. The desired number of buses during different time interval are given in the following table:

Time intervals	Minimum number of buses required
5 AM–9 AM	5
9 AM–1 PM	13
1 PM–5 PM	11
5 PM–9 PM	14
9 PM–1 AM	4

The depo keeps in view the reduction of air pollution and smog problem. It is required to determine the number of buses to operate during different shifts that will meet the minimum requirement while minimizing the total number of daily buses in operation.

Let x_i be the number of buses starting at the beginning of the ith period, $i = 1$ to 5. Note that each bus operates during two consecutive shifts. Buses which join the crew at 5 AM and 9 AM will be in operation between 9 AM and 1 PM. As the minimum number of buses required in this interval is 13, we have $x_1 + x_2 \geq 13$, and similarly others.

The LPP formulation is

$$\min \quad z = x_1 + x_2 + x_3 + x_4 + x_5$$
$$\text{s.t.} \quad x_1 + x_2 \geq 13$$
$$x_2 + x_3 \geq 11$$
$$x_3 + x_4 \geq 14$$
$$x_4 + x_5 \geq 4$$
$$x_5 + x_1 \geq 5$$
$$x_1, x_2, x_3, x_4, x_5 \geq 0 \text{ and are integers.}$$

The warehousing problem. A warehouse has a capacity of 2000 units. The manager of the warehouse buys and sells the stock of potatoes over a period of 6 weeks to make profit. Assume that in the jth week the same unit price p_j holds for both purchase and sale. In addition, there is unit cost \$15 as weekly expenses for holding stock. The warehouse is empty at the beginning and is required to be empty after the sixth week. How should the manager operate?

The major activities involve buying, selling, and holding the stock for a week. Define the variables

x_j = the level of the stock at the beginning of the jth week;

y_j = the amount bought during the jth week;

z_j = the amount sold during the jth week.

Then the manager tries to maximize

$$\sum_{j=1}^{6} p_j(z_j - y_j) - 15x_j$$

subject to the stock balance constraints

$$x_{j+1} = x_j + y_j - z_j, \quad j = 1, 2, \ldots, 5$$

the warehouse capacity constraints

$$x_j \leq 2000, \ j = 1, 2, \ldots, 6$$

the boundary conditions

$$x_1 = 0, \quad x_6 + y_6 - z_6 = 0$$

and the nonnegative restrictions

$$x_j \geq 0, \ y_j \geq 0, \ z_j \geq 0, \quad j = 1, 2, \ldots, 6.$$

Caterer problem. TIET has to organize its annual cultural festival continuously for next five days. There is an arrangement of dinner for every invited team. The requirement of napkins during these five days is

Days	1	2	3	4	5
Napkins required	80	50	100	80	150.

Accordingly, a caterer has been requested to supply the napkins according to the above schedule. After the festival is over caterer has no use of napkins. A new napkin costs \$2. The washing charges for a used napkin is \$0.5 by ordinary services and \$1, if express service is used. A napkin given for washing by ordinary service is returned third day, while under express service it is return next day. How the caterer should meet the requirement of the festival organizers so that the total cost is minimized.

Define the decision variables as

x_i = number of napkins purchased on the ith day, $i = 1$ to 5.

y_j = number of napkins given for washing on jth day under express service, $j = 1$ to 4.

z_k = number of napkins given for washing on kth day under ordinary service, $k = 1$ to 3.

v_ℓ = number of napkins left in the stock on ℓth day after the napkins have been given for washing, $\ell = 1$ to 5.

The data is tabulated as

Type	Number of napkins required on days				
	1	2	3	4	5
New napkins	x_1	x_2	x_3	x_4	x_5
Express service	—	y_1	y_2	y_3	y_4
Ordinary service	—	—	z_1	z_2	z_3
Napkins required	80	50	100	80	150

We have to minimize

$$2(x_1 + x_2 + x_3 + x_4 + x_5) + y_1 + y_2 + y_3 + y_4 + 0.5(z_1 + z_2 + z_3).$$

From the table:

$x_1 = 80$, $x_2 + y_1 = 50$, $x_3 + y_2 + z_1 = 100$, $x_4 + y_3 + z_2 = 80$, $x_5 + y_4 + z_3 = 150$.

Also, there is another set of constraints which shows the total number of napkins which may be given for washing and some napkins which were not given for washing just on the day these have been used. These constraints are: $y_1 + z_1 + v_1 = 80$, $y_2 + z_2 + v_2 = 50 + v_1$, $y_3 + z_3 + v_3 = 100 + v_2$, $y_4 + v_4 = 80 + v_3$, $v_5 = 150 + v_4$.

Thus, the desired LPP model is

$$\min \quad z = 160 + 2(x_2 + x_3 + x_4 + x_5) + y_1 + y_2 + y_3 + y_4$$
$$+ 0.5(z_1 + z_2 + z_3)$$

$$\text{s.t.} \quad x_2 + y_1 = 50$$
$$x_3 + y_2 + z_1 = 100$$
$$x_4 + y_3 + z_2 = 80$$
$$x_5 + y_4 + z_3 = 150$$
$$y_1 + z_1 + v_1 = 80$$
$$y_2 + z_2 + v_2 - v_1 = 50$$
$$y_3 + z_3 + v_3 - v_2 = 100$$
$$y_4 + v_4 - v_3 = 80$$
$$v_5 - v_4 = 150$$
$$\text{all var } \geq 0.$$

Trim-loss problem. Paper cutting machines are available to cut standard news print rolls into the subrolls. Each standard roll is of 180 cm width and a number of them must be cut to produce smaller subrolls at the current orders for 30 of width 70 cm, 60 of width 50 cm and 40 of width 30 cm. Formulate the problem so as to minimize the amount of wastes. Ignoring the recycling or other uses for the trim, assume that the length of each required subroll is the same as that of the standard roll.

A standard roll may be cut according to the following patterns.

Widths ordered in cm	Number of subrolls cut on different patterns							
	p_1	p_2	p_3	p_4	p_5	p_6	p_7	p_8
30	6	4	3	2	2	1	1	0
50	0	1	0	1	2	0	3	2
70	0	0	1	1	0	2	0	1
Trim loss	0	10	20	0	20	10	0	10

Let x_i be the number of the standard news print rolls pieces to cut on the pattern p_i, $i = 1, 2, \ldots, 8$. Thus, the required LPP is

$$
\begin{aligned}
\min \quad & z = 10x_2 + 20x_3 + 20x_5 + 10x_6 + 10x_8 \\
\text{s.t.} \quad & 6x_1 + 4x_2 + 3x_3 + 2x_4 + 2x_5 + x_6 + x_7 = 40 \\
& x_2 + x_4 + 2x_5 + 3x_7 + 2x_8 = 60 \\
& x_3 + x_4 + 2x_6 + x_7 + x_8 = 30 \\
& x_i \geq 0, \ i = 1, 2, \ldots, 8 \text{ and are integers.}
\end{aligned}
$$

Here, in the constraints the equality is desired due to the fact any thing left is of no use.

Example 4. Two alloys, A and B are made from four different metals, I, II, III, and IV, according to the following specifications:

Alloy	Specifications	Selling price ($)/ton
A	at most 80% of I	200
	at least 30% of II	
	at least 50% of IV	
B	between 40% & 60% of II	300
	at least 30% of III	
	at most 70% of IV	

The four metals, in turn, are extracted from three different ores with the following data:

Ore	Max. Quantity (tons)	I	II	III	IV	others	Purchase Price ($)/ton
1	1000	20	10	30	30	10	30
2	2000	10	20	30	30	10	40
3	3000	5	5	70	20	0	50

How much of each alloy should be produced to maximize the profit. Formulate the problem as a LP model.

Define

x_{ij} = tons of ore i allocated to alloy j; $i = 1, 2, 3$; $j = $ A, B

w_j = tons of alloy j produced

$$\max z = 200w_A + 300w_B - 30(x_{1A} + x_{1B}) - 40(x_{2A} + x_{2B}) - 50(x_{3A} + x_{3B})$$

Specification constraints:

$$0.2x_{1A} + 0.1x_{2A} + 0.05x_{3A} \leq 0.8w_A$$
$$0.1x_{1A} + 0.2x_{2A} + 0.05x_{3A} \geq 0.3w_A$$
$$0.3x_{1A} + 0.3x_{2A} + 0.2x_{3A} \geq 0.5w_A$$
$$0.1x_{1B} + 0.2x_{2B} + 0.05x_{3B} \geq 0.4w_B$$
$$0.1x_{1B} + 0.2x_{2B} + 0.05x_{3B} \leq 0.6w_B$$
$$0.3x_{1B} + 0.3x_{2B} + 0.7x_{3B} \geq 0.3w_B$$
$$0.3x_{1B} + 0.3x_{2B} + 0.2x_{3B} \leq 0.7w_B$$

Ore constraints:

$$x_{1A} + x_{1B} \leq 1000$$
$$x_{2A} + x_{2B} \leq 2000$$
$$x_{3A} + x_{3B} \leq 3000$$

Alloy constraints:

$$x_{1A} + x_{2A} + x_{3A} \geq w_A$$
$$x_{1B} + x_{2B} + x_{3B} \geq w_B$$
$$x_{iA} \geq 0, x_{iB}, w_j \geq 0 \quad i = 1, 2, 3, \ j = \text{A, B}.$$

Nonlinear models. The formulation of nonlinear problems requires little more efforts in comparison to the linear models. In this section, we formulate some nonlinear programming problems.

Gambler problem. A gambler has $24,000 to play a game. In the game there are three places for stake. He divides his total money among three choices. There are three outcomes in the game. The return per unit deposited at each choice can be read from the table:

Outcomes	Gain or loss per rupee at choice		
	1	2	3
1	-5	1	1
2	-7	6	10
3	13	-2	6

The probabilities of different outcomes are not known. The gambler wants least risk as far as loss is concerned. He decides to divide his money among three choices in such a way if there is any loss, then it is least. Any way he maximizes the minimum return.

Suppose that x_1, x_2, x_3 dollars are invested by the gambler on the choices 1,2,3, respectively. Then the returns depending upon outcomes 1,2,3 are

$$-5x_1 + x_2 + x_3, \qquad -7x_1 + 6x_2 + 10x_3, \qquad 13x_1 - 2x_2 + 6x_3.$$

The problem is formulated as

max $\quad z = \min\{-5x_1 + x_2 + x_3, \ -7x_1 + 6x_2 + 10x_3, \ 13x_1 - 2x_2 + 6x_3\}$

s.t. $\quad x_1 + x_2 + x_3 = 24000$

$\quad x_1, x_2, x_3 \geq 0$ and are integers.

Remark. This is a nonlinear programming problem (NLPP). However, it may be converted into an LPP as follows:

Let $y = \min\{-5x_1 + x_2 + x_3, \ -7x_1 + 6x_2 + 10x_3, \ 13x_1 - 2x_2 + 6x_3\}$. Then

$$y \leq -5x_1 + x_2 + x_3$$
$$y \leq -7x_1 + 6x_2 + 10x_3$$
$$y \leq 13x_1 - 2x_2 + 6x_3$$

The required LPP is

max $\quad z = y$

s.t. $\quad 5x_1 - x_2 - x_3 + y \leq 0$

$\quad 7x_1 - 6x_2 - 10x_3 + y \leq 0$

$\quad -13x_1 + 2x_2 - x_3 + y \leq 0$

$\quad x_1 + x_2 + x_3 = 24000$

$\quad x_1, x_2, x_3, y \geq 0$ and are integers.

Production planning problem. The municipal corporation of Patiala decides to clean the water system of open wells in urban areas, and for this purpose a medicated product is made by mixing two parts of potassium permagnate and three parts of bleaching powder. These products are processed in departments of Chemical, Chemistry and Biotechnology operating in Ranbaxy laboratories. Departments have limited number of production hours available, viz.,, 100, 150 and 200, respectively. The production rate of potassium permagnate and bleaching powder in each department is given in the following table

| Department | Production rate(no. of units/hour) | |
	Potassium permagnate	Bleaching powder
Chemical	20	25
Chemistry	25	20
Biotechnology	20	5

The objective is to determine the number of hours to be assigned to each department to maximize the completed units of the medicated product. Formulate the appropriate model.

Writing the given data in the tabular form as

| Department | Production rate(no. of units/hour) | | Limited hours |
	Potassium permagnate	Bleaching powder	
Chemical	20	25	100
Chemistry	25	20	150
Biotechnology	20	5	200

Let x_{ij} be the number of hours assigned to ith department for jth part, $i = 1, 2, 3$ and $j = a, b$, where suffixes $1 =$ Chemical, $2 =$ Chemistry, $3 =$ Biotechnology and $a =$ Potassium permagnate and $b =$ bleaching powder.

Total number of parts of a manufactured $= 20x_{1a} + 25x_{2a} + 20x_{3a}$.

Total number of parts of b manufactured $= 25x_{1b} + 20x_{2b} + 5x_{3b}$.

Complete units of the final product are

$$y = \min \left\{ \frac{20x_{1a} + 25x_{2a} + 20x_{3a}}{2}, \frac{25x_{1b} + 20x_{2b} + 5x_{3b}}{3} \right\}.$$

Thus,

$$\begin{aligned}
\max \quad & z = y \\
\text{s.t.} \quad & x_{1a} + x_{1b} \le 100 \\
& x_{2a} + x_{2b} \le 150 \\
& x_{3a} + x_{3b} \le 200 \\
& x_{ij} \ge 0, \ i = 1, 2, 3; \ j = a, b, \text{ and are integers.}
\end{aligned}$$

The problem is formulated as a nonlinear programming problem.

Remark. The formulation of this problem can be done as LPP, see Problem 19, Problem set 1.

Example 5. A firm produces two products A and B using two limited resources. The maximum amount of Resource 1 available per week is 3000, while for Resource 2 is 2500. The production of one unit of A requires 3 units of Resource 1 and 1 unit of of Resource 2, and the production of B requires 2 units of Resource 1 and 2 units of Resource 2. The unit cost of Resource 1 is $(1.5 - .001u_1)$, where u_1 is the number of units of Resource 1 used. The unit cost of Resource 2 is $(2 - .004u_2)$, where u_2 is the number of units of Resource 2 is used. The selling price per unit of A and B are fixed as

$$p_A = 8 - .001x_A - .005x_B,$$

$$p_B = 9 - .002x_A - .004x_B,$$

where x_A and x_B are the number of units sold for product A and B, respectively. Assuming that how much has been manufactured is disposed off, formulate the above problem to maximize the profit over a week.

Let x_A and x_B be number of units of the products A and B produced per week. The requirement of Resource 1 per week is $(3x_A + 2x_B)$, while that of Resource 2 is $(x_A + 2x_B)$ and the constraints on the resources availability are $3x_A + 2x_B \le 3000$ and $x_A + 2x_B \le 2500$.

The total cost of Resource 1 and 2 per week is

$$(3x_A + 2x_B)[1.5 - .001(3x_A + 2x_B)] + (x_A + 2x_B)[2 - .004(x_A + 2x_B)]$$

and the total return per week from selling of A and B is

$$x_A(8 - .001x_A - .005x_B) + x_B(9 - .002x_A - .004x_B).$$

As the total profit is the difference of total cost and total return, the formulation of model is

$$
\begin{aligned}
\max \quad & z = .012x_A^2 + .016x_B^2 + .021x_Ax_B + 1.5x_A + 2x_B \\
\text{s.t.} \quad & 3x_A + 2x_B \le 3000 \\
& x_A + 2x_B \le 2500 \\
& x_A, x_B \ge 0
\end{aligned}
$$

This is a quadratic programming problem.

Problem Set 1

1. Write the following LPP into the standard form:

$$
\begin{aligned}
\text{opt} \quad & z = 2x_1 + x_2 - x_3 \\
\text{s.t.} \quad & 2x_1 + x_2 - x_3 \le 5 \\
& -3x_1 + 2x_2 + 3x_3 \ge -3 \\
& x_1 - 3x_2 + 4x_3 \ge 2 \\
& x_1 + x_2 + x_3 = 4 \\
& x_1 \ge 0, x_2 \ge 1 \text{ and } x_3 \text{ is unrestricted in sign.}
\end{aligned}
$$

2. Write the standard form of the LPP

$$
\begin{aligned}
\max \quad & z = 2x_1 + x_2 + x_3 \\
\text{s.t.} \quad & x_1 - x_2 + 2x_3 \ge 2 \\
& |2x_1 + x_2 - x_3| \le 4 \\
& 3x_1 - 2x_2 - 7x_3 \le 3 \\
& x_1, x_3 \ge 0, x_2 \le 0.
\end{aligned}
$$

3. Write the following linear programme into a standard form:

$$
\begin{aligned}
\text{opt} \quad & z = x_1 + 2x_2 - x_3 \\
\text{s.t.} \quad & x_1 + x_2 - x_3 \le 5 \\
& -x_1 + 2x_2 + 3x_3 \ge -4 \\
& 2x_1 + 3x_2 - 4x_3 \ge 3 \\
& x_1 + x_2 + x_3 = 2 \\
& x_1 \ge 0, \ x_2 \ge p \text{ and } x_3 \text{ is unrestricted in sign.}
\end{aligned}
$$

Mention the range of p in the standard LPP.

4. Consider the following optimization problem:

$$\min \quad z = |x_1| + 2|x_2| - x_3$$
$$\text{s.t.} \quad x_1 + x_2 - x_3 \leq 9$$
$$x_1 - 2x_2 + 3x_3 = 11$$
$$x_3 \geq 0$$

(a) Is this a linear programming problem?

(b) Can you convert it into LPP? If yes, write the standard form.

5. Convert the following problem into a standard linear programme by using only three nonnegative variables in place of x_1 and x_2 and the objective function must be free of the constant term.

$$\min \quad z = x_1 + x_2 - 1$$
$$\text{s.t.} \quad x_1 + x_2 \leq 7$$
$$x_1 - 2x_2 \geq 4.$$

Suggestion. At first, replace x_1 by $x_1 + 1$ and then use the proposition.

6. Convert the following problem into an equivalent linear model

$$\max \quad \frac{-3 + 2x_1 + 4x_2 - 5x_3}{6 + 3x_1 - x_2}$$
$$\text{s.t.} \quad x_1 - x_2 \geq 0$$
$$7x_1 + 9x_2 + 10x_3 \leq 30$$
$$x_1 \geq 0, x_2 \geq 1, x_3 \geq 0$$

Suggestion. This is a linear fractional programming problem. To avoid various possibilities, assume $r = (6 + 3x_1 - x_2)^{-1} > 0$, and define $rx_j = y_j$, $j = 1, 2, 3$.

7. Suppose n different food items are available at the market and the selling price for the jth food is c_j per unit. Moreover, there are m basic nutritional ingredients for the human body and minimum b_i units of the ith ingredient are required to achieve a balanced diet for good health. In addition, a study shows that each unit of the jth food contains a_{ij} units of the ith ingredients. A dietitian of a large group may face a problem of determining the most economical diet that satisfies the basic nutritional requirement for good health. Formulate the problem so that problem of dietitian is solved.

8. A small manufacturing plant produces two products, A and B. Each product must be worked on by a bank of CNC lathe machines and then, in succession, by a group of CNC milling machines. Product A requires 1 hr on CNC lathe machines and 3 hrs on CNC milling machines. Product B requires 5 hrs on CNC lathe machines and 1 hr on CNC milling machines. A total of 10000 hrs is available per week on CNC lathe machines and 7000 hrs on CNC milling machines. The net profit is $5 per unit for product A and $10 per unit for product B. Formulate the problem so as to maximize the weekly profit. Assume that all the quantities manufactured are disposed off.

9. A company makes two kinds of leather belts. Belt A is a high quality belt, and belt B is of lower quality. Each belt of type A requires twice as much time as a belt of type B, and if all belts were of type B, the company could make 1500 per day. The supply of leather is sufficient for only 1000 belts per day (both A and B combined). Belt A requires a fancy buckle, and only 500 per day are available. There are only 800 buckles a day available for belt B. The profits in belt A and B are $3 and $2 per belt, respectively. Formulate the linear programming problem to maximize the profit.

10. The New Delhi Milk Corporation (NDMC) has two plants each of which produces and supplies two products: Milk and Butter. Plants can each work up to 16 hours a day. In Plant-I, it takes 3 hours to prepare from powder and pack 1000 liters of milk and 1 hour to prepare and pack 100 kg of butter. In Plant-II, it takes 2 hours to prepare and pack 1000 liters of milk and 1.5 hours to prepare and pack 100 kg of butter. In Plant-I, it costs $15,000 to prepare and pack 1000 liters of milk and $28,000 to prepare and pack 100 kg of butter, whereas these costs are $18,000 and $26,000, respectively for Plant-II. The NDMC is obliged to produce daily at least 10,000 liters of milk and 800 kg of butter. Formulate this as LPP to find as to how should the company organize its production so that the required amount of the products be obtained at minimum cost.

   Suggestion. Let m_i = units of milk produced in ith plant per day and b_i = units of butter produced in ith plant per day, $i = 1, 2$. 1000 liters = one unit and 100 kg = one unit. Note that one unit of milk is produced by Plant-I in 3 hours and hence m_1 is produced in $3m_1$ hours and so on.

11. A farmer has to plant two kinds of trees, say A and B on a land with 4400 sq m area. Each A tree requires at least 25 sq m of land, and B requires 40 sq m. The annual water requirement of tree A is 30 units and that of B is 15 units, while at most 3300 units water is available. It is estimated that the ratio of the number of B trees to the number of A trees should not be less than 6/19 and not be more that 17/8. The return from one B tree is $50, while from one A tree is one and a half times that of return from B. Describe the plantation project of the farmer in terms of LPP so that the return is maximum.

12. A metal slitting company cuts master rolls with width 200 cm into subrolls of small width. Customers specify that they need subrolls of different widths given in the following table

Width of subrolls (in cm)	Number required
35	200
80	90
90	350
120	850

The objective is to use a minimum number of master rolls to satisfy a set of customers' orders. Formulate the problem as LPP.

13. The Materials Science Division of TIET needs circular metallic plates of diameters 3 cm and 6 cm to perform experiments on heat treatment studies, and requirement of these plates are 2500 and 1500, respectively. These are to be cut from parent metallic sheets of dimension 6×15 cm^2. Formulate the problem as a linear programming problem so that the minimum number of parent metallic sheets are used.

14. Martin furniture company manufactures tables and chairs using wood and labour only. Wood required for one table is 30 units and for one chair is 20 units, and the labour spent on table is 10 units and for chair is 5 units. Total units of wood available are 381 and of labour are 117. The unit profit for table is $9 and for chair is $6. How many tables and chairs should be made to get maximum profit?

15. IBM produces two kinds of memory chips (Chip-1 and Chip-2) for memory usage. The unit selling price is \$15 for Chip-1 and \$25 for Chip-2. To make one Chip-1, IBM has to invest 3 hours of skilled labour, 2 hours of unskilled labour and 1 unit of raw material. To make one Chip-2, it takes 4 hours of skilled labour, 3 hours of unskilled labour, and 2 units of raw material. The company has 100 hours of skilled labour, 70 hours of unskilled labour and 30 units of raw material available, and is interested to utilize the full potential of skilled labour. The sales contract signed by IBM requires that at least 3 units of chip-2 have to be produced and any fractional quantity is acceptable. Formulate a linear programme to help IBM determine its optimal product mix.

16. A manufacturer produces three models (I, II and III) of a certain product. He uses two types of raw material (A and B) of which 2000 and 3000 units are available, respectively. The raw material requirement can be read from the following table

Raw material	Requirement per unit of given model		
	I	II	III
A	2	3	5
B	4	2	7

The labour time for each unit of model I is twice that of model II and three times that of model III. The entire labour force can produce the equivalent of 700 units of model I. A market survey indicates that the minimum demand of three models are 200, 200 and 150 units, respectively. Formulate the LPP to determine the number of units of each product which will maximize the profit. Assume that the profit per unit of models I, II, III are \$30, \$20, and \$60, respectively.

17. There are m machines and n products, and the time a_{ij} is required to process one unit of product j on machine i. The x_{ij} is the number of units of product j produced on machine i and c_{ij} is the respective cost of processing them. The b_i is the total time available on machine i whereas d_j is the number of units of product j which must be processed. Formulate the problem with an objective of minimizing the total cost.

18. A ship has three cargo loads, forward, after and centre; the capacity limits are

Placement	Weight (tonnes)	Capacity in m^3
Forward	2000	100,000
Centre	3000	135,000
After	1500	30,000

The following cargoes are offered, the ship owner may accept all or any part of each commodity:

Commodity	Weight (tonnes)	Volume per ton (m^3)	Profit per ton($)
A	6000	60	60
B	4000	50	80
C	2000	25	50

In order to preserve the trim of the ship, the weight in each load must be proportional to the capacity. The objective is to maximize the profit. Formulate the linear programming model for this problem.

19. Convert the nonlinear problem obtained in the production planning problem of Section 1.3 into a linear programming problem.

20 Reformulate the LPP of Problem 16 with the modification: "The labour time for each unit of model-I is twice that of model-II and labour time for each unit of model-II is thrice that of model-III". The remaining data is same as given in Problem 16.

21. A company manufactures a product which consists of n type of ingredients that are being produced in m departments. Each department has limited number of production hours, viz.,, the ith $(i = 1, 2, \ldots, m)$ department has b_i hours available. The production rate of jth $(j = 1, 2, \ldots, n)$ ingredient is a_{ij} units per hour in the ith department. The final product is made just by mixing one part of the first ingredient, two parts of the second ingredient, and so on n parts of the nth ingredients. The objective is to determine the number of hours of each department to be assigned to each ingredient to maximize the completed units of the product. Formulate the problem as NLPP.

Suggestion. x_{ij} = the number of hours assigned to ith department for the production of jth ingredient. This is a generalization of the production planning problem of Section 1.3.

22. A canteen of an institute which remains functional only for five days in a week has to recruit waiters. A waiter has to work continuously for three days and have two days off. The minimum number of waiters required on individual days are

Days	1	2	3	4	5
Number of waiters required	25	35	40	30	20

Not more than 30 waiters can be recruited on any day. Formulate the LPP model to minimize the number of waiters recruited.

23. A transporter company assigns three type of buses to four cities Bombay, Calcutta, Chennai and Delhi for tourists according to the following data

Bus type	Capacity of passengers	No. of buses
1	100	5
2	70	8
3	50	10

Bus type	No. of weekly trips to			
	Bombay	Calcutta	Chennai	Delhi
1	3	2	2	1
2	4	3	3	2
3	5	5	4	2
Number of tourists	1000	2000	900	1200

The associated costs, including the penalties for losing customers because of space limitation, are

	Operating cost($) per trip for cities			
Bus type	Bombay	Calcutta	Chennai	Delhi
1	1000	1100	1200	1500
2	900	1000	1100	1200
3	700	900	900	1000
Penalty($) per lost customer	40	50	45	70

Formulate the LPP model that determines the optimum allocation of the buses to different cities and the associated number of trips.

24. A retired employee wants to invest $200,000 which he received as provident fund. He was made acquainted with two schemes. In scheme-A he is ensured that for each rupee invested will earn $0.6 a year, and in scheme-B each rupee will earn $1.4 after two years. In scheme-A investments can be made annually, while in scheme-B investments are allowed for periods that are multiples of two years only. How should the employee invest his hard earn money to maximize the earnings at the end of three years? Formulate the LP model for the problem.

25. A factory is to produce two products P_1 and P_2. The product requires machining on two machines M_1 and M_2. Product P_1 requires 5 hours on machine M_1 and 3 hours on machine M_2. Product P_2 requires 4 hours on machine M_1 and 6 hours on machine M_2. Machine M_1 is available for 120 hours per week during regular working hours and 50 hours on overtime. Weekly machine hours on M_2 are limited to 150 hours on regular working hours and 40 hours on overtime. Product P_1 earns a unit profit of $8 if produced on regular time and $6, if produced on regular time on M_1 and on overtime on M_2, and $4 if produced on overtime on both the machines. Product P_2 earns a unit profit of $10 if produced on regular time and $9, if produced on regular time on M_1 and on overtime on M_2, and $8 if produced on overtime

on both the machines. Formulate the LPP model for designing an optimum production schedule for maximizing the profit.

Suggestion. Define the variables as

x_1 = number of units of P_1 made on regular time

x_2 = number of units of P_2 made on regular time

x_3 = number of units of P_1 made on overtime

x_4 = number of units of P_2 made on overtime

x_5 = number of units of P_1 made on regular time on M_1 and overtime on M_2

x_6 = number of units of P_2 made on regular time on M_1 and overtime on M_2

The objective function is $8x_1 + 10x_2 + 4x_3 + 8x_4 + 6x_5 + 9x_6$ and to find the constraints, construct the table:

Machine type	Product type		Available time	
	Time for P_1	Time for P_2	Regular time	Overtime
M_1	5	4	120	50
M_2	3	6	150	40

$5x_1 + 4x_2 + 5x_5 + 4x_6 \le 120$ (regular time of M_1)

$5x_3 + 4x_4 \le 50$ (overtime of M_1)

$3x_1 + 6x_2 \le 150$ (regular time on M_2)

$3x_3 + 6x_4 + 3x_5 + 6x_6 \le 40$ (overtime on M_2)

$x_j \ge 0, \ j = 1, 2, \ldots, 6.$

26. A chemical company has been requested by its state government to install and employ antipollution devices. The company makes two products; for each of these products, the manufacturing process yields excessive amount of irritant gases and particulates (airborne solids). The table shows the daily emission, in pounds, of each pollutant for every 1000 gallons of product manufactured. The company is prohibited from emitting more than G_1, G_2 and P_1 pounds of gas CM, gas SD, and Particulates,

respectively. The profit for each thousand gallons of Products 1 and 2 manufactured per day is p_1 and p_2, respectively.

Type of	Pounds of pollution emitted	
Pollutant	Per 1000 gallons of Product 1	Per 1000 gallons of Product 2
Gas CM	24	36
Gas SD	8	13
Particulates	100	50

The production manager has approved the installations of two antipollution devices. The first device removes 0.75 of gas CM, 0.5 of gas SD and 0.9 of the Particulates, regardless of the product made. The second device removes 0.33 of gas CM, none of gas SD, and 0.6 of the Particulates for Product 2. The first device reduces profit per thousand gallons manufactured daily by c_1, regardless of the product; similarly, the second device reduces profit by c_2 per thousand gallons manufactured, regardless of the product. Sales commitments dictate that at least R_1 thousand gallons of Product 1 be produced per day, and R_2 thousand of gallons of Product 2. Formulate the appropriate optimization model.

Suggestion. Define the decision variables as $x_1 = 1000$ gallons of Product 1 made per day without using any control device; $x_{11} = 1000$ gallons of Product 1 made per day using the first control device; $x_{12} = 1000$ gallons of Product 1 made per day using the second device. Define the similar variables y_1, y_{11}, y_{12} for Product 2.

27. A company produces two products P_1 and P_2. The sales volume for P_1 is at least 40% of the total sales of both P_1 and P_2. The market survey ensures that it is not possible to sell more than 100 unit of P_1 per day. Both product use one raw material whose availability to 120 lb a day. The usage rates of the raw material are 1 lb per unit for P_1 and 2 lb for per unit for P_2. The unit prices for P_1 and P_2 are \$10 and \$30, respectively. Formulate the LPP model to optimize the product mix for the company.

28. A farming organization operates three farms of comparable productivity. The output of each farm is limited both by the usable

acreage and the water available for irrigation. The data for the upcoming season are

Farm	Usable acreage	Water available (acre feet)
1	400	1500
2	600	2000
3	300	900

The organization is interested in three crops for planting which differ primarily in their expected profit per acre and their consumption of water. Furthermore, the total acreage that can be devoted to each of the crops is limited by the amount of appropriate harvesting equipment available

Crop	Maximum acreage	Water consumption in acre feet	Expected profit per acre ($)
A	700	5	4000
B	800	4	3000
C	3000	3	1000

In order to maintain a uniform workload among farms, the policy of the organization is that the percentage of the usable acreage planted be the same at each farm. However, any combination of the crops may be grown at any of the farms. The organization wishes to know how much each crop should be planted at the respective farms to maximize the expected profit.

Suggestion. x_{ij} = number of acres of $i$the farm to be allotted to jth crop, $i = 1, 2, 3$ and $j = A, B, C$.

29. Weapons of three types are to be assigned to 8 different targets. Upper limits on available weapons and lower limits on weapons to be assigned are specified. The characteristics of the three weapons type are as follows

(a) W_1 : Fighter bombers

(b) W_2 : Medium-range ballistic missiles

(c) W_3 : Intercontinental ballistic missiles

Table 1 gives the values of the parameters needed for the model: probabilities that target will be undamaged by weapons, total number of weapons available, minimum number of weapons to be assigned (b_j), and military value of targets (u_j).

Table 1a

Targets	T_1	T_2	T_3	T_4	T_5	T_6	T_7	T_8
W_1	1.00	0.90	1.00	0.95	1.00	0.90	1.00	0.95
W_2	0.85	0.85	0.90	1.00	0.95	1.00	0.95	0.90
W_3	0.95	1.00	0.95	0.90	0.90	0.95	0.85	1.00
b_j	30	100	50	40	60	70	50	10
u_j	60	50	75	80	40	200	100	150

Table 1b

Targets	Weapons Available
W_1	200
W_2	100
W_3	300

Formulate the model for maximizing the expected target damage value.

Chapter 2

Geometry of Linear Programming

The intent of this chapter is to provide a geometric interpretation of linear programming problems. To conceive fundamental concepts and validity of different algorithms encountered in optimization, convexity theory is considered the key of this subject. The last section is on the graphical method of solving linear programming problems.

2.1 Geometric Interpretation

Let $\mathbb{R}^n$ denote the n-dimensional vector space (Euclidean) defined over the field of reals. Suppose $X, Y \in \mathbb{R}^n$. For $X = (x_1, x_2, \ldots, x_n)^T$ and $Y = (y_1, y_2, \ldots, y_n)^T$ we define the distance between X and Y as

$$|X - Y| = \left((x_1 - y_1)^2 + (x_2 - y_2)^2 + \cdots + (x_n - y_n)^2\right)^{1/2}.$$

Neighbourhood. Let X_0 be a point in $\mathbb{R}^n$. Then δ-neighbourhood of X_0, denoted by $N_\delta(X_0)$ is defined as the set of points satisfying

$$N_\delta(X_0) = \{X \in \mathbb{R}^n : |X - X_0| < \delta,\ \delta > 0\}.$$

$N_\delta(X_0) \setminus X_0 = \{X \in \mathbb{R}^n : 0 < |X - X_0| < \delta\}$ will be termed as deleted neighbourhood of X_0.

In $\mathbb{R}^2$, $N_\delta(X_0)$ is a circle without circumference, and in $\mathbb{R}^3$, $N_\delta(X_0)$ is sphere without boundary, and for $\mathbb{R}$, an open interval on the real line. For $n > 3$, figures are hypothetical.

Let $S \subset \mathbb{R}^n$. We give few elementary definitions.

Boundary point. A point X_0 is called a boundary point of S if each deleted neighbourhood of X_0 intersects S and its compliment S^c.

Interior point. A point $X_0 \in S$ is said to be an interior point of S, if there exists a neighbourhood of X_0 which is contained in S.

Open set. A set S is said to be open if for each $X \in S$ there exists a neighbourhood of X which is contained in S.

For example, $S = \{X \in \mathbb{R}^n : |X - X_0| < 2\}$ is an open set. The well known results: (i) A set is open $\Longleftrightarrow$ it contains all its interior points, and (ii) The union of any number of open sets is an open set, are left as exercises for the reader.

Close set. A set S is closed if its compliment S^c is open.

For example, $S = \{X \in \mathbb{R}^n : |X - X_0| \leq 3\}$ is a closed set. Again a useful result arises: intersection of any number of closed sets is closed.

A set S in $\mathbb{R}^n$ is bounded if there exists a constant $M > 0$ such that $|X| \leq M$ for all X in S.

Definition 1. A line joining X_1 and X_2 in $\mathbb{R}^n$ is a set of points given by the linear combination

$$L = \{X \in \mathbb{R}^n : X = \alpha_1 X_1 + \alpha_2 X_2, \ \alpha_1 + \alpha_2 = 1\}.$$

Obviously,

$$L^+ = \{X : X = \alpha_1 X_1 + \alpha_2 X_2, \ \alpha_1 + \alpha_2 = 1 \, , \alpha_2 \geq 0\}$$

is a half-line originating from X_1 in the direction of X_2 as, for $\alpha_2 = 0$, $X = X_1$ and $\alpha_2 = 1$, $X = X_2$.

Similarly,

$$L^- = \{X : X = \alpha_1 X_1 + \alpha_2 X_2, \ \alpha_1 + \alpha_2 = 1 \, , \alpha_1 \geq 0\}$$

is a half-line emanating from X_2 in the direction of X_1 as, for $\alpha_1 = 0$, $X = X_2$ and $\alpha_1 = 1$, $X = X_1$.

Definition 2. A point $X \in \mathbb{R}^n$ is called a convex linear combination (clc) of two points X_1 and X_2, if it can be expressed as

$$X = \alpha_1 X_1 + \alpha_2 X_2, \ \alpha_1, \alpha_2 \geq 0, \ \alpha_1 + \alpha_2 = 1.$$

Geometrically, speaking convex linear combination of any points X_1 and X_2 is a line segment joining X_1 and X_2.

For example, let $X_1 = (1, 2)$ and $X_2 = (3, 7)$. Then,

$$X = \left(\frac{1}{3}\right)(1, 2) + \left(\frac{2}{3}\right)(3, 7) = \left(\frac{1}{3}, \frac{2}{3}\right) + \left(2, \frac{14}{3}\right) = \left(\frac{7}{3}, \frac{16}{3}\right)$$

is a point lying on the line joining X_1 and X_2.

Convex set. A set S is said to be convex if clc of any two points of S belongs to S, i.e.,

$$X = \alpha_1 X_1 + \alpha_2 X_2 \in S, \quad \alpha_1 + \alpha_2 = 1, \quad \alpha_1, \alpha_2 \geq 0 \ \forall \ X_1, X_2 \in S.$$

Geometrically, this definition may be interpreted as the line segment joining every pair of points X_1, X_2 of S lies entirely in S. For more illustration, see Fig. 2.1.

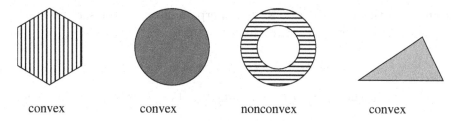

| convex | convex | nonconvex | convex |

Figure 2.1

By convention empty set is convex. Every singleton set is convex. A straight line is a convex set, and a plane in $\mathbb{R}^3$ is also a convex set. Convex sets have many pleasant properties that give strong mathematical back ground to the optimization theory.

Proposition 1. Intersection of two convex sets is a convex set.

Proof. Let S_1 and S_2 be two convex sets. We have to show that $S_1 \cap S_2$ is a convex set. If this intersection is empty or singleton there is nothing to prove.

Let X_1 and X_2 be two arbitrary points in $S_1 \cap S_2$. Then $X_1, X_2 \in S_1$ and $X_1, X_2 \in S_2$. Since S_1 and S_2 are convex, we have

$$\alpha_1 X_1 + \alpha_2 X_2 \in S_1 \text{ and } \alpha_1 X_1 + \alpha_2 X_2 \in S_2, \ \alpha_1, \alpha_2 \geq 0, \ \alpha_1 + \alpha_2 = 1.$$

Thus,

$$\alpha_1 X_1 + \alpha_2 X_2 \in S_1 \cap S_2,$$

and hence $S_1 \cap S_2$ is convex.

Remarks. 1. Moreover, it can be shown that intersection of any number of convex sets is a convex set, see Problem 3.

2. The union of two or more convex sets may not be convex. As an example the sets $S_1 = \{(x_1, 0) : x_1 \in \mathbb{R}\}$ and $S_2 = \{(0, x_2) : x_2 \in \mathbb{R}\}$ are convex in xy-plane, but their union $S_1 \cup S_2 = \{(x_1, 0), (0, x_2) : x_1, x_2 \in \mathbb{R}\}$ is not convex, since $(2, 0), (0, 2) \in S_1 \cup S_2$, but their clc,

$$\left(\frac{1}{2}\right)(2, 0) + \left(\frac{1}{2}\right)(0, 2) = (1, 1) \notin S_1 \cup S_2.$$

Hyperplanes and Half-spaces. A plane is $\mathbb{R}^3$ is termed as a hyperplane. The equation of hyperplane in $\mathbb{R}^3$ is the set of points $(x_1, x_2, x_3)^T$ satisfying

$$a_1 x_1 + a_2 x_2 + a_3 x_3 = \beta.$$

Extending the above idea to $\mathbb{R}^n$, a hyperplane in $\mathbb{R}^n$ is the set of points $(x_1, x_2, \ldots, x_n)^T$ satisfying the linear equation

$$a_1 x_1 + a_2 x_2 + \cdots + a_n x_n = \beta$$

or $a^T X = \beta$, where $a = (a_1, a_2, \ldots, a_n)^T$. Thus, a hyperplane in $\mathbb{R}^n$ is the set

$$H = \{X \in \mathbb{R}^n : a^T X = \beta\}. \tag{2.1}$$

A hyperplane separates the whole space into two closed half-spaces

$$H_L = \{X \in \mathbb{R}^n : a^T X \leq \beta\}, \quad H_U = \{X \in \mathbb{R}^n : a^T X \geq \beta\}.$$

Removing H results in two disjoint open half-spaces

$$H_L' = \{X \in \mathbb{R}^n : a^T X < \beta\}, \quad H_U' = \{X \in \mathbb{R}^n : a^T X > \beta\}.$$

From (2.1), it is clear that the defining vector a of hyperplane H is orthogonal to H. Since, for any two vectors X_1 and $X_2 \in H$

$$a^T(X_1 - X_2) = a^T X_1 - a^T X_2 = \beta - \beta = 0.$$

Moreover, for each vector $X \in H$ and $W \in H_L'$,

$$a^T(W - X) = a^T W - a^T X < \beta - \beta = 0.$$

This shows that the normal vector a makes an obtuse angle with any vector that points from the hyperplane toward the interior of H_L. In

$$\{X \in \mathbb{R}^n : a^T X = \beta\}$$

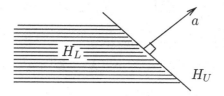

Figure 2.2

other words, a is directed toward the exterior of H_L. Fig. 2.2 illustrates the geometry.

Proposition 2. A hyperplane in $\mathbb{R}^n$ is a closed convex set.

Proof. The hyperplane in $\mathbb{R}^n$ is the set $S = \{X \in \mathbb{R}^n : a^T X = \alpha\}$. We prove that S is closed convex set. First, we show that S is closed. To do this we prove S^c is open, where

$$S^c = \{X \in \mathbb{R}^n : a^T X < \alpha\} \cup \{X \in \mathbb{R}^n : a^T X > \alpha\} = S_1 \cup S_2.$$

Let $X_0 \in S^c$. Then $X_0 \notin S$. This implies

$$a^T X_0 < \alpha \quad \text{or} \quad a^T X_0 > \alpha.$$

Suppose $a^T X_0 < \alpha$. Let $a^T X_0 = \beta < \alpha$. Define

$$N_\delta(X_0) = \left\{X \in \mathbb{R}^n : |X - X_0| < \delta,\ \delta = \frac{\alpha - \beta}{|a|}\right\}. \qquad (2.2)$$

If $X_1 \in N_\delta(X_0)$, then in view of (2.2),

$$a^T X_1 - a^T X_0 \leq |a^T X_1 - a^T X_0| = |a^T(X_1 - X_0)|$$
$$= |a^T| |X_1 - X_0| < \alpha - \beta.$$

But $a^T X_0 = \beta$. This implies $a^T X_1 < \alpha$ and hence $X_1 \in S_1$. Since X_1 is arbitrary, we conclude that $N_\delta(X_0) \subset S_1$. This implies S_1 is open.

Similarly, it can be shown that $S_2 = \{X : a^T X_0 > \alpha\}$ is open. Now, $S^c = S_1 \cup S_2$ is open (being union of open sets) which proves that S is closed.

Let $X_1, X_2 \in S$. Then $a^T X_1 = \alpha$ and $a^T X_2 = \alpha$, and consider

$$X = \beta_1 X_1 + \beta_2 X_2,\ \beta_1, \beta_2 \geq 0,\ \beta_1 + \beta_2 = 1$$

and operating a^T, note that

$$a^T X = \beta_1 a^T X_1 + \beta_2 a^T X_2 = \beta_1 \beta + \beta_2 \beta = \beta(\beta_1 + \beta_2) = \beta.$$

Thus, $X \in S$ and hence S is convex.

Proposition 3. A half-space $S = \{X \in \mathbb{R}^n : a^T X \le \alpha\}$ is a closed convex set.

Proof. Let $S = \{X \in \mathbb{R}^n : a^T X \le \alpha\}$. Suppose $X_0 \in S^c$. Then $a^T X_0 > \alpha$. Now, $a^T X_0 = \beta > \alpha$. Consider the neighbourhood $N_\delta(X_0)$ defined by

$$N_\delta(X_0) = \left\{ X \in \mathbb{R}^n : |X - X_0| < \delta, \ \delta = \frac{\beta - \alpha}{|a|} \right\}.$$

Let X_1 be an arbitrary point in $N_\delta(X_0)$. Then

$$a^T X_0 - a^T X_1 \le |a^T X_1 - a^T X_0| = |a^T| \, |X_1 - X_0| < \beta - \alpha$$

Since $a^T X_0 = \beta$, we have

$$-a^T X_1 < -\alpha \Rightarrow a^T X_1 > \alpha \Rightarrow X_1 \in S^c \Rightarrow N_\delta(X_0) \subset S^c.$$

This implies S^c is open and hence S is closed.

Take $X_1, X_2 \in S$. Hence $a^T X_1 \le \alpha$, $a^T X_2 \le \alpha$. For

$$X = \alpha_1 X_1 + \alpha_2 X_2, \ \alpha_1, \alpha_2 \ge 0, \ \alpha_1 + \alpha_2 = 1.$$

note that

$$a^T X = a^T (\alpha_1 X_1 + \alpha_2 X_2) = \alpha_1 a^T X_1 + \alpha_2 a^T X_2$$
$$\le \alpha_1 \alpha + \alpha_2 \alpha = \alpha(\alpha_1 + \alpha_2) = \alpha.$$

This implies $X \in S$, and hence S is convex.

Polyhedral set. A set formed by the intersection of finite number of closed half-spaces is termed as polyhedron or polyhedral.

If the intersection is nonempty and bounded, it is called a polytope. For a linear programme in standard form, we have m hyperplanes

$$H_i = \left\{ X \in \mathbb{R}^n : a_i^T X = b_i, \ X \ge 0, \ b_i \ge 0, \quad i = 1, 2, \ldots, m \right\},$$

where $a_i^T = (a_{i1}, a_{i2}, \ldots, a_{in})$ is the ith row of the constraint matrix A, b_i is the ith element of the right-hand vector b.

Moreover, for a linear program in standard form, the hyperplanes $H = \{X \in \mathbb{R}^n : C^T X = \beta, \ \beta \in \mathbb{R}\}$ depict the contours of the linear objective function, and cost vector C^T becomes the normal of its contour hyperplanes.

Set of feasible solutions. The set of all feasible solutions forms a feasible region, generally denoted by P_F, and this is the intersection of hyperplanes H_i, i=1,2,...,m and the first octant of $\mathbb{R}^n$.

Note that each hyperplane is intersection of two closed half-spaces H_L and H_U, and the first octant of $\mathbb{R}^n$ is the intersection of n closed half-spaces $\{x_i \in \mathbb{R} : x_i \geq 0\}$. Hence the feasible region is a polyhedral set, and is given by

$$P_F = \{X \in \mathbb{R}^n : AX = b, \ X \geq 0, \ b \geq 0\}.$$

When P_F is not empty, the linear programme is said to be consistent. For a consistent linear programme with a feasible solution $X^* \in P_F$, if $C^T X^*$ attains the minimum or maximum value of the objective function $C^T X$ over the feasible region P_F, then we say X^* is an optimal solution to the linear programme.

Moreover, we say a linear programme has a bounded feasible region, if there exists a positive constant M such that for every $X \in P_F$, we have $|X| \leq M$. On the other hand for minimization problem, if there exists a constant K such that $C^T X \geq K$ for all $X \in P_F$, then we say linear programme is bounded below. Similarly, we can define bounded linear programme for maximization problem.

Remarks. 1. In this context, it is worth mentioning that a linear programme with bounded feasible region is bounded, but the converse may not be true, i.e., a bounded LPP need not to have a bounded feasible region, see Problem 2.

2. In $\mathbb{R}^3$, a polytope has prism like shape.

Converting equalities to inequalities. To study the geometric properties of a LPP we consider the LPP in the form, where the constraints are of the type $\leq$ or $\geq$, i.e.,

$$\text{opt} \quad z = C^T X$$
$$\text{s.t.} \quad g_i(X) \leq \text{ or } \geq 0, \ i = 1, 2, \ldots, m$$
$$X \geq 0.$$

In case there is equality constraint like $x_1 + x_2 - 2x_3 = 5$, we can write this as (equivalently): $x_1 + x_2 - 2x_3 \leq 5$ and $x_1 + x_2 - 2x_3 \geq 5$.

This tells us that m equality constraints will give rise to $2m$ inequality constraints. However, we can reduce the system with m equality constraints to an equivalent system which has $m + 1$ inequality constraints.

Example 1. Show that the system having m equality constraints

$$\sum_{j=1}^{n} a_{ij}x_j = b_i, \quad i = 1, 2, \ldots, m$$

is equivalent to the system with $m + 1$ inequality constraints.

To motivate the idea, note that $x = 1$ is equivalent to the combination $x \le 1$ and $x \ge 1$. As we can check graphically, the equations $x = 1$ and $y = 2$ are equivalent to the combinations $x \le 1$, $y \le 2$, $x + y \ge 3$. The other way to write equivalent system is $x \ge 1$, $y \ge 2$, $x + y \le 3$. This idea can further be generalized to m equations. Consider the system of m equations

$$\sum_{j=1}^{n} a_{ij}x_j = b_i, \quad i = 1, 2, \ldots, m.$$

This system has the equivalent form:

$$\sum_{j=1}^{n} a_{ij}x_j \le b_i \quad \text{and} \quad \sum_{j=1}^{n} a_{ij}x_j \ge b_i$$

or

$$\sum_{j=1}^{n} a_{ij}x_j \le b_i \quad \text{and} \quad \sum_{j=1}^{n} \left(\sum_{i=1}^{m} a_{ij} \right) x_j \ge \sum_{i=1}^{m} b_i.$$

If we look at the second combination, then the above system is equivalent to

$$\sum_{j=1}^{n} a_{ij}x_j \ge b_i, \quad \text{and} \quad \sum_{j=1}^{n} \left(\sum_{i=1}^{m} a_{ij} \right) x_j \le \sum_{i=1}^{m} b_i.$$

Consider the constraints and nonnegative restrictions of a LPP in its standard form

$$x_1 + x_2 + s_1 = 1$$
$$-x_1 + 2x_2 + s_2 = 1$$
$$x_1, x_2, s_1, s_2 \ge 0.$$

Although it has four variables, the feasible reason P_F can be represented as a two dimensional graph. Write the basic variables s_1 and s_2 in terms of nonbasic variables, see Problem 28, and use the conditions $s_1 \geq 0$, $s_2 \geq 0$ to have

$$x_1 + x_2 \leq 1$$
$$- x_1 + 2x_2 \leq 1$$
$$x_1, x_2 \geq 0$$

and is shown in Fig. 2.3.

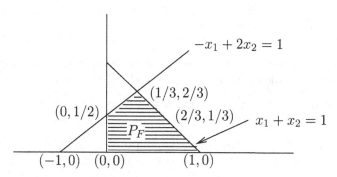

Figure 2.3

Remark. If a linear programming problem in its standard form has n variables and $n - 2$ nonredundant constraints, then the LPP has a two-dimensional representation. Why?, see Problem 28.

Proposition 4. The set of all feasible solutions of a LPP (feasible region P_F) is a closed convex set.

Proof. By definition $P_F = \{X : AX = b, X \geq 0\}$. Let X_1 and X_2 be two points of P_F. This means that $AX_1 = b$, $AX_2 = b$, $X_1 \geq 0$, $X_2 \geq 0$. Consider

$$Z = \alpha X_1 + (1 - \alpha)X_2, \ 0 \leq \alpha \leq 1.$$

Clearly, $Z \geq 0$ and $AZ = \alpha AX_1 + (1 - \alpha)AX_2 = \alpha b + (1 - \alpha)b = b$. Thus, $Z \in P_F$, and hence P_F is convex.

Remark. Note that in above proposition b may be negative, i.e., the LPP may not be in standard form. The only thing is that we have equality constraint.

Alternative proof. Each constraint $a_i^T X = b_i$, $i = 1, 2, \ldots m$ is closed (being a hyperplane), and hence intersection of these m hyperplanes $(AX = b)$ is closed. Further, each nonnegative restriction $x_i \geq 0$

is a closed (being closed half-spaces) is closed, and hence their intersection $X \geq 0$ is closed. Again, intersection $P_F = \{AX = b, X \geq 0\}$ is closed. This concludes that P_F is a closed convex set.

Clc's of two or more points. Convex linear combination of two points gives a line segment The studies on different regions need the clc of more than two points. This motivates the idea of extending the concept.

Definition 3. The point X is called a clc of m points $X_1, X_2, \ldots, X_m$ in $\mathbb{R}^n$, if there exist scalars α_i, $i = 1, 2, \ldots, m$ such that

$$X = \alpha_1 X_1 + \alpha_2 X_2 + \cdots + \alpha_m X_m, \ \alpha_i \geq 0, \ \alpha_1 + \alpha_2 + \cdots + \alpha_m = 1.$$

Remark. This definition includes clc of two points also. Henceforth, whenever we talk about clc, it means clc of two or more points.

Theorem 1. A set S is convex $\iff$ every clc of points in S belongs to S.

Proof. ($\Leftarrow$) Given that every clc of points in S belongs to S includes the assertion that every clc of two points belongs to S. Hence S is convex.

($\Rightarrow$) Suppose S is convex. we prove the result by induction. S is convex $\Rightarrow$ clc of every two points in S belongs to S. Hence the proposition is true for clc of two points. Assume that proposition is true for clc of n points. we must show that it is true for $n + 1$ points. Consider

$$X = \beta_1 X_1 + \beta_2 X_2 + \cdots + \beta_n X_n + \beta_{n+1} X_{n+1}$$

such that

$$\beta_1 + \beta_2 + \cdots + \beta_{n+1} = 1, \ \beta_i \geq 0.$$

If $\beta_{n+1} = 0$, then X, being clc of n points belongs to S (by assumption). If $\beta_{n+1} = 1$, then $\beta_1 = \beta_2 = \cdots = \beta_n = 0$ and $X = 1.X_{n+1}$, the proposition is trivially true. Assume, $\beta_{n+1} \neq 0$ or 1, i.e., $0 < \beta_{n+1} < 1$. Now, $\beta_1 + \beta_2 + \cdots + \beta_n \neq 0$.

$$X = \frac{\beta_1 + \beta_2 + \cdots + \beta_n}{\beta_1 + \beta_2 + \cdots + \beta_n} (\beta_1 X_1 + \beta_2 X_2 + \cdots + \beta_n X_n) + \beta_{n+1} X_{n+1}$$

or

$$X = (\beta_1 + \beta_2 + \cdots + \beta_n)(\alpha_1 X_1 + \alpha_2 X_2 + \cdots + \alpha_n X_n) + \beta_{n+1} X_{n+1},$$

where, $\alpha_i = \beta_i/(\beta_1 + \beta_2 + \cdots + \beta_n)$, $i = 1, 2, \ldots, n$. Clearly, $\alpha_i \geq 0$ and $\alpha_1 + \alpha_2 + \cdots + \alpha_n = 1$. Hence, by assumption $\alpha_1 X_1 + \alpha_2 X_2 + \cdots + \alpha_n X_n = Y$ (say) belongs to S. Again,

$$X = (\beta_1 + \beta_2 + \cdots + \beta_n)Y + \beta_{n+1}X_{n+1}$$

such that

$$\beta_1 + \beta_2 + \cdots + \beta_n \geq 0, \ \beta_{n+1} \geq 0, \ \sum_{i=1}^{n+1} \beta_i = 1.$$

Thus, X is the clc of two points and hence belongs to S.

Convex hull. Let S be a nonempty set. Then convex hull of S, denoted by $[S]$ is defined as all clc's of points of S,

$$[S] = \{X \in \mathbb{R}^n : X \text{ is clc of points in } S\}.$$

Remarks. 1. By convention $[\emptyset] = \{0\}$.

2. The above discussion reveals that the convex hull of finite number of points $X_1, X_2, \ldots, X_m$ is the convex combination of the m points. This is the convex set having at most m vertices. Here, at most means some points may be interior points. Moreover, convex hull generated in this way is a closed convex set.

3. The convex hull of m points is given a special name as convex polyhedron.

Theorem 2. Let S be a nonempty set. Then the convex hull $[S]$ is the smallest convex set containing S.

Proof. Let $X, Y \in [S]$. Then

$$X = \alpha_1 X_1 + \alpha_2 X_2 + \cdots + \alpha_n X_n, \ \alpha_i \geq 0, \ \sum_{i=1}^{n} \alpha_i = 1,$$

$$Y = \beta_1 Y_1 + \beta_2 Y_2 + \cdots + \beta_m Y_m, \ \beta_j \geq 0, \ \sum_{j=1}^{m} \beta_j = 1.$$

Consider the linear combination $\alpha X + \beta Y$, $\alpha, \beta \geq 0$, $\alpha + \beta = 1$, and note that

$$\alpha X + \beta Y = \alpha(\alpha_1 X_1 + \alpha_2 X_2 + \cdots + \alpha_n X_n)$$

$$+ \beta(\beta_1 Y_1 + \beta_2 Y_2 + \cdots + \beta_m Y_m)$$

$$= (\alpha\alpha_1)X_1 + \cdots + (\alpha\alpha_n)X_n + (\beta\beta_1)Y_1 + \cdots + (\beta\beta_m)Y_m.$$

Now, each $\alpha\alpha_i \geq 0$, $\beta\beta_j \geq 0$ and

$$\sum_{i=1}^{n}(\alpha\alpha_i) + \sum_{j=1}^{m}(\beta\beta_j) = \alpha\sum_{i=1}^{n}\alpha_i + \beta\sum_{j=1}^{m}\beta_j = \alpha + \beta = 1,$$

i.e., $\alpha X + \beta Y$ is a clc of points $X_1, X_2, \ldots, X_n, Y_1, Y_2, \ldots, Y_m$. This implies that $\alpha X + \beta Y \in [S]$ and hence $[S]$ is convex.

Clearly, it contains S because each $X \in S$ can be written as $X = 1.X + 0.Y$, i.e., clc of itself. To prove that $[S]$ is the smallest convex set containing S, we show that if there exists another convex set T containing S, then $[S] \subset T$.

Suppose T is a convex set which contains S. Take any element $X \in [S]$. Then

$$X = \alpha_1 X_1 + \alpha_2 X_2 + \cdots + \alpha_n X_n, \; \alpha_i \geq 0, \; \sum_{i=1}^{n}\alpha_i = 1 \; \forall \; X_1, X_2, \ldots, X_n \in S.$$

Since $S \subset T$, it follows that $X_1, X_2, \ldots, X_n \in T$ and, moreover convexity of T ensures that

$$\alpha_1 X_1 + \alpha_2 X_2 + \cdots + \alpha_n X_n = X \in T.$$

Hence $[S] \subset T$.

Remark. If S is convex, then $S = [S]$.

For convex set $S \subset \mathbb{R}^n$, a key geometric figure is due to the following separation theorem. The proof is beyond the scope of the book.

Theorem 3(Separation Theorem). Let $S \subset R^n$ and X be a boundary point of S. Then there is a hyperplane H containing X with S contained either in lower half-plane or upper half-plane.

Based on this theorem we can define a supporting hyperplane H to be the hyperplane such that (i) the intersection of H and S is nonempty; (ii) lower half-plane contains S, see Fig. 2.4.

One very important fact to point out here is that the intersection set of the polyhedral set and the supporting hyperplane with negative cost vector C^T as its normal provides optimal solution of a LPP. This is the key idea of solving solving linear programming problems by the graphical method.

To verify this fact, let us take

$$\min x_0 = -x_1 - 2x_2$$

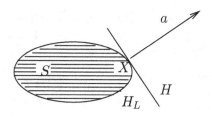

Figure 2.4

as the objective function for the LPP whose feasible region is shown in Fig. 2.3. Note that $-x_1 - 2x_2 = -80$ is the hyperplane passing through $(0, 40)$ and the vector $-C^T = (1, 2)$ is normal to this plane. This is a supporting hyperplane passing through $(0, 40)$, since $H_L = \{(x_1, x_2) : x_1 + 2x_2 \leq 80\}$ contains P_F and is satisfied by the points $(20, 20)$ and $(30, 0)$.

However, the hyperplane passing through $(20, 20)$ which is normal to $-C^T = (1, 2)$ is given by $-x_1 - 2x_2 = -60$. This is not a supporting hyperplane as point $(0, 40)$ is not in $\{(x_1, x_2) : x_1 + 2x_2 \leq 60\}$. Similarly it can be shown that hyperplane at $(3, 0)$ which is normal to $-C^T$ is also not a supporting hyperplane. This implies that $x_1 = 0, x_2 = 40$ is the optimal solution.

2.2 Extreme Points and Basic Feasible Solutions

Definition 1. A point X of a convex set S is said to be an extreme (vertex) of S if X is not a clc of any other two distinct points of S, i.e., X can not be expressed as

$$X = \alpha_1 X_1 + \alpha_2 X_2, \ \alpha_1, \alpha_2 > 0, \ \alpha_1 + \alpha_2 = 1.$$

In other words, a vertex is a point that does not lie strictly within the line segment connecting two other points of the convex set.

From the pictures of convex polyhedron sets, especially in lower dimensional spaces it is clear to see the vertices of a convex polyhedron, Analyze the set P_F as depicted in Fig. 2.3. Further, we note the following observations:

(a) $A(0, 0), B(1, 0), C(1/3, 2/3)$ and $D(0, 1/2)$ are vertices and moreover, these are boundary points also. But every boundary point need not be a vertex. In Fig. 2.3, as E is boundary point of the feasible region P_F but not vertex, since it can be written as clc of distinct points of the set P_F as

$$\left(\frac{2}{3},\frac{1}{3}\right) = \left(\frac{1}{2}\right)(1,0) + \left(\frac{1}{2}\right)\left(\frac{1}{3},\frac{2}{3}\right).$$

(b) All boundary points of $\{(x,y) : x^2+y^2 \leq 9\}$ are vertices. Hence, vertices of a bounded closed set may be infinite. However, in a LPP, if P_F is bounded and closed, then it contains finite number of vertices, see Proposition 5.

(c) Needless to mention whenever we talk about vertex of a set S, it is implied that S is convex.

(d) If S is unbounded, then it may not have a vertex, e.g., , $S = \mathbb{R}^2$.

(e) If S is not closed, then it may not have vertex, e.g., , $S = \{(x,y) : 0 < x < 1, 2 < y < 3\}$ has no vertex.

Remark. Here extreme points are in reference to convex sets. However, extreme points of a function will be defined in Chapter 13.

To characterize the vertices of a feasible region $P_F = \{X \in \mathbb{R}^n : AX = b,\ X \geq 0\}$ of a given LPP in standard form, we may assume A is an $m \times n$ matrix with $m < n$ and also denote the jth column of the coefficient matrix A by A_j, $j = 1, 2, \ldots, n$. Then, for each $X = (x_1, x_2, \ldots, x_n)^T \in P_F$, we have

$$x_1 A_1 + x_2 A_2 + \cdots + x_n A_n = b.$$

Therefore, A_j is the column of A corresponding to the jth component x_j of X.

Theorem 4. A point X of feasible region P_F is a vertex of P_F $\Longleftrightarrow$ the columns of A corresponding to the positive components of X are linearly independent.

Proof. Without loss of generality, we may assume that the components of X are zero except for the first p components, namely

$$X = \begin{bmatrix} \overline{X} \\ \hline 0 \end{bmatrix}, \quad \overline{X} = (x_1, x_2, \ldots, x_p)^T > 0.$$

We also denote the first p columns of matrix A by $\overline{A}$. Hence $AX = \overline{A}\,\overline{X} = b$.

($\Rightarrow$) Suppose that the columns of $\overline{A}$ are not linearly independent, then there exists nonzero vector $\overline{w}$ (at least one of the p components is nonzero) such that $\overline{A}\overline{w} = 0$. Now, define

$$Y = \overline{X} + \delta\overline{w} \quad \text{and} \quad Z = \overline{X} - \delta\overline{w}$$

For sufficiently small $\delta > 0$, we note that $Y, Z \geq 0$ and

$$\overline{A}Y = \overline{A}Z = \overline{A}\,\overline{X} = b.$$

We further define

$$Y_1 = \begin{bmatrix} \overline{Y} \\ 0 \end{bmatrix} \quad \text{and} \quad Z_1 = \begin{bmatrix} \overline{Z} \\ 0 \end{bmatrix}$$

Note that $Y_1, Z_1 \in P_F$ and $X = (1/2)Y_1 + (1/2)Z_1$. In other words X is not an vertex of P_F.

($\Leftarrow$) Suppose that X is not vertex of P_F, then $X = \alpha Y_1 + (1 - \alpha)Z_1$, $\alpha \geq 0$ for some distinct $Y_1, Z_1 \in P_F$. Since $Y_1, Z_1 \geq 0$ and $0 \leq \alpha \leq 1$, the last $n - p$ components of Y_1 must be zero, as

$$0 = \alpha y_j + (1 - \alpha)z_j, \quad j = p+1, p+2, \ldots, n.$$

Consequently, we have a nonzero vector $w = X - Y_1$ ($X \neq Y_1$) such that

$$\overline{A}w = Aw = AX - AY_1 = b - b = 0.$$

This shows that columns of $\overline{A}$ are linearly dependent.

Consider the LPP in a standard form, $AX = b$, suppose we have, $n = $ number of unknowns; $m = $ number of equations. Assume $m < n$ (otherwise the problem is over-specified). However, after introducing the slack and surplus variables, generally this assumption remains valid. Let $r(A)$ and $r(A, b)$ be the ranks of matrix A and augmented matrix (A, b), respectively.

(i) $r(A) = r(A, b)$ guarantees the consistency, i.e., $AX = b$ has at least one solution.

(ii) $r(A) \neq r(A, b)$ the system is inconsistent,, i.e., $AX = b$ has no solution. For example

$$x_1 + x_2 + x_3 = 1$$
$$4x_1 + 2x_2 - x_3 = 5$$
$$9x_1 + 5x_2 - x_3 = 11$$

has no solution and hence inconsistent.

For consistent systems we have difficulty when $r(A) = r(A, b) < m =$ number of equations. It means that all m rows are not linearly independent. Hence, any row may be written as linear combination of other rows. We consider this row (constraint) as redundant. For example

$$x_1 - x_2 + 2x_3 = 4$$
$$2x_1 + x_2 - x_3 = 3$$
$$5x_1 + x_2 = 10$$
$$x_1, x_2, s_1, s_2 \geq 0$$

In this example $r(A) = r(A, b) = 2$ which is less than number of equations. The third constraint is the sum of the first constraint and two times of the second constraint. Hence, the third constraint is redundant.

Another type of redundancy happens when $r(A) = r(A, b)$, but some of the constraint does not contribute any thing to find the optimal solution. However, in this case $r(A) = r(A, b) = m$ number of equations. Such type of cases we shall deal in Chapter 3. The following simple example illustrates the fact.

$$x_1 + x_2 + s_1 = 1$$
$$x_1 + 2x_2 + s_2 = 2$$
$$x_1, x_2, s_1, s_2 \geq 0$$

Here $r(A) = r(A, b) = 2$. The vertices of the feasible region are $(0, 0)$, $(1, 0)$ and $(0, 1)$. The second constraint is redundant as it contributes only $(0, 1)$ which is already given by the first constraint.

It is advisable to delete redundant constraints, if any, before a LPP is solved to find its optimal solution, otherwise computational difficulty may arise.

Basic solutions. Let $AX = b$ be a system of m simultaneous linear equations in n unknowns $(m < n)$ with $r(A) = m$. This means that that there exist m linearly independent column vectors. In this case group these linearly independent column vectors to form a basis B and leave the remaining $n - m$ columns as nonbasis N. In other words, we can rearrange $A = [B|N]$.

We can also rearrange the components of any *solution* vector X in the corresponding order, namely

$$X = \left[\frac{X_B}{X_N}\right]$$

For a component in X_B, its corresponding column is in the basis B. Similarly, components in X_N correspond to nonbasis matrix N.

If all $n - m$ variables X_N which are not associated with columns of B are equated to zero, then the solution of the resulting system $BX_B = b$ is called a basic solution of $AX = b$. Out of n columns, m columns can be selected in $n!/m!(n-m)!$ ways. The m variables (left after putting $n - m$ variables equal to zero) are called basic variables and remaining $n - m$ variables as nonbasic variables. The matrix corresponding to basic variables is termed as the basis matrix.

Basic feasible solution. A basic solution with nonnegative restrictions is called a basic feasible solution.

Nondegenerate BFS. If all the m basic variables in a BFS are positive than it is called nondegenerate basic feasible solution.

The following result is a direct consequences of Theorem 4.

Corollary 1. A point $X \in P_F$ is an vertex of $P_F \iff X$ is a basic feasible solution corresponding to some basis B.

Proof. By Theorem 4, we have

$$X \in P_F \text{ is vertex} \iff \text{columns } A_i \text{ for } x_i > 0 \ (i = 1 \text{ to } m) \text{ are}$$
$$\text{linearly independent}$$
$$\iff B = [A_1, A_2, \ldots, A_m] \text{ is a nonsingular}$$
$$\text{matrix of } X$$
$$\iff X \text{ is a basic feasible solution.}$$

Degenerate BFS. A BFS which is not nondegenerate is called degenerate basic feasible solution, i.e., at least one of the basic variable is at zero level in the BFS.

Remarks. 1. This corollary reveals that there exists a one-one correspondence between the set of basic feasible solutions and set of vertices of P_F only in the absence of degeneracy. Actually, in case degeneracy exists, then a vertex may correspond to many degenerate basic feasible solutions. The following examples will make the remark more clear and justified.

2. When we select m variables out of n variables to define a basic solution it is must that the matrix B formed with the coefficients of m variables must be nonsingular, otherwise we may get no solution or infinity of solutions (which are not basic), see Problem 11, Problem set 2.

Example 2. Without sketching P_F find the vertices for the system

$$-x_1 + x_2 \leq 1$$
$$2x_1 + x_2 \leq 2$$
$$x_1, x_2 \geq 0.$$

First, write the system in standard form

$$-x_1 + x_2 + s_1 = 1$$
$$2x_1 + x_2 + s_2 = 2.$$

Here $n = 4, m = 2 \Rightarrow 4!/2!2! = 6$ as basic solutions. To find all basic solutions we take any of the two variables as basic variables from the set $\{x_1, x_2, s_1, s_2\}$ to have

$$\left(\frac{1}{3}, \frac{4}{3}, 0, 0\right), \ (1, 0, 2, 0), \ (-1, 0, 0, 4), \ (0, 2, -1, 0), \ (0, 1, 0, 1), \ (0, 0, 1, 2).$$

The system has 6 basic solutions, and out of these 4 are basic feasible solutions. The solution set is nondegenerate and hence there exists one-one correspondence between BFS and vertices, i.e., the feasible region P_F has 4 vertices.

Note that $(-1, 0, 0, 4)$ and $(0, 2, -1, 0)$ are basic solutions but not feasible.

Example 3. The system $AX = b, \ X \geq 0$ is given by

$$x_1 + x_2 - 8x_3 + 3x_4 = 2$$
$$-x_1 + x_2 + x_3 - 2x_4 = 2.$$

Find (i) a nonbasic feasible solution; (ii) a basic solution which is not feasible; (iii) a vertex which corresponds to two different basic feasible solutions.

Here $X = (x_1, x_2, x_3, x_4)^T$, $b = (2, -2)^T$ and the coefficient matrix A and the augmented matrix (A, b) are written as

$$A = \begin{bmatrix} 1 & 1 & -8 & 3 \\ -1 & 1 & 1 & -2 \end{bmatrix}; \quad (A, b) = \begin{bmatrix} 1 & 1 & -8 & 3 & 2 \\ -1 & 1 & 1 & -2 & 2 \end{bmatrix}$$

We make the following observations:

(i) Since $r(A) = r(A, b) = 2 <$ number of unknowns, the system is consistent and has infinity of solutions with two degrees of freedom. Reduction of the matrix (A, b) in row reduced echelon yields

$$\begin{bmatrix} 1 & 0 & -9/2 & 5/2 & 0 \\ 0 & 1 & -7/2 & -1/2 & 2 \end{bmatrix}$$

This gives

$$\begin{aligned} x_1 &= \tfrac{9}{2}x_3 - \tfrac{5}{2}x_4 \\ x_2 &= \tfrac{7}{2}x_3 - \tfrac{1}{2}x_4 + 2 \end{aligned} \tag{2.3}$$

Let us assign $x_3 = x_4 = 2$, i.e., nonzero values to x_3 and x_4 to have $x_1 = 4$ and $x_2 = 8$. Thus, one of the feasible solution is $(4, 8, 2, 2)$ but not a basic feasible solution, since at least two variables must be at zero level.

(ii) If x_1 and x_4 are chosen basic variables, then the system reduces to

$$x_1 + 3x_4 = 2$$

$$-x_1 - 2x_4 = 2$$

with solution $x_1 = -10$ and $x_4 = 4$. Thus, $(-10, 0, 0, 4)$ is basic solution which is not feasible. This can also be obtained by pivoting at x_1 and x_4 to get row reduced echelon form.

(iii) Further, for x_1, x_2 as basic variables,

$$(0, 2, 0, 0) \text{ is a BFS with basis matrix } \begin{bmatrix} 1 & 1 \\ -1 & 1 \end{bmatrix}$$

while, for x_2, x_4 as basic variables,

$$(0, 2, 0, 0) \text{ is a BFS with basis matrix } \begin{bmatrix} 1 & 3 \\ 1 & -2 \end{bmatrix}$$

Both BFS given by $(0, 2, 0, 0)$ seem to be same but these are different as their basis matrices are different. However, both BFS correspond to the same vertex $(0, 2)$. The vertex $(0, 2)$ can be identified by writing the equivalent form of LPP as two-dimensional graph in $x_1 - x_2$ plane.

Remarks. 1. Note that system (2.3) can be written as

$$1x_1 + 0x_2 - \tfrac{9}{2}x_3 + \tfrac{5}{2}x_4 = 0$$

$$0x_1 + 1x_2 - \tfrac{7}{2}x_3 + \tfrac{1}{2}x_4 = 2$$

This is defined as canonical form of constraint equations. Further, note that

$$\begin{bmatrix} -8 \\ 1 \end{bmatrix} = -\frac{9}{2}\begin{bmatrix} 1 \\ 1 \end{bmatrix} - \frac{7}{2}\begin{bmatrix} 1 \\ 1 \end{bmatrix}$$

This extracts a good inference that if x_1 and x_2 are basic variables and are pivoted, then coefficients of x_3 and x_4 are coordinate vectors of A_3 and A_4 column vectors of A with respect to the basic vectors A_1 and A_2. This phenomenon will be used in simplex method to be discussed in next chapter.

2. All basic solutions can also be obtained by pivoting at any two variables and assigning zero value to the remaining variables. This can be done in 6 ways and hence will give six different basic solutions.

3. For the existence of all basic solutions it is necessary that column vectors of the coefficient matrix A must be linearly independent. It is also possible that after keeping requisite number of variables at zero level the remaining system may have infinity of solutions, see Problem 12(c). This happens when at least two columns are not linearly independent. In Problem 11, A_3 and A_4 are not linearly independent. We do not term these infinity solutions as as basic solutions because for a basis solution the coefficient matrix formed with the coefficients of basic variables (in order) must be nonsingular.

Proposition 5. The set of all feasible solutions P_F of a LPP has finite number of vertices.

Proof. Let P_F be the set of all feasible solutions of the LPP, where constraints are written in standard form, $AX = b, X \geq 0$. If rank of A is m, then the system $AX = b$ has

$$\binom{n}{m} = \frac{n!}{m!(n-m)!}$$

basic solutions. Again, by note the fact that every basic feasible solution is a basic solution.

With inclusion of $X \geq 0$, we can say $AX = b$ has at most $n!/m!(n-m)!$ basic feasible solutions. As there exists one to one correspondence between BFS set and set of all vertices, provided nondegeneracy persists in the problem we conclude that set of vertices may have at most $n!/m!(n-m)!$ elements and hence number of vertices is finite.

In case of degeneracy more than one BFS may correspond to the same vertex. Hence, in this situation number of vertices will be less than $n!/m!(n-m)!$, and again the vertices are finite.

Note that two basic feasible solutions (vertices) are adjacent, if they use $m-1$ basic variables in common to form basis. For example, in Figure 2.3, it is easy to verify that $(0, 1/2)$ is adjacent to $(0, 0)$ but not adjacent to $(1, 0)$ since $(0, 1/2)$ takes x_2 and s_1 as basic variables, while $(0, 0)$ takes s_1 and s_2 and $(1, 0)$ takes x_1 and s_2. Under the nondegeneracy assumption, since each of the $n-m$ nonbasic variables could replace one current basic variable in a given basic feasible solution, we know that every BFS (vertex) has $n-m$ neighbours. Actually, each BFS can be reached by increasing the value of one nonbasic from zero to positive and decreasing the value of one basic variable from positive to zero. This is the basic concept of pivoting in simplex method to be discussed in next chapter.

Suppose the feasible region P_F is bounded, in other words it is a polytope. From Fig. 2.5 it is easy to observe that each point of P_F can be represented as a convex combination of finite number of vertices of P_F. From the figure it is clear that X can be written as convex linear combination of the vertices X_1, X_3, X_4.

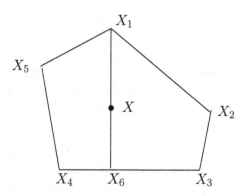

Figure 2.5

This idea of convex resolution can be verified for a general polyhedron (may be unbounded) with the help of following definition:

Definition 2. An extremal direction of a polyhedron set is a nonzero vector $d \in \mathbb{R}^n$ such that for each $X_0 \in P_F$, the ray $\{X \in \mathbb{R}^n : X = X_0 + \alpha d, \ \alpha \geq 0\}$ is contained in P_F.

Remark. From the definition of feasible region, we see that a nonzero vector $d \in \mathbb{R}^n$ is an extremal direction of $P_F \iff Ad = 0$ and $d \geq 0$. Also, P_F is unbounded $\iff P_F$ has an extremal direction.

Using vertices and extremal directions every point in P_F can be represented by the following useful result known as the resolution theorem.

Theorem 5 (Resolution theorem). Let $B = \{V_i \in \mathbb{R}^n : i \in \mathbb{Z}\}$ be the set of all vertices of P_F with a finite index set $\mathbb{Z}$. Then, for each $X \in P_F$, we have

$$X = \sum_{i \in B} \alpha_i V_i + d, \quad \sum_{i \in \mathbb{Z}} \alpha_i = 1, \quad \alpha_i \geq 0, \tag{2.4}$$

where d is either the zero vector or an extremal direction of P_F.

Proof. To prove the theorem by the induction, we let p be the number of positive components of $X \in P_F$. When $p = 0$, $X = (0, 0, \ldots, 0)$ is obviously a vertex. Assume that the theorem holds for $p = 0, 1, \ldots, k$ and X has $k+1$ positive components. If X is a vertex there is nothing to prove. If X is not a vertex, we let $X^T = (x_1, x_2, \ldots, x_{k+1}, 0, \ldots, 0) \in \mathbb{R}^n$ such that $(x_1, x_2, \ldots, x_{k+1}) > 0$ and $A = [\overline{A}|N]$, $\overline{A}$ is the matrix corresponding to positive components of X. Then, by Theorem 4, the columns of $\overline{A}$ are linearly independent, in other words there exists a nonzero vector $\overline{w} \in \mathbb{R}^{k+1}$ such that $\overline{A}\overline{w} = 0$. We define $w = (\overline{w}, 0, \ldots, 0) \in \mathbb{R}^n$, then $w \neq 0$ and $Aw = \overline{A}\overline{w} = 0$. There are three possibilities: $w \geq 0$, $w < 0$ and w has both positive and negative components. For $w \geq 0$, consider $X(\theta) = X + \theta w$ and pick θ^* to be the largest value of θ such that $X^* = X(\theta^*)$ has at least one more zero component than X. Then follow the induction hypothesis to show that theorem holds. Similarly, show that in the remaining two cases, the theorem still holds.

The direct consequences of the resolution theorem are:

Corollary 2. If P_F is a bounded feasible region (polytope), then each point $X \in P_F$ is a convex linear combination of its vertices.

Proof. Since P_F is bounded, by the remark following Definition 2 the extremal direction d is zero, and application of (2.4) ensures that X is a clc of vertices.

Corollary 3. If P_F is nonempty, then it has at least one vertex.

Example 4. Consider the set P_F : $x_1 + x_2 \leq 1$, $-x_1 + 2x_2 \leq 1$, $x_1, x_2 \geq 0$ and show that a point of P_F may be clc of different vertices.

Take the point $(1/3, 1/6) \in P_F$. Now

$$\left(\frac{1}{3}, \frac{1}{6}\right) = \frac{1}{3}(0,0) + \frac{1}{3}(1,0) + \frac{1}{3}\left(0, \frac{1}{2}\right) + 0\left(\frac{1}{3}, \frac{2}{3}\right)$$

or,

$$\left(\frac{1}{3}, \frac{1}{6}\right) = \frac{3}{8}(0,0) + \frac{5}{16}(1,0) + \frac{1}{16}\left(\frac{1}{3}, \frac{2}{3}\right) + \frac{1}{4}\left(0, \frac{1}{2}\right).$$

Thus, an additional information is that a point of P_F may have different clc's of its vertices.

2.3 Fundamental Theorem of Linear Programming

Theorem 6. The maximum of the objective function f(X) of a LPP occurs at least at one vertex of P_F, provided P_F is bounded.

Proof. Given that the LPP is a maximization problem. Suppose that maximum of $f(X)$ occurs at some point X_0 in feasible region P_F. Thus,

$$f(X) \leq f(X_0) \ \forall \ X \in P_F.$$

We show that this X_0 is nothing but some vertex of of P_F. Since P_F is bounded and problem is LPP, it contains finite number of vertices $X_1, X_2, \ldots, X_n$. Hence,

$$f(X_i) \leq f(X_0), \ i = 1, 2, \ldots, n. \tag{2.5}$$

By Corollary 2,

$$X_0 = \alpha_1 X_1 + \alpha_2 X_2 + \cdots + \alpha_n X_n, \ \alpha_i \geq 0, \ \sum_{i=1}^{n} \alpha_i = 1.$$

Using linearity of f, we have

$$f(X_0) = \alpha_1 f(X_1) + \alpha_2 f(X_2) + \cdots + \alpha_n f(X_n).$$

Let
$$f(X_k) = \max \{f(X_1), f(X_2), \ldots, f(X_n)\},$$
where $f(X_k)$ is one of the values of $f(X_1), f(X_2), \ldots, f(X_n)$. Then

$$f(X_0) \leq \alpha_1 f(X_k) + \alpha_2 f(X_k) + \cdots + \alpha_n f(X_k) = f(X_k). \qquad (2.6)$$

Combining (2.5) and (2.6), we have $f(X_0) = f(X_k)$. This implies that X_0 is the vertex X_k and hence the result.

The minimization case can be treated on parallel lines just by reversing the inequalities. Thus, we have proved that the optimum of a LPP occurs at some vertex of P_F, provided P_F is bounded.

Remark. Theorem 6 does not rule out the possibility of having an optimal solution at a point which is not vertex. It simply says among all optimal solutions to a LPP at least one of them is a vertex. The following proposition further strengthens Theorem 6.

Proposition 6. In a LPP, if the objective function $f(X)$ attains its maximum at an interior point of P_F, then f is constant, provided P_F is bounded.

Proof. Given that the problem is maximization, and let X_0 be an interior point of P_F, where maximum occurs, i.e.,

$$f(X) \leq f(X_0) \ \forall \ X \in P_F.$$

Assume contrary that $F(X)$ is not constant. Thus, we have $X_1 \in P_F$ such that
$$f(X_1) \neq f(X_0), \ \ f(X_1) < f(X_0).$$

Since P_F is nonempty bounded closed convex set, it follows that X_0 can be written as a clc of two points X_1 and X_2 of P_F

$$X_0 = \alpha X_1 + (1 - \alpha)X_2, \ 0 < \alpha < 1.$$

Using linearity of f, we get

$$f(X_0) = \alpha f(X_1) + (1 - \alpha)f(X_2) \Rightarrow f(X_0) < \alpha f(X_0) + (1 - \alpha)f(X_2).$$

Thus, $f(X_0) < f(X_2)$. This is a contradiction and hence the theorem.

2.4   Graphical Method

This method is convenient in case of two variables. By Theorem 6 (see also Problem 13), the optimum value of the objective function occurs at one of the vertices of P_F. We exploit this result to find an optimal solution of any LPP. First we sketch the feasible region and identify its vertices. Compute value of the objective function at each vertex, and take largest of these values to decide optimal value of the objective function, and the vertex at which this largest value occurs is the optimal solution. For the minimization problem we consider the smallest value.

Example 5. Solve the following LPP by the graphical method

$$\text{max} \quad z = x_1 + 5x_2$$
$$\text{s.t.} \quad -x_1 + 3x_2 \leq 10$$
$$x_1 + x_2 \leq 6$$
$$x_1 - x_2 \leq 2$$
$$x_1, x_2 \geq 0.$$

Rewrite each constraint in the forms:

$$-\frac{x_1}{10} + \frac{x_2}{10/3} \leq 1$$
$$\frac{x_1}{6} + \frac{x_2}{6} \leq 1$$
$$\frac{x_1}{2} - \frac{x_2}{2} \leq 1$$

Draw the each constraint first by treating as linear equation. Then use the inequality condition to decide the feasible region. The feasible region and vertices are shown in Fig. 2.6.

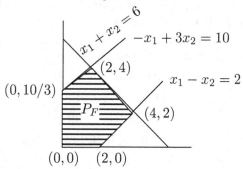

Figure 2.6

The vertices are $(0,0)$, $(2,0)$, $(4,2)$, $(2,4)$, $(0,10/3)$. The values of the objective function is computed at these points are

$$z = 0 \qquad \text{at} \quad (0,0)$$
$$z = 2 \qquad \text{at} \quad (2,0)$$
$$z = 14 \qquad \text{at} \quad (4,2)$$
$$z = 22 \qquad \text{at} \quad (2,4)$$
$$z = 50/3 \quad \text{at} \quad (0,10/3)$$

Obviously, the maximum occurs at vertex $(2,4)$ with maximum value 22. Hence,

$$\text{optimal solution: } x_1 = 2, \; x_2 = 4, \quad z = 22.$$

Example 6. A machine component requires a drill machine operation followed by welding and assembly into a larger subassembly. Two versions of the component are produced: one for ordinary service and other for heavy-duty operation. A single unit of the ordinary design requires 10 min of drill machine time, 5 min of seam welding, and 15 min for assembly. The profit for each unit is \$100. Each heavy-duty unit requires 5 min of screw machine time, 15 min for welding and 5 min for assembly. The profit for each unit is \$150. The total capacity of the machine shop is 1500 min; that of the welding shop is 1000 min; that of assembly is 2000 min. What is the optimum mix between ordinary service and heavy-duty components to maximize the total profit?

Let x_1 and x_2 be number of ordinary service and heavy-duty components. The LPP formulation is

$$\begin{aligned}
\max \quad & z = 100x_1 + 150x_2 \\
\text{s.t.} \quad & 10x_1 + 5x_2 \leq 1500 \\
& 5x_1 + 15x_2 \leq 1000 \\
& 15x_1 + 5x_2 \leq 2000 \\
& x_1, x_2 \geq 0 \text{ and are integers.}
\end{aligned}$$

Draw the feasible region by taking all constraints in the format as given in Example 5 and determine all the vertices. The vertices are $(0,0)$, $(400/3, 0)$, $(125, 25)$, $(0, 200/3)$. The optimal solution exists at the vertex $x_1 = 125$, $x_2 = 25$ and the maximum value: $z = 16250$.

Problem Set 2

1. Which of the following sets are convex

   (a) $\{(x_1, x_2) : x_1 x_2 \leq 1\}$;

   (b) $\{(x_1, x_2) : x_1^2 + x_2^2 < 1\}$;

   (c) $\{(x_1, x_2) : x_1^2 + x_2^2 \geq 3\}$;

   (d) $\{(x_1, x_2) : 4x_1 \geq x_2^2\}$;

   (e) $\{(x_1, x_2) : 0 < x_1^2 + x_2^2 \leq 4\}$;

   (f) $\{(x_1, x_2) : x_2 - 3 \geq -x_1^2,$
   $$x_1, x_2 \geq 0\}$$

2. Prove that a linear program with bounded feasible region must be bounded, and give a counterexample to show that the converse need not be true.

3. Prove that arbitrary intersection of convex sets is convex.

4. Prove that the half-space $\{X \in \mathbb{R}^n : a^T X \geq \alpha\}$ is a closed convex set.

5. Show that the convex set in $\mathbb{R}^n$ satisfy the following relations:
   (a) If S is a convex set and β is a real number, the set

   $$\beta S = \{\beta X : X \in S\}$$

   is convex;

   (b) If S_1 and S_2 are convex sets in $\mathbb{R}^n$, then the set

   $$S_1 \pm S_2 = \{X_1 \pm X_2 : X_1 \in S_1, \ X_2 \in S_2\}$$

   is convex.

6. A point X_v in S is a vertex of $S \iff S \setminus \{X_v\}$ is convex.

7. Write the system

   $$x_1 + x_2 = 1$$
   $$2x_1 - 4x_3 = -5$$

   into its equivalent system which contains only three inequality constraints.

8. Define the convex hull of a set S as

$$[S] = \{\cap_{i\in\Lambda}A_i : A_i \supset S \text{ and } A_i \text{ is convex}\}.$$

Show that this definition and the definition of convex hull in Section 2.1 are equivalent.

9. Using the definition of convex hull in Problem 7, show that $[S]$ is the smallest convex set containing S.

10. Find the convex hull of the following sets

(a) $\{(1,1),(1,2),(2,0),(0,-1)\}$;   (b) $\{(x_1,x_2) : x_1^2 + x_2^2 > 3\}$;

(c) $\{(x_1,x_2) : x_1^2 + x_2^2 = 1\}$;   (d) $\{(0,0),(1,0),(0,1),(1,1)\}$.

11. Prove that convex linear combinations of finite number of points is a closed convex set.

Suggestion. For convexity see Theorem 1.

12. Consider the following constraints of a LPP written in standard form:

$$x_1 + x_2 + 4x_3 + 2x_4 + 3x_5 = 8$$

$$4x_1 + 2x_2 + 2x_3 + x_4 + 6x_5 = 4$$

$$x_1, x_2, x_3, x_4 \geq 0$$

Identify (a) a basic feasible solution; (b) basic infeasible solution; (c) infinity of solutions; (d) nonexisting solution (infeasible solution).

13. Find all vertices in which at least x_1 or x_2 is positive from the set of all feasible solutions of the system

$$x_1 + x_2 + x_3 \geq 3$$

$$-2x_1 + 2x_2 - x_3 \leq 2$$

$$x_1, x_2, x_3 \geq 0$$

14. Use the resolution theorem to prove the following generalization of Theorem 6.

For a consistent linear program in its standard form with a feasible region P_F, the maximum objective value of $z = C^T X$ over P_F is either unbounded or is achievable at least at one vertex of P_F.

15. **Prove** Theorem 6 and Proposition 6 for the minimization case.

16. **Prove** that if optimal value of a LPP occurs at more than one vertex of P_F, then it also occurs at clc of these vertices.

17. **Consider** the above problem and mention whether the point other than vertices where optimal solution exists is a basic solution of the LPP.

18. **Show** that set of all optimal solutions of a LPP is a closed convex set.

19. **Consider** the system $AX = b$, $X \geq 0$, $b \geq 0$ (with m equations and n unknowns). Let X be a basic feasible solution with $p < m$ components positive. How many different bases will correspond to X due to degeneracy in the system.

20. **In** view of Problem 19, the BFS $(0, 2, 0, 0)$ of Example 3 has one more different basis. Find this basis.

21. **Write** a solution of the constraint equations in Example 3 which is neither basic nor feasible.

22. **If** X_0 is any optimal solution of the LPP min $x_0 = C^T X$, subject to $AX = b$, $X \geq 0$ in standard form and X^* is any optimal solution when C is replaced by C^*, then prove that

$$(C^* - C)^T (X^* - X_0) \geq 0.$$

23. **To** make the graphical method work, prove that the intersection set of the feasible domain P_F and the supporting hyperplane whose normal is given by the negative cost vector $-C^T$ provides the optimal solution to a given linear programming problem.

24. **Find** the solution of the following linear programming problems using the graphical method

(a) min $z = -x_1 + 2x_2$
    s.t. $-x_1 + 3x_2 \leq 10$
          $x_1 + x_2 \leq 6$
          $x_1 - x_2 \leq 2$
          $x_1, x_2 \geq 0$

(b) max $z = 3x_1 + 4x_2$
    s.t. $x_1 - 2x_2 \leq -1$
          $-x_1 + 2x_2 \geq 0$
          $x_1, x_2 \geq 0$

25. What difficulty arises if all the constraints are taken as strict inequalities?

26. Show that by properly choosing c_i's in objective function of a LPP, every vertex can be made optimal.

27. Let X be basic solution of the system $AX = b$ having both positive and negative variables. How can X be reduced to a BFS.

28. Prove that a LPP in standard form with n variables and $n - 2$ nonredundant constraints can be represented by two dimensional graph.

29. Let us take the following LPP

$$\begin{aligned} \min \quad & z = -2x_1 - x_2 + x_3 + x_4 + 2x_5 \\ \text{s.t.} \quad & -2x_1 + x_2 + x_3 + x_4 + x_5 = 12 \\ & -x_1 + 2x_2 + x_4 - x_5 = 5 \\ & x_1 - 3x_2 + x_3 + 4x_5 = 11 \\ & x_1, x_2, x_3, x_4, x_4 \geq 0 \end{aligned}$$

Using preceding problem represent this LPP as a two-dimensional problem and then find its optimal solution by graphical method.

Suggestion. Do pivoting at x_3, x_4, x_5.

30. Consider the following constraints:

$$\begin{aligned} x_1 + x_2 + 4x_3 + 2x_4 + 3x_5 &= 8 \\ 4x_1 + 2x_2 + 2x_3 + x_4 + 6x_5 &= 4 \end{aligned}$$

Identify all basic feasible solutions of the above system.

Suggestion. Note that $(0, 7/2, 0, -3/2)$ is one of the basic feasible solutions. Why?

Chapter 3

The Simplex Algorithm

This chapter develops the theory of simplex method to solve linear programming problems. The other methods are slight variants of the simplex method known as big-M method and two phase method. In the end, we explain some difficulties which are encountered when simplex iterations are executed.

3.1 Introduction

Simplex means a polytope having finite number of vertices. To find the optimal solution by simplex method one starts from some convenient BFS (vertex), and goes to another adjacent BFS (vertex) so that value of the objective function is improved. In this way, after few steps the desired basic feasible solution (vertex) is reached where the optimal solution exists.

The theory of simplex method is developed in stages with induction of some analysis. This is an iterative method in the sense that rules and conditions of previous step work invariably to the last. The iterative scheme consists of the three steps:

(i) Start from somewhere;

(ii) Check if the aim is met;

(iii) Move to a place closer to the aim.

3.2   The Simplex Algorithm

Consider the LPP in standard form as the basic feasible solution (BFS) is calculated after writing the problem in this format

$$\text{opt}\quad z = C^T X$$
$$\text{s.t.}\quad AX = b$$
$$X \geq 0,\ b \geq 0,$$

where A, X, C, b have already been defined in Chapter 1.

The above LPP can also be written in the form

$$\text{opt}\quad z = c_1 x_1 + c_2 x_2 + \cdots + c_n x_n$$
$$\text{s.t.}\quad x_1 A_1 + x_2 A_2 + \cdots + x_n A_n = b \qquad (3.1)$$
$$x_1, x_2, \ldots, x_n \geq 0,\ b \geq 0,$$

where $A = [A_1, A_2, \ldots, A_n]$ is a m by n matrix and A_1, A_2 and A_n are the first, second and nth columns of A, respectively.

Let m be the rank of A, and every set of m column vectors is linearly independent. The number of equations in the system $AX = b$ are m. This determines the number of m basic variables and $n - m$ nonbasic variables. The total number of basic feasible solutions of LPP can not exceed $n!/m!(n-m)!$, see Proposition 5, Chapter 2.

Suppose a BFS $x_1, x_2, \ldots, x_m$ is available at our disposal. This implies

$$x_{m+1} = x_{m+2} = \cdots = x_n = 0 \qquad (3.2)$$

are left as nonbasic variables. Thus, $X_B = (x_1, x_2, \ldots, x_m)^T$ is a basic vector with basis matrix given as

$$B = \begin{bmatrix} a_{11} & a_{12} & \cdots & a_{1m} \\ a_{21} & a_{22} & \cdots & a_{2m} \\ \vdots & & & \\ a_{m1} & a_{m2} & \cdots & a_{mm} \end{bmatrix}$$

Using (3.2) in (3.1), we have $x_1A_1 + x_2A_2 + \cdots + x_mA_m = b$, or

$$(A_1, A_2, \ldots, A_m) \begin{bmatrix} x_1 \\ x_2 \\ \vdots \\ x_m \end{bmatrix} = b.$$

Since, $(A_1, A_2, \ldots, A_m) = B$, this has the compact form:

$$BX_B = b \iff X_B = B^{-1}b. \tag{3.3}$$

All column vectors of B are linearly independent (by assumption), and hence, B is a nonsingular matrix, and moreover, all column vectors of B generate $\mathbb{R}^m$, an m-dimensional real linear space. This ensures that each A_j, $j = 1, 2, \ldots, n$ can uniquely be expressed as a linear combination of the elements of the ordered basis $\{A_1, A_2, \ldots, A_m\}$,

$$A_{m+1} = \alpha_1^{m+1}A_1 + \alpha_2^{m+1}A_2 + \cdots + \alpha_m^{m+1}A_m$$
$$A_{m+2} = \alpha_1^{m+2}A_1 + \alpha_2^{m+2}A_2 + \cdots + \alpha_m^{m+2}A_m$$
$$\vdots =$$
$$A_n = \alpha_1^nA_1 + \alpha_2^nA_2 + \cdots + \alpha_m^nA_m.$$

The compact form of the above system is

$$A_j = \alpha_1^jA_1 + \alpha_2^jA_2 + \cdots + \alpha_m^jA_m, \quad j = m+1, m+2, \ldots, n$$

or

$$A_j = (A_1, A_2, \ldots, A_m) \begin{bmatrix} \alpha_1^j \\ \alpha_2^j \\ \vdots \\ \alpha_m^j \end{bmatrix} \Rightarrow A_j = B\alpha^j,$$

where $\alpha^j = (\alpha_1^j, \alpha_2^j, \ldots, \alpha_m^j)^T$ is the coordinate vector of A_j. Thus, we get the relation

$$\alpha^j = B^{-1}A_j. \tag{3.4}$$

Also, in view of (3.3), one has

$$f(X_B) = c_1x_1 + c_2x_2 + \cdots + c_mx_m = C_B^TX_B = C_B^TB^{-1}b$$

is the value of objective function at X_B, where $C_B^T = (c_1, c_2, \ldots, c_m)$ is the cost of basic vector X_B. Now, we prove an important result.

Proposition 1. If $z_j = C_B^T \alpha^j$, $j = 1, 2, \ldots, n$, then $z_j - c_j = 0$ for all basic variables.

Proof. Since $A_1, A_2, \ldots, A_m$ is an ordered basis, the coordinate vector α^j of A_j, $j = 1, 2, \ldots, m$ is $(0, 0, \ldots, 1, \ldots, 0)$, here 1 is at jth place. Hence

$$z_j = C_B^T \alpha^j = (c_1, c_2, \ldots, c_j, \ldots, c_m) \begin{bmatrix} 0 \\ \vdots \\ 1 \leftarrow \\ \vdots \\ 0 \end{bmatrix} = c_j.$$

Here arrow toward 1 indicates that it is at jth place. From the above relation, $z_j - c_j = 0$.

Remark. Note that $z_j - c_j$ may or may not be zero for a nonbasic variable. In our subsequent discussion $z_j - c_j$ will be termed as relative cost of jth variable.

In order to introduce the simplex method in algebraic terms, let us introduce some notations here. For a given basic feasible solution X^*, we can always denote it by

$$X^* = \begin{bmatrix} X_B^* \\ \hline X_N^* \end{bmatrix}$$

where the elements of vector X_B^* represent the basic variables and the elements of vector X_N^* represent nonbasic variables. Needless to mention $X_B^* \geq 0$ and $X_N^* = 0$ for the basic feasible solution.

Also, for a given cost vector C (column vector) and the coefficient matrix A, we can always denote these as

$$C = \begin{bmatrix} C_B \\ \hline C_N \end{bmatrix} \quad \text{and} \quad A = \begin{bmatrix} B \,|\, N \end{bmatrix},$$

where B is a $m \times m$ nonsingular matrix that is referred to as basis and N is referred to as nonbasis with dimensionality $m \times (n - m)$.

Proposition 2. Let the LPP be max $z = C^T X$, subject to $AX = b$, $X \geq 0$. If, for any BFS, X_B^* all $z_j - c_j \geq 0$, then X_B is the optimal solution of the problem.

Proof. Once a basis is known, every feasible solution $X \in P_F$ arranged in a order as

$$X = \begin{bmatrix} X_B \\ \overline{X_N} \end{bmatrix}$$

Note that both X_B and X_N are nonnegative. Hence the LPP becomes

$$\max \quad z = C_B^T X_B + C_N^T X_N \tag{3.5a}$$
$$\text{s.t.} \quad B X_B + N X_N = b \tag{3.5b}$$
$$X_B \geq 0, \; X_N \geq 0 \tag{3.5c}$$

Equation (3.5b) implies that

$$X_B = B^{-1} b - B^{-1} N X_N \tag{3.6}$$

Substituting (3.6) back into (3.5a) results in

$$\begin{aligned} z &= C_B^T \left(B^{-1} b - B^{-1} N X_N \right) + C_N^T X_N \\ &= C_B^T B^{-1} b + \left(C_N^T - C_B^T B^{-1} N \right) X_N \\ &= C_B^T B^{-1} b - r^T \begin{bmatrix} X_B \\ \overline{X_N} \end{bmatrix} \end{aligned} \tag{3.7}$$

where

$$r = \begin{bmatrix} 0 \\ \overline{(B^{-1} N)^T C_B - C_N} \end{bmatrix}$$

Observe that r is an n-dimensional vector. Its first m components, corresponding to basic variables are set to be zero (see Proposition 1) and the remaining $n - m$ components correspond to nonbasic variables. Also, note that objective value z^* at current basic feasible solution X^* is $C_B^T B^{-1} b$, since $X_B^* = B^{-1} b$ and $X_N^* = 0$. Consequently (3.7) becomes

$$z^* - z = r^T \begin{bmatrix} X_B \\ \overline{X_N} \end{bmatrix} \quad \text{for each } X \in P_F$$

It is apparent that $r^T \geq 0$, i.e., every component of $C_B^T B^{-1} N - C_N^T$ is nonnegative, then $z^* - z \geq 0$ for each feasible solution $X \in P_F$. Hence z^* is optimal value and X^* is optimal solution.

Equivalently, for optimal solution to exist we write

$$r_j = z_j - c_j = C_B^T B^{-1} A_j - c_j \geq 0$$

where j runs over nonbasic variables as $z_j - c_j = 0$ for all basic variables.

This proves the proposition.

Remark. For a minimization problem all $z_j - c_j \leq 0$ in the last iteration table is the desired condition for any BFS, X_B^* to be optimal.

To develop further, the theory of simplex method for solving linear programming problems, we give a brief exposition of linear algebra tools which are very much needed at different stages.

Proposition 3. Let $B = \{X_1, X_2, \ldots, X_n\}$ be a basis in a linear space $\mathbb{R}^n$ and let $X \notin B$ such that $X = \alpha_1 X_1 + \alpha_2 X_2 + \cdots + \alpha_n X_n$. If $\alpha_i = 0$ then the vector X_i can not be replaced by X to form a new basis of $\mathbb{R}^n$.

Proof. Given that $B = \{X_1, X_2, \ldots, X_n\}$ is a basis of $\mathbb{R}^n$. For any vector $X \notin B$, we have the unique representation

$$X = \alpha_1 X_1 + \alpha_2 X_2 + \cdots + \alpha_n X_n. \tag{3.8}$$

If $\alpha_i = 0$, then (3.8) becomes

$$X = \alpha_1 X_1 + \alpha_2 X_2 + \cdots + \alpha_{i-1} X_{i-1} + \alpha_{i+1} X_{i+1} + \cdots + \alpha_n X_n.$$

By using commutativity and associativity repeatedly in above, we have

$$\alpha_1 X_1 + \alpha_2 X_2 + \cdots + \alpha_{i-1} X_{i-1} + (-1)X + \alpha_{i+1} X_{i+1} + \cdots + \alpha_n X_n = 0. \tag{3.9}$$

From (3.9), we conclude that $\{X_1, X_2, \ldots, X_{i-1}, X, X_{i+1}, \ldots, X_n\}$ is linearly dependent, and hence, it is not a basis. Thus, we have established that the vector whose coefficient is zero in linear representation for X can not be replaced by X to form a new basis.

Example 1. Let $\{(1,0,0)^T, (0,1,1)^T, (0,0,1)^T\}$ be a basis in $\mathbb{R}^3$. Which vector can not be replaced by $(2,3,3)$ in this basis to form a new basis.

Since

$$\begin{bmatrix} 2 \\ 3 \\ 3 \end{bmatrix} = 2 \begin{bmatrix} 1 \\ 0 \\ 0 \end{bmatrix} + 3 \begin{bmatrix} 0 \\ 1 \\ 1 \end{bmatrix} + 0 \begin{bmatrix} 0 \\ 0 \\ 1 \end{bmatrix}$$

in view of the above proposition the vector $(0, 0, 1)^T$ can not be replaced by $(2, 3, 3)^T$ to form a new basis.

Rules for entering and leaving variables. Suppose we are considering maximization problem and the basic vector $X_B = (x_1, x_2, \ldots, x_m)^T$ is at our disposal. With the help of Proposition 2, we can check whether this BFS is optimal or not. If not, our next step is to move to some other BFS (vertex) so that the current value of the objective function improves (increases) or optimal BFS is obtained. The basis matrix associated with X_B is

$$B = (A_1, A_2, \ldots, A_m)$$

To go to other vertex one of the basic variables from $x_1, x_2, \ldots, x_m$ is supposed to leave and other nonbasic variable will occupy its position. This is equivalent to saying that some of the column vector from B will leave the basis and a column vector from $A_{m+1}, A_{m+2}, \ldots, A_n$ occupies its place to form a new basis (see Fig. 3.1).

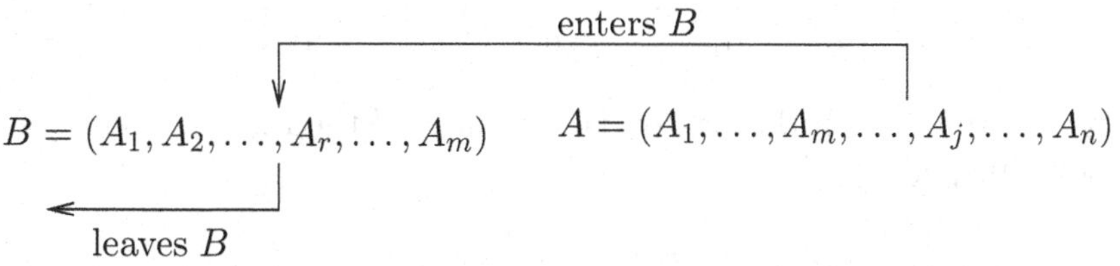

Figure 3.1

$$A_j = \alpha_1^j A_1 + \alpha_2^j A_2 + \cdots + \alpha_r^j A_r + \cdots + \alpha_m^j A_m. \qquad (3.10)$$

Proposition 3 ensures that $A_1, A_2, \ldots, A_j, \ldots, A_m$ forms a new basis provided $\alpha_r^j \neq 0$. From the original BFS, we have

$$x_1 A_1 + x_2 A_2 + \cdots + x_r A_r + \cdots + x_m A_m = b.$$

We use equation (3.10) to replace A_r in above equation by A_j

$$\left(x_1 - x_r \frac{\alpha_1^j}{\alpha_r^j} \right) A_1 + \left(x_2 - x_r \frac{\alpha_2^j}{\alpha_r^j} \right) A_2 + \cdots + \frac{x_r}{\alpha_r^j} A_j$$

$$+ \cdots + \left(x_m - x_r \frac{\alpha_m^j}{\alpha_r^j} \right) A_m = b.$$

The position of A_r has been occupied by A_j, and

$$x_1 - x_r \frac{\alpha_1^j}{\alpha_r^j}, \quad x_2 - x_r \frac{\alpha_2^j}{\alpha_r^j}, \quad \cdots, \quad \frac{x_r}{\alpha_r^j}, \quad \cdots, \quad x_m - x_r \frac{\alpha_m^j}{\alpha_r^j} \qquad (3.11)$$

is a new basic solution, i.e., it satisfies $AX = b$ when $n - m$ variables are kept at zero level, and (3.11) gives the new set of basic variables. Now, we have to choose A_r such that the set in (3.11) defines a feasible solution, i.e., each term must be nonnegative. For this, the first requirement is that in rth term of (3.11), α_r^j must be positive (this is nonzero by assumption). Note that the first basic variable ≥ 0 if $\alpha_1^j < 0$. To ensure that each term in (3.11) is nonnegative, i.e.,

$$x_1 - x_r \frac{\alpha_1^j}{\alpha_r^j} \geq 0, \ x_2 - x_r \frac{\alpha_2^j}{\alpha_r^j} \geq 0, \ \cdots, \ \frac{x_r}{\alpha_r^j} \geq 0, \ \cdots, \ x_m - x_r \frac{\alpha_m^j}{\alpha_r^j} \geq 0$$

or

$$\frac{x_1}{\alpha_1^j} \geq \frac{x_r}{\alpha_r^j}, \ \frac{x_2}{\alpha_2^j} \geq \frac{x_r}{\alpha_r^j}, \ \cdots, \ \frac{x_r}{\alpha_r^j} \geq 0, \ \cdots, \ \frac{x_m}{\alpha_m^j} \geq \frac{x_r}{\alpha_r^j}$$

choose $r(1$ to $m)$ such that x_r/α_r^j is minimum of the right side entries in above inequalities, i.e.,

$$\frac{x_r}{\alpha_r^j} = \min_i \left\{ \frac{x_i}{\alpha_i^j}, \ \alpha_i^j > 0 \right\} = \theta_j, \ i = 1, 2, \ldots, m. \qquad (3.12)$$

This determines r, i.e., which variable x_r or column A_r leaves the basis, when its position is to be occupied by A_j (j fixed) so that the resulting set $A_1, A_2, \ldots, A_j, \ldots, A_m$ forms a basis and the set in (3.11) forms a basic feasible solution.

In the above analysis A_j is arbitrary but fixed. Now our main purpose is the selection of proper A_j, i.e., the entering variable so that there is maximum improvement in the value of the objective function as compared its value on the earlier BFS. Imposing condition (3.12), the new BFS is

$$\hat{x}_1 = x_1 - \theta_j \alpha_1^j, \ \ \hat{x}_2 = x_2 - \theta_j \alpha_2^j, \ \ \ldots, \ \ \hat{x}_r = \theta_j, \ \ \ldots, \ \ \hat{x}_m = x_m - \theta_j \alpha_m^j.$$

The value of the objective functions at this new BFS is

$$\hat{x}_0 = c_1 \hat{x}_1 + c_2 \hat{x}_2 + \cdots + c_j \hat{x}_r + \cdots + c_m \hat{x}_m.$$

Here c_r is changed to c_j. Insert the values of $\hat{x}_1, \hat{x}_2, \ldots, \hat{x}_m$, we have

$$\hat{x}_0 = c_1 x_1 - \frac{x_r}{\alpha_r^j} c_1 \alpha_1^j + c_2 x_2 - \frac{x_r}{\alpha_r^j} c_2 \alpha_2^j + \cdots + c_j \frac{x_r}{\alpha_r^j}$$
$$+ \cdots + c_m x_m - \frac{x_r}{\alpha_r^j} c_m \alpha_m^j + c_r x_r - \frac{x_r}{\alpha_r^j} c_r \alpha_r^j$$

Here we added and subtracted $c_r x_r$. Since $x_0 = c_1 x_1 + x_2 x_2 + \cdots + c_r x_r + \cdots + c_m x_m$ and $z_j = C_B^T \alpha^j = c_1 \alpha_1^j + c_2 \alpha_2^j + \cdots + c_r \alpha_r^j + \cdots + c_m \alpha_m^j$, it follows that

$$\hat{x}_0 - x_0 = -\theta_j (z_j - c_j).$$

Since $x_r / \alpha_r^j = \theta_j \geq 0$, for maximum increase in x_0, we select that A_j as the entering variable for which

$$\theta_j (z_j - c_j)$$

is the most negative.

Thus, theoretically we permit the entry of the nonbasic variable determined by the most negative (for maximization) or most positive (for minimization) nature of $\theta_j (z_j - c_j)$. But this is not convenient because we have to compute θ_j for all nonbasic variables. This will make our task time consuming. The only thing we do is to see the most negative or the most positive nature of $z_j - c_j$ for deciding the entering variable. This serves our purpose well to a large extent. In an arbitrary selection the only possibility persists is that the number of iterations may increase.

While there appears a tie between entering variables, any nonbasic variable may be chosen to enter the basis. In case more than two nonbasic variables desire to enter the basis based upon the observation of $z_j - c_j$ values, it is profitable to compute $\theta_j (z_j - c_j)$ for all nonbasic variables to decide the appropriate entering variable. Such type of problems usually come in the way when Games theory problems are solved using simplex method.

Now, we solve a maximization problem using the theoretical details developed so far, and some more concepts are introduced during the analysis.

Note. From now onwards, in every simplex table the abbreviations B V and Soln stand for 'Basic Variables' and 'Solution', respectively.

Example 2. Solve the following LPP by the simplex method.

$$\max \quad z = -x_1 + 3x_2 - 3x_3$$
$$\text{s.t.} \quad 3x_1 - x_2 + x_3 \le 7$$
$$-x_1 + 2x_2 \le 6$$
$$-4x_1 + 3x_2 + 8x_3 \le 10$$
$$x_1, x_2, x_3 \ge 0$$

Write the LPP in standard form (including slack variables at zero cost level in objective function)

$$\max \quad z = -x_1 + 3x_2 - 3x_3 + 0s_1 + 0s_2 + 0s_3$$
$$\text{s.t.} \quad 3x_1 - x_2 + x_3 + s_1 = 7$$
$$-x_1 + 2x_2 + s_2 = 6$$
$$-4x_1 + 3x_2 + 8x_3 + s_3 = 10$$
$$x_1, x_2, x_3, s_1, s_2, s_3 \ge 0$$

Here $C = (-1, 3, -3, 0, 0, 0)^T$, $b = (7, 6, 10)^T$ and the coefficient matrix is

$$
\begin{array}{cccccc}
A_1 & A_2 & A_3 & A_4 & A_5 & A_6
\end{array}
$$
$$
A = \begin{bmatrix}
3 & -1 & 1 & 1 & 0 & 0 \\
-1 & 2 & 0 & 0 & 1 & 0 \\
-4 & 3 & 8 & 0 & 0 & 1
\end{bmatrix}
$$

The rank of A is 3. We choose the simplest basic variables as s_1, s_2, s_3, i.e., basic vector $X_B = (s_1, s_2, s_3)^T$. Thus, starting BFS represents the origin. The corresponding basis matrix is

$$
B = \begin{bmatrix}
1 & 0 & 0 \\
0 & 1 & 0 \\
0 & 0 & 1
\end{bmatrix}
$$

1. Our starting requirement is that the coefficient matrix A must contain an identity submatrix of order m, the rank of A. First, we construct the starting simplex format. Write the objective function in the form

$$z + 1x_1 - 3x_2 + 3x_3 - 0s_1 - 0s_2 - 0s_3 = 0.$$

The coefficients of the basic and nonbasic variables constitute the z-row (for starting table only). Note that the entries of z-row are the relative costs $z_j - c_j$ and can also be computed using $z_j = C_B^T \alpha^j = C_B^T B^{-1} A_j$, $j = 1, 2, \ldots, 6$. Below solution in z-row is $f(X_B) = C_B^T X_B = (0, 0, 0)(7, 6, 10)^T = 0$, which is nothing but the right-hand entry of the above equation.

2. Each of the bottom row (below z-row) represents a constraint equation, while the right-hand vector is inserted in the 'Solution' column. These entries can also be computed, using $\alpha^j = B^{-1} A_j$, $j = 1$ to 6. Since $B^{-1} = I$ (by our choice of starting BFS), α^j and b are written as it is. The entries in front of basic variables (except z-row and solution column) constitute the body matrix. Thus, the elements of the body matrix are coordinate vectors (column-wise) α^j, $j = 1$ to 6, see Table 0.

Table 0

B V	x_1	x_2	x_3	s_1	s_2	s_3	Soln
z	1	-3	3	0	0	0	0
s_1	3	-1	1	1	0	0	7
s_2	-1	2	0	0	1	0	6
s_3	-4	3	8	0	0	1	10

3. The LPP we are solving is a maximization problem and hence to have the optimal solution all entries in z-row must be ≥ 0, see Proposition 2. This is not satisfied in Table 0 and hence $(0, 0, 0, 7, 6, 10)^T$ is not an optimal BFS. Now, search for another basic feasible solution so that the value of objective function improves. Hence, one of the nonbasic variable from x_1, x_2, x_3 is to be converted into basic variable, i.e., to take to positive level (equivalently, A_1 or A_2 or A_3 will enter the basis). Note that the relative cost of x_2 is the most negative than relative cost of x_1 and x_3, this suggests that it will be profitable to choose x_2 as the entering variable. Thus, we reach at the agreement that for starting the first iteration in maximization problem we choose the most negative variable as the entering variable, see Table 1.

4. It is natural when one nonbasic has entered the basis, certainly, one has to leave the basis as the number of basic variables in

each BFS remains fixed. To decide this we take the ratios of the entries of the solution column with the *positive entries*, see equation (3.12), of the column below the entering variable. This is the *minimum ratio rule*. Find the minimum ratio. The basic variable in B. V. column corresponding to this minimum ratio will leave the basic vector. Here, in this example

$$\min\left\{\frac{6}{2}, \frac{10}{3}\right\} = 3.$$

Since the above minimum corresponds to the s_2-row, this implies that s_2 should leave the BFS, see Table 1. The variable at the intersection of entering column and leaving row becomes pivot element which has been boxed.

Table 1

B V	x_1	$x_2 \downarrow$	x_3	s_1	s_2	s_3	Soln
z	1	-3	3	0	0	0	0
s_1	3	-1	1	1	0	0	7
$\leftarrow s_2$	-1	$\boxed{2}$	0	0	1	0	6
s_3	-4	3	8	0	0	1	10

5. Now, the problem is how to execute the next iteration table. The entering variable is x_2 and the leaving variable is s_2, i.e., new BFS is $(s_1, x_2, s_3)^T$. The coordinate vector (α_1^2, α_2^2) will be $(0, 1, 0)^T$. Why? This means that in x_2-column of body matrix we must now have $(0, 1, 0)^T$ instead of $(-1, 2, 3)^T$. Now, x_2 is a basic variable and hence in z-row below this one must have $z_j - c_j = 0$, see Proposition 1. Any way below x_2 the entries must be $(0; 0, 1, 0)^T$ instead of $(-3; -1, 2, 3)^T$. In a simple manner this can be achieved by elementary row operations. Divide s_2-row by 2 (pivot element), and write this new s_2-row as the x_2-row (pivot row) in the next table, see Table 2. With the help of pivot row bring zeros in x_2-column. What we do is as follows

z-row (Table 1) $-(-3)$ times of pivot row (Table 2)=z-row

of Table 2;

s_1-row (Table 1) + pivot row (Table 2)=s_1-row of Table 2;

s_3-row (table 1) -3 times of pivot row (Table 2)=s_3-row

of Table 2.

This gives Table 2.

Table 2

B V	$x_1 \downarrow$	x_2	x_3	s_1	s_2	s_3	Soln
z	$-1/2$	0	3	0	3/2	0	9
$\leftarrow s_1$	5/2	0	1	1	1/2	0	10
x_2	$-1/2$	1	0	0	1/2	0	3
s_3	$-5/2$	0	8	0	$-3/2$	1	1

6. Still Table 2 is not optimal table because all $z_j - c_j$ are not nonnegative. Further, x_1 is the entering variable and s_1 is the leaving variable. We repeat Steps 3-6 to have Table 3.

Table 3

B V	x_1	x_2	x_3	s_1	s_2	s_3	Soln
z	0	0	16/5	1/5	8/5	0	11
x_1	1	0	2/5	2/5	1/5	0	4
x_2	0	1	1/5	1/5	3/5	0	5
s_3	0	0	9	1	-1	1	11

Since all $z_j - c_j \geq 0$ in z-row of Table 3, it follows that the optimality is reached. This table is now the optimal table. The optimal solution can be read directly from the table.

The optimal solution is $x_1 = 4$, $x_2 = 5$, $x_3 = 0$, max value $z = 11$.

Remark. It seems in Table 0 that we have filled up the coefficient matrix in s_1, s_2 and s_3 rows. In fact we have not done this. Notice that in every simplex table these entries are coordinate vectors α^j. However, for Table 0, the special construction of basis matrix B as the identity submatrix ensures that column vectors are nothing but the coordinate vectors with respect to the standard basis $\{(1,0,0),(0,1,0),(0,0,1)\}$, and hence we can fill up the coefficient matrix. This may not happen in other tables as the basis may not be standard basis.

What we have learned to start with simplex method is

1. Write the LPP in standard form.

2. The coefficient matrix A must contain the identity submatrix. The variables constituting the identity submatrix give the starting basis (B V), and the solution is b.

3. The objective function must be expressed in terms of nonbasic variables.

After the above all is done, we summarize here how to complete all simplex iterations to reach at optimality for a maximization problem.

(i) The variable (column) with the most negative coefficient will enter as basic variable (basis). This ensures largest possible increase in objective function.

(ii) The leaving variable is decided by

$$\min\left\{\frac{\text{entries of the solution column}}{\text{corresponding entering column entries} > 0}\right\}.$$

This ensures feasibility.

(iii) If all the entries in the z-row are ≥ 0, the optimality is reached, and the optimal solution can be read from the table.

The minimization problem can be solved by converting into maximization problem as

$$\min f(X) = -\max\{-f(X)\}.$$

Therefore, we simply multiply the cost coefficients by -1 to convert to a minimization problem into maximization. But once the maximum of the new problem is found, remember to multiply the maximum by -1 for the original minimum.

Similarly,
$$\max f(X) = -\min\{-f(X)\}.$$

We can also solve minimization problem directly (without converting into maximization problem). The rules for minimization problem are

(i) The variable (column) with the most positive coefficient will enter as basic variable. This ensures largest possible decrease in objective function.

(ii) The leaving variable is decided by the same rule as for maximization problem which ensures feasibility.

(iii) If all the entries in the z-row are ≤ 0, the optimality is reached, and the optimal solution can be read from the table.

Now, we solve a minimization problem without converting into maximization problem.

Example 3. Solve the following LPP by simplex method

$$\begin{aligned} \min \quad & z = x_1 - 2x_2 + x_3 \\ \text{s.t.} \quad & x_1 + 2x_2 - 2x_3 \leq 4 \\ & x_1 - x_3 \leq 3 \\ & 2x_1 - x_2 + 2x_3 \leq 2 \\ & x_1, x_2, x_3 \geq 0. \end{aligned}$$

Write the above problem in standard form

$$\begin{aligned} \min \quad & z = x_1 - 2x_2 - x_3 \\ \text{s.t.} \quad & x_1 + 2x_2 - 2x_3 + s_1 = 4 \\ & x_1 - x_3 + s_2 = 3 \\ & 2x_1 - x_2 + 2x_3 + s_3 = 2 \\ & x_i, s_i \geq 0, \ i = 1, 2, 3 \end{aligned}$$

Obviously, the coefficient matrix A of standard form of the LPP contains identity submatrix and hence simplex method can be applied. The starting BFS is s_1, s_2, s_3. Construct the starting table as explained in Steps 1 and 2 of Example 2.

This is a minimization problem, and hence, the nonbasic variable with the most positive relative cost $z_j - c_j$ in z-row will enter the BFS.

The leaving variable is decided as for maximization problem. The complete working is shown in the following table.

B V	x_1	$x_2 \downarrow$	x_3	s_1	s_2	s_3	Soln
z	-1	2	-1	0	0	0	0
$\leftarrow s_1$	1	$\boxed{2}$	-2	1	0	0	4
s_2	1	0	-1	0	1	0	3
s_3	2	-1	2	0	0	1	2
z	-2	0	$1 \downarrow$	-1	0	0	-4
x_2	$1/2$	1	-1	$1/2$	0	0	2
s_2	1	0	-1	0	1	0	3
$\leftarrow s_3$	$5/2$	0	$\boxed{1}$	$1/2$	0	1	4
z	$-9/2$	0	0	$-3/2$	0	-1	-8
x_2			0				6
s_2			0				7
x_3	$5/2$	0	1	$1/2$	0	1	4

The last table is optimal because all $z_j - c_j \leq 0$ (as this is a minimization problem). The optimal solution can be read from this table as

optimal solution: $x_1 = 0$, $x_2 = 6$, $x_3 = 4$, min value $z = -8$.

Example 4. Consider the LPP

$$\begin{aligned} \max \quad & z = 4x_1 + x_2 + 3x_3 + 5x_4 \\ \text{s.t.} \quad & 4x_1 - 6x_2 - 5x_3 + 4x_4 \geq -20 \\ & 3x_1 - 2x_2 + 4x_3 + x_4 \leq 11 \\ & 8x_1 - 3x_2 + 3x_3 + 2x_4 \leq 23 \\ & x_1, x_2, x_3, x_4 \geq 0. \end{aligned}$$

One of the simplex iterations table of the above LPP is given below.

Without performing simplex iterations, find the missing entries.

B V	x_1	x_2	x_3	x_4	s_1	s_2	s_3	Soln
z	17	0	-35	0	0	-17	*	*
s_1	*	0	*	*	1	0	2	*
x_4	*	0	*	*	0	-3	2	*
x_2	*	1	*	*	0	-2	*	*

Writing the problem in standard form

$$\begin{aligned}
\max \quad & z = 4x_1 + x_2 + 3x_3 + 5x_4 \\
\text{s.t.} \quad & -4x_1 + 6x_2 + 5x_3 - 4x_4 + s_1 = 20 \\
& 3x_1 - 2x_2 + 4x_3 + x_4 + s_2 = 11 \\
& 8x_1 - 3x_2 + 3x_3 + 2x_4 + s_3 = 23 \\
& \text{all var} \geq 0.
\end{aligned}$$

First of all we write simplex format by putting 1 at the intersection of each column and row corresponding to basic variables and elsewhere 0 in that column. Thus, $\alpha^5 = (1,0,0)^T$, $\alpha^4 = (0,1,0)^T$ and $\alpha^2 = (0,0,1)^T$ are determined.

Further, write the basis matrix B corresponding to the basic vector (s_1, x_4, x_2) and find its inverse to have

$$B = \begin{bmatrix} 1 & -4 & 6 \\ 0 & 1 & -2 \\ 0 & 2 & -3 \end{bmatrix} ; \quad B^{-1} = \begin{bmatrix} 1 & 0 & 2 \\ 0 & -3 & 2 \\ 0 & -2 & 1 \end{bmatrix}$$

Always B^{-1} for each BFS is kept below starting basis. This places $(2,2,1)^T$ below s_3 and hence $\alpha_3^7 = 1$.

Now, the solution column is

$$B^{-1}b = \begin{bmatrix} 1 & 0 & 2 \\ 0 & -3 & 2 \\ 0 & -2 & 1 \end{bmatrix} \begin{bmatrix} 20 \\ 11 \\ 23 \end{bmatrix} = \begin{bmatrix} 66 \\ 13 \\ 1 \end{bmatrix}$$

For coordinate vectors of x_1 and x_3 in body matrix of the table, use the formula $\alpha^j = B^{-1}A_j$ to have

$$\alpha^1 = \begin{bmatrix} 1 & 0 & 2 \\ 0 & -3 & 2 \\ 0 & -2 & 1 \end{bmatrix} \begin{bmatrix} -4 \\ 3 \\ 8 \end{bmatrix} = \begin{bmatrix} 12 \\ 7 \\ 2 \end{bmatrix} \; ; \quad \alpha^3 = \begin{bmatrix} 1 & 0 & 2 \\ 0 & -3 & 2 \\ 0 & -2 & 1 \end{bmatrix} \begin{bmatrix} 5 \\ 4 \\ 3 \end{bmatrix} = \begin{bmatrix} 11 \\ -6 \\ -5 \end{bmatrix}$$

The cost vector is $(0, 5, 1)^T$ and $X_B = (66, 13, 1)^T$. Thus, we compute the value of the objective function at X_B as

$$f(X_B) = C_B^T X_B = (0, 5, 1)(66, 13, 1)^T = 66.$$

The entry (relative cost) in z-row below s_3 is given by

$$z_7 - c_7 = (0, 5, 1)(2, 2, 1)^T - 0 = 11.$$

After inserting all missing entries, the complete simplex table is

B V	x_1	x_2	x_3	x_4	s_1	s_2	s_3	Soln
z	17	0	-35	0	0	-17	11	66
s_1	12	0	11	0	1	0	2	66
x_4	7	0	-6	1	0	-3	2	13
x_2	16	1	-5	0	0	-2	1	1

3.3   The Big-M Method

After the LPP is written in standard form and the coefficient matrix A in $X = b$ does not contain identity submatrix then we extend the idea of solving the problem by using Big-M method or Two phase method. The first is preferable for theoretical investigations, while latter is very much useful for computational purposes.

For any equation i that does not have the slack variable, we augment an artificial variable $R_i(\geq 0)$. With the induction of artificial variables the matrix A is modified and now, it contains identity submatrix. The artificial variables then become part of the starting basic feasible solution. Because artificials are extraneous in LP model, we assign penalties to them in objective function to force to come to zero level at later simplex iterations. As M is sufficiently large positive number, the variable R_i is penalized in the objective function using $-MR_i$ in maximization problem, while by $+MR_i$ in minimization. This is the reason that sometimes the big M-method is called *penalty method*.

Example 5. Solve the following LPP by using the big M-method

$$\max \quad z = -x_1 + 3x_2$$
$$\text{s.t.} \quad x_1 + 2x_2 \geq 2$$
$$3x_1 + x_2 \leq 3$$
$$x_1 \leq 4$$
$$x_1, x_2 \geq 0$$

The standard form of the LPP is written as

$$\max \quad z = -x_1 + 3x_2$$
$$\text{s.t.} \quad x_1 + 2x_2 - s_1 = 2$$
$$3x_1 + x_2 + s_2 = 3$$
$$x_1 + s_3 = 4$$
$$x_1, x_2, s_1, s_2, s_3 \geq 0$$

The coefficient matrix

$$A = \begin{bmatrix} 1 & 2 & -1 & 0 & 0 \\ 3 & 1 & 0 & 1 & 0 \\ 1 & 0 & 0 & 0 & 1 \end{bmatrix}$$

does not have identity submatrix. This is because first equation does not have slack variable, we thus use artificial variable R_1 in the equation and penalize in the objective function by $-MR_1$.

The resulting LPP is given as

$$\max \quad z = -x_1 + 3x_2 - MR_1$$
$$\text{s.t.} \quad x_1 + 2x_2 - s_1 + R_1 = 2$$
$$3x_1 + x_2 + s_2 = 3$$

$$x_1 + s_3 = 4$$
$$x_1, x_2, s_1, s_2, s_3 \geq 0, \quad M > 0 \text{ (big number)}$$

The above form of the LPP is said to be in *standard form -II*, whereas *standard form-I* means just standard form (excluding artificial variables). Here, the modified coefficient matrix

$$\begin{bmatrix} 1 & 2 & -1 & 1 & 0 & 0 \\ 3 & 1 & 0 & 0 & 1 & 0 \\ 1 & 0 & 0 & 0 & 0 & 1 \end{bmatrix}$$

has the identity matrix as submatrix. As usual, the starting table is

B.V.	x_1	x_2	s_1	R_1	s_2	s_3	Soln
z	1	-3	0	M	0	0	0
R_1	1	2	-1	1	0	0	2
s_2	3	1	0	0	1	0	3
s_3	1	0	0	0	0	1	4

But this is not a simplex table because R_1 is in BFS, its relative cost must be zero. Hence we perform row operation to reduce to zero, and for this purpose subtract M times of R_1-row from z-row. This operation converts the starting table into simplex format. The next table is

B.V.	x_1	$x_2 \downarrow$	s_1	R_1	s_2	s_3	Soln
z	$1-M$	$-3-2M$	M	0	0	0	$-2M$
$\leftarrow R_1$	1	$\boxed{2}$	-1	1	0	0	2
s_2	3	1	0	0	1	0	3
s_3	1	0	0	0	0	1	4

The above table is in simplex format. Now, the computational procedure just involves simplex iterations. x_2 enters and R_1 leaves to give the following table

B.V.	x_1	$x_2 \downarrow$	s_1	R_1	s_2	s_3	Soln
z	10	0	$-3/2 \downarrow$	$\times$	0	0	3
x_2	1/2	1	$-1/2$	$\times$	0	0	1
$\leftarrow s_2$	5/2	0	$\boxed{1/2}$	$\times$	1	0	2
s_3	1	0	0	$\times$	0	1	4

In next iteration s_1 enters and s_2 leaves to give the next table

B.V.	x_1	x_2	s_1	R_1	s_2	s_3	Soln
z	5	0	0	×	3	0	9
x_2	3	1	0	×	1	0	3
s_1	5	0	1	×	2	0	4
s_3	1	0	0	×	0	1	4

This is optimal table, since all $z_j - c_j \geq 0$, and

optimal solution: $x_1 = 0$, $x_2 = 3$, max value $z = 9$.

Remarks. 1. Once any artificial basis has left the basis it should not be permitted for re-entry.

2. The question arises how much big M should be selected when the computer code is to be executed. Any way M is positive finite and large, but it should not be too large to produce accumulation of round off errors during simplex iterations. To justify the choice we quote an example

$$\max \quad z = 0.2x_1 + 0.5x_2$$
$$\text{s.t.} \quad 3x_1 + 2x_2 \geq 6$$
$$x_1 + 2x_2 \leq 4$$
$$x_1, x_2 \geq 0$$

With $M = 10$, computer gives the optimal solution $x_1 = 1, x_2 = 1.5$, while with $M = 999999$ the optimal solution is $x_1 = 4, x_2 = 0$. Note that first solution is correct. This is the reason we avoid to implement big M-method on computer codes.

3.4 Two Phase Method

As usual we write the LPP in standard form and seek for the presence of identity submatrix in coefficient matrix A. When A does not contain the identity submatrix, the addition of artificial variable is used to do so. Instead of big-M method, we may also use two phase method to solve such problems. This method is computationally efficient.

During Phase-I, we find a basic feasible solution of the system of constraints with the help of an auxiliary objective function to be minimized by using simplex iterations.

Once Phase-I is done, we go for Phase-II that tests whether the BFS obtained in Phase-I is optimal in reference to the main objective function. In case this BFS is not optimal, we continue further simplex iterations to reach at optimality.

We solve a problem to demonstrate the complete working of the method.

Example 6. Solve the following LPP using two phase method

$$\min \quad z = 4x_1 + 6x_2 + 5x_3$$
$$\text{s.t.} \quad 2x_1 + 4x_2 + 3x_3 \geq 32$$
$$x_1 + 2x_2 + 4x_3 \geq 28$$
$$x_1, x_2, x_3 \geq 0$$

Write the problem in standard form

$$\min \quad z = 4x_1 + 6x_2 + 5x_3$$
$$\text{s.t.} \quad 2x_1 + 4x_2 + 3x_3 - s_1 = 32$$
$$x_1 + 2x_2 + 4x_3 - s_2 = 28$$
$$x_1, x_2, x_3 \geq 0$$

This is very much obvious that the coefficient matrix A does not contain the identity submatrix. Hence artificial variables are introduced for the purpose of developing identity submatrix as

$$\min \quad z = 4x_1 + 6x_2 + 5x_3$$
$$\text{s.t.} \quad 2x_1 + 4x_2 + 3x_3 - s_1 + R_1 = 32$$
$$x_1 + 2x_2 + 4x_3 - s_2 + R_2 = 28$$
$$x_1, x_2, s_1, s_2, R_1, R_2 \geq 0.$$

This is standard form-II in which the coefficient matrix of the LPP contains identity submatrix. The LPP is now solved in two phases as shown in Tables 1 and 2.

Phase-I. We construct an auxiliary LPP (always a minimization problem) and its objective function is the sum of artificial variables which have been used to create identity submatrix in A. The optimal solution of the auxiliary will give a basic feasible solution of the constraint system of the original problem. In case there are more than two extreme points the auxiliary LPP will have alternate optimal solutions, see Section 3.5. However, we need only one optimal solution of the auxiliary LPP to move to Phase-II iterations.

$$\begin{aligned}
\min \quad & r = R_1 + R_2 \\
\text{s.t.} \quad & 2x_1 + 4x_2 + 3x_3 - s_1 + R_1 = 32 \\
& x_1 + 2x_2 + 4x_3 - s_2 + R_2 = 28 \\
& x_1, x_2, s_1, s_2, R_1, R_2 \geq 0.
\end{aligned}$$

Obviously, the minimum of the auxiliary problem is $r = 0$, and this occurs when $R_1 = R_2 = 0$. Thus the optimal solution of the above LPP gives BFS of of the constraint system. We solve the LPP by using the simplex method, and the solution is given in Table 4.

Table 4

B V	x_1	x_2	$x_3 \downarrow$	s_1	s_2	R_1	R_2	Soln
r	3	6	7	-1	-1	0	0	60
	0	0	0	0	0	-1	-1	0
R_1	2	4	3	-1	0	1	0	32
$\leftarrow R_2$	1	2	$\boxed{4}$	0	-1	0	1	28
r	5/4	5/2 $\downarrow$	0	-1	3/4	0	$-7/4$	11
$\leftarrow R_1$	5/4	$\boxed{5/2}$	0	-1	3/4	1	$-3/4$	11
x_3	1/4	1/2	1	0	$-1/4$	0	1/4	7
r	0	0	0	0	0	-1	-1	0
x_2	1/2	1	0	$-2/5$	3/10	2/5	$-3/10$	22/5
x_3	0	0	1	1/5	$-2/5$	$-1/5$	2/5	24/5

Remark. In Phase-I of two phase method, if any artificial variable remains at zero level in the basis, then degenerate BFS is available. Suppose the kth artificial variable $R_k = 0$ in BFS. Then two possibilities are there.

(i) If $e_k^T B^{-1} A_j \neq 0$ for any nonbasic variable x_j, then we can bring x_j to the current basis to replace R_k. In this case, the optimal solution to the Phase-I problem provides a starting basis without any artificial variable in it for the original problem.

(ii) If $e_k^T B^{-1} A_j = 0$ for every nonbasic variable x_j, then it can be verified that the kth constraint of the system $AX = b$ is redundant. In this case we can remove the redundant row from the original constraints and restart the Phase-I problem.

Here e_k^T is the m-dimensional vector having 1 at kth place and elsewhere zero.

Phase-II. In second phase we use the basic feasible solution available from the Phase I to find the optimal solution of the original problem. To construct next table, we first write body matrix ignoring artificial variables and all nonbasic variables which have negative relative cost in the optimal table of Phase-I. Also, write the entries of solution column of Table 1.

Now, insert the objective function coefficients in z-row, i.e., the coefficients from the left hand side of the equation

$$z - 4x_1 - 6x_2 - 5x_3 - 0s_1 - 0s_2 = 0,$$

and right hand side of this equation below solution. This disturbs simplex format in z-row as x_2 and x_3 are the basic variables but in z-row we have inserted -6 and -5, respectively. Restore simplex format by adding six times of x_2-row and five times of x_3-row in corresponding entries of z-row. This gives simplex format, see Table 5.

Table 5

B V	x_1	x_2	x_3	s_1	s_2	Soln
z	-1	0	0	$-7/5$	$-1/5$	$252/5$
	-4	-6	-5	0	0	0
x_2	$1/2$	1	0	$-2/5$	$3/10$	$22/5$
x_3	0	0	1	$1/5$	$-2/5$	$24/5$

Now, check Table 5 for optimality. Since all $z_j - c_j \leq 0$, Table 5 is optimal with

optimal solution: $x_1 = 0$, $x_2 = 22/5$, $x_3 = 24/5$, $z = 252/5$.

Note that while simplex format is restored and the optimality is not visible in z-row, then continue further with simplex iterations to reach at optimality.

Remarks. 1. While going to Phase II, delete all nonbasic variables for which the relative cost $z_j - c_j < 0$ in optimal table of the Phase I. This is the *Elimination rule* which will be well placed in 'Goal Programming' at later stage. This is due to the fact that nonbasic variables with negative relative cost in optimal table of Phase-I will never enter the basis in Phase-II, otherwise the solution of auxiliary LPP will become infeasible.

For instance, if $x_1, x_2, \ldots, x_m$ are basic variables and $x_{m+1}, x_{m+2}, \ldots, x_n$ and R_i are nonbasic variables then objective function row in optimal table of Phase-I can be written as

$$r = 0 \cdot x_1 + 0 \cdot x_2 + \cdots + 0 \cdot x_m + k_1 x_{m+1} + k_2 x_{m+1} + \cdots + k_n x_n$$
$$+ r_1 R_1 + \cdots + r_m R_m,$$

where k_i and r_i are nonnegative. If any of the nonbasic variables from $x_{m+1}, x_{m+2}, \ldots, x_n$ enters in basis at positive level in Phase-II, then r will have positive value which results infeasibility in auxiliary LPP, since set of constraints set is same for both phases.

2. In case any LPP contains an equation which has a variable (may be decision variable) with unity coefficient and this variable is not contained in other constraints, then take this variable in the starting BFS to facilitate the computations.

3. In Phase I, the auxiliary LPP is always a minimization problem.

4. We suggest a good variant of two phase method which does not require the addition of artificial variables to the constraints to develop identity submatrix inside the coefficient matrix of a LPP in standard form.

Let us workout the example of this section again. Write the LPP in standard form. Find its BFS by reducing the coefficient matrix in row reduced echelon form (canonical form of equations) maintaining feasibility in the last column.

$$\begin{bmatrix} 2 & 4 & 3 & -1 & 0 & 32 \\ 1 & 2 & 4 & 0 & -1 & 28 \end{bmatrix} \simeq \begin{bmatrix} 1 & 2 & -1 & -1 & 1 & 4 \\ 1 & 2 & 4 & 0 & -1 & 28 \end{bmatrix} \simeq$$

$$
\begin{bmatrix} 1 & 2 & -1 & -1 & 1 & 4 \\ 0 & 0 & 5 & 1 & -2 & 24 \end{bmatrix} \simeq \begin{bmatrix} 1 & 2 & -1 & -1 & 1 & 4 \\ 0 & 0 & 1 & 1/5 & -2/5 & 24/5 \end{bmatrix} \simeq
$$

$$
\begin{bmatrix} 1 & 2 & 0 & -4/5 & 3/5 & 44/5 \\ 0 & 0 & 1 & 1/5 & -2/5 & 24/5 \end{bmatrix}
$$

Thus, x_1 and x_3 function as starting basic variables. On inserting the objective function in z-row, we get

B V	x_1	x_2	x_3	s_1	s_2	Soln
z	−4	−6	−5	0	0	0
x_1	1	2	0	−4/5	3/5	44/5
x_3	0	0	1	1/5	−2/5	24/5

Note that after insertion of the objective function the simplex format is disturbed as there must be zero in z-row below basic variables x_1 and x_3. To restore simplex format, add four times of the first row and five times of second row to the z-row. The next table is in simplex format

B V	x_1	$x_2 \downarrow$	x_3	s_1	s_2	Soln
z	0	2	0	−11/5	2/5	296/5
← x_1	1	$\boxed{2}$	0	−4/5	3/5	44/5
x_3	0	0	1	1/5	−2/5	24/5

x_2 enters and x_1 leaves the basis, and we get

B V	x_1	x_2	x_3	s_1	s_2	Soln
z	−4	0	0	−7/5	−1/5	252/5
x_2	1/2	1	0	−2/5	3/10	22/5
x_3	0	0	1	1/5	−2/5	24/5

This is the optimal table and optimal solution is the same as computed by two phase method.

3.5 Exceptional Cases in LPP

Here, we discuss some situations which are encountered during simplex iterations.

1. Non-existing feasible solution. This situation arises when no point satisfies all the constraints, i.e., $P_F = \emptyset$. If the standard form of LPP amenable to simplex iterations does not involve artificial variable then this situation does not arise. The following two situations characterize the occurrence of infeasible solutions.

(i) If, in the optimal table of big-M method, at least one of the artificial is nonzero; or

(ii) In the optimal table of Phase I of two phase method, if at least one artificial variable is nonzero, then the problem has infeasible solution.

For example consider the problem

$$\begin{aligned}
\min \quad & z = x_1 - x_2 \\
\text{s.t.} \quad & x_1 + x_2 \leq 1 \\
& 3x_1 + 4x_2 \geq 6 \\
& x_1, x_2, \geq 0.
\end{aligned}$$

Let us perform Phase-I to find initial basic feasible solution. The auxiliary problem is

$$\begin{aligned}
\min \quad & r = R_2 \\
\text{s.t.} \quad & x_1 + x_2 + s_1 = 1 \\
& 3x_1 + 4x_2 - s_2 + R_2 = 6 \\
& \text{all var} \geq 0.
\end{aligned}$$

B V	x_1	$x_2 \downarrow$	s_2	s_1	R_2	Soln
z	3	4	−1	0	0	6
	0	0	0	0	−1	0
$\leftarrow x_1$	1	$\boxed{1}$	0	1	0	1
R_2	3	4	−1	0	1	6
z	−1	0	−1	−4	0	2
x_2	1	1	0	1	0	1
R_2	−1	0	−1	−4	1	2

Since in the last table we can not proceed for further iterations and even then the artificial variable $R_2 > 0$, this indicates that the problem has infeasible solution.

2. Unbounded solution. This may happen when the feasible region is unbounded. The feasible region is unbounded if while applying simplex method, it is observed that all entries of the column in body matrix corresponding to some nonbasic variable are nonpositive. In other words, if in any simplex iteration, the minimum ratio rule fails, the LPP has unbounded solution, i.e., appropriate nonbasic variable desires to enter the basis but $\alpha^j \leq 0$ do not permit its entry. The solution becomes unbounded because the entering variable can enter the basis at an arbitrary level. The following LPP gives unbounded solution.

$$\begin{aligned} \max \quad & z = 4x_1 + x_2 \\ \text{s.t.} \quad & x_1 - x_2 \leq 1 \\ & -2x_1 + x_2 \leq 2 \\ & x_1, x_2, \geq 0. \end{aligned}$$

The standard form of the LPP is

$$\begin{aligned} \max \quad & z = 4x_1 + x_2 \\ \text{s.t.} \quad & x_1 - x_2 + s_1 = 1 \\ & -2x_1 + x_2 + s_2 = 2 \\ & x_1, x_2, s_1, s_2 \geq 0. \end{aligned}$$

The simplex iterations are

B V	$x_1 \downarrow$	x_2	s_1	s_2	Soln
z	-4	-1	0	0	0
$\leftarrow s_1$	$\boxed{1}$	-1	1	0	1
s_2	-2	1	0	1	2
z	0	$-5 \downarrow$	4	0	4
x_1	1	-1	1	0	1
s_2	0	-1	2	1	4

After first iteration we do not get optimal table and hence second iteration is performed. In second iteration the variable x_2 with the most negative relative cost desires to be a basic variable but $\alpha_1^2 < 0, \alpha_2^2 < 0$ prohibit its entry into basic vector. Thus, we conclude that the problem has an unbounded solution.

Remark. The condition of unboundedness of feasible region is necessary but not sufficient. There are linear programming problems in which the feasible region is unbounded even though the solution is finite. One such example is

$$\min \quad z = -4x_1 + x_2$$
$$\text{s.t.} \quad x_1 - 2x_2 \leq 2$$
$$- 2x_1 + x_2 \leq 2$$
$$x_1, x_2, \geq 0.$$

By sketching the feasible region, we observe that minimum value occurs at the vertex $(2, 0)$. However, the feasible region is unbounded.

3. Alternative optimal solution. If in the optimal table (obtained from any method), the relative cost $z_j - c_j = 0$ for at least one of the nonbasic variables, then alternate optimal solution exits provided P_F is bounded. Bring this nonbasic variable into basis and find a new optimal solution. Again, if the optimality occurs at two or more vertices, then it also occurs at clc of these vertices, see Problem 15, Problem set 2. However, an important fact is pointed out here that that clc of optimal basic feasible solutions may not be basic solution, see Problem 16, Problem set 2. We consider one example in which it is shown how alternative optimal solution is computed. Consider the LPP

$$\max \quad z = 4x_1 + 10x_2$$
$$\text{s.t.} \quad 2x_1 + x_2 \leq 10$$
$$2x_1 + 5x_2 \leq 20$$
$$2x_1 + 3x_2 \leq 18$$
$$x_1, x_2 \geq 0$$

The LPP is amenable to solution by simplex method. We take up only the optimal table (outcome of the last iteration).

BV	$x_1 \downarrow$	x_2	s_1	s_2	s_3	Soln
z	0	0	0	2	0	40
$\leftarrow s_1$	8/5	0	1	−1/5	0	6
x_2	2/5	1	0	1/5	0	4
s_3	4/5	0	0	−3/5	1	6

Optimal solution: $x_1 = 0$, $x_2 = 4$, max value $z = 40$.

From the above table, note that x_1, s_2 are nonbasic variables and relative cost of x_1 is zero and this implies that alternative optimal solution exists. Choose x_1 as the entering variable, and consequently minimum ratio rule permits s_1 to leave. The remaining computations are done as usual to have the alternative optimal table as

BV	x_1	x_2	s_1	s_2	s_3	Soln
z	0	0	0	2	0	40
x_1	1	0	5/8	−1/8	0	15/4
x_2	0	1	−1/4	1/4	0	5/2
s_3	0	0	−1/2	−1/2	1	3

Optimal solution: $x_1 = 15/4$, $x_2 = 5/2$, max value $z = 40$.

Thus, both the optimal solutions are different giving the same maximum objective value. Note that every convex linear combination of these solutions will again be an optimal solution. However, all the optimal solutions obtained by taking clc will be non-basic feasible optimal solutions. For instance, in this case,

$$X = \frac{1}{2}(0, 4, 6, 6) + \frac{1}{2}\left(\frac{15}{4}, \frac{5}{2}, 0, 0, 3\right)$$

$$= \left(\frac{15}{4}, \frac{9}{2}, 0, 3, \frac{9}{2}\right)$$

is the nonbasic feasible optimal solution.

4. Degeneracy. One of the reasons for degeneracy may be due to presence of some redundant constraint. The system is redundant if one or more constraints in a LPP are not at all essential to find the optimal solution. For example the system

$$3x_1 + 2x_2 \leq 6$$

$$2x_1 + x_2 \geq 4$$

$$x_1 + x_2 \geq 3$$

$$x_1, x_2 \geq 0$$

has last constraint as redundant. It is suggested to eliminate such constraints before we proceed for simplex iterations, otherwise degeneracy will occur in simplex iterations. Phase-I may be helpful to detect redundancy, see Remark just after Phase-I in Section 3.4.

However, in absence of redundant constraint the degeneracy may occur in the LPP. If in any simplex table, there is tie between two or more leaving variables we can select any one of them to leave the basis but the new solution thus obtained will have remaining such variables (tied) at zero level in the next simplex table. It means new BFS will be degenerate.

Suppose at some stage of simplex iterations s_2 enters, and the minimum ratios for two basic variables are same, i.e., two basic variables are candidates for leaving at the same time. This will give next BFS as degenerate. The following table is self explanatory in this regard.

B V	x_1	x_2	s_1	$s_2 \downarrow$	s_3	Soln
z						
$\leftarrow x_1$				α		α'
$\leftarrow x_2$				β		β'
s_1				γ		γ'
z				0		
s_2				1		α'/α
x_2				0		$\beta' - \beta\alpha'/\alpha = 0$
s_1				0		$\gamma' - \gamma\alpha'/\alpha$

We summarize the above discussion as

Degeneracy (due to tie among leaving variables) causes three possibilities:

(i) Temporary degeneracy. After some iteration degeneracy disappears and nondegenerate optimal solution is obtained.

(ii) Permanent degeneracy. Degenerate optimal solution is obtained.

(iii) Cyclic degeneracy. Simplex table starts repeating after some iterations.

In the case of permanent and temporary degeneracy the nonbasic variables which have a tie to leave the basis can be chosen at random as the leaving variable. However, it can not be done when cycle is detected. Cycling can be detected at early stage by noting the fact that for tied variables

$$\min_i \left\{ \frac{x_i}{\alpha_i^j}, \ \alpha_i^j > 0 \right\} = 0.$$

Remark. Whatever type of degeneracy occurs in a LPP, we suggest that Charne's perturbation method should be applied to solve the problem. Consider the problem:

$$\text{opt} \quad z = C^T X$$
$$\text{s.t.} \quad AX = b$$
$$X \geq 0, \ b \geq 0$$

The requirement vector b, perturbed to $b(\varepsilon)$ is given by

$$b(\varepsilon) = b + \sum_{j=1}^{n} \varepsilon^j A_j + \sum_{i=1}^{n+s} \varepsilon^{n+1} q_i,$$

where q_i is the ith column corresponding to ith artificial variable. Since $X_B = B^{-1}b$, we have

$$X_{B(\varepsilon)} = B^{-1}b + \sum_{j=1}^{n} \varepsilon^j B^{-1} A_j + \sum_{i=1}^{n+s} \varepsilon^{n+1} B^{-1} q_i$$

or

$$X_{B(\varepsilon)} = X_B + \sum_{j=1}^{n+s} \varepsilon^j \alpha^j.$$

Now, take the kth component of $X_{B(\varepsilon)}$ as

$$X_{B_k(\varepsilon)} = X_{B_k} + \sum_{i=1}^{n+s} \varepsilon^j \alpha_k^j.$$

Let $x_k(A_k)$ enter the basis. The variable $X_{B_r(\varepsilon)}$ will leave the basis, if

$$\frac{X_{B_r(\varepsilon)}}{\alpha_r^k} = \min_i \left\{ \frac{X_{B_i(\varepsilon)}}{\alpha_i^k}, \ \alpha_i^k > 0 \right\}$$

and

$$Q_k(\varepsilon) = \min_i \left\{ \frac{X_{B_i(\varepsilon)}}{\alpha_i^k} + \sum_{j=1}^{n+s} \varepsilon^j \frac{\alpha_i^j}{\alpha_i^k}, \ \alpha_i^k > 0 \right\}$$

$$= \min_i \left\{ Q_k + \sum_{j=1}^{n+s} \varepsilon^j \frac{\alpha_i^j}{\alpha_i^k}, \ \alpha_i^k > 0 \right\}.$$

If Q_k is not unique then cycling may occur. To avoid this, examine

$$\min\left\{\frac{\alpha_i^1}{\alpha_i^k},\ \alpha_i^k > 0\right\} \qquad \text{(i)}$$

$$\min\left\{\frac{\alpha_i^2}{\alpha_i^k},\ \alpha_i^k > 0\right\} \qquad \text{(ii)}$$

$$\vdots$$

$$\min\left\{\frac{\alpha_i^{n+s}}{\alpha_i^k},\ \alpha_i^k > 0\right\} \qquad \text{(n+s)}$$

If (i) is unique, then stop and we have decided the leaving variable. In case (i) is not unique then proceed to (ii) and so on until uniqueness is achieved. Charnes has claimed that in proceeding like this way the uniqueness is necessarily obtained.

Example 7. Solve the following LPP by the simplex method.

$$\begin{aligned}
\max \quad & z = 2x_1 + x_2 \\
\text{s.t.} \quad & 4x_1 + 3x_2 \leq 12 \\
& 4x_1 + x_2 \leq 8 \\
& 4x_1 - x_2 \leq 8 \\
& x_1, x_2, \geq 0.
\end{aligned}$$

Procedure. The standard form of the LPP is

$$\begin{aligned}
\max \quad & z = 2x_1 + x_2 \\
\text{s.t.} \quad & 4x_1 + 3x_2 + s_1 = 12 \\
& 4x_1 + x_2 + s_2 = 8 \\
& 4x_1 - x_2 + s_3 = 8 \\
& x_1, x_2, s_1, s_2, s_3 \geq 0.
\end{aligned}$$

Obviously, the coefficient matrix A contains identity submatrix. Rewrite the matrix A as A' (including columns of artificial variables, if any) in such a way that the first column of A' is the first column of identity matrix contained in A, second column of A' is the second column of identity submatrix contained in A' and so on till identity submatrix columns are exhausted. In the last, we write columns of the decision variables.

Note that when simplex method is applied, there is a tie between s_2 and s_3 to leave the basis, see the table, and hence degeneracy occurs. To remove degeneracy find the basic variable for which

$$\frac{\text{Elements of the first column (facing tied variables) of unit matrix}}{\text{corresponding elements of entering column with key element}},$$

or

$$\frac{\text{Elements of the second column (facing tied variables) of unit matrix}}{\text{corresponding elements of entering column with key element}}$$

or so on (till all basic variables are exhausted), is unique. This basic variable will leave the basis. For tied variables s_2 and s_3, using above formula, we have $\min\{0/4, 0/4\} = 0$ is not unique. This informs that cycling may happen. Move to the next column of the unit matrix and note that $\min\{1/4, 0/4\} = 0$ is unique, and this is for s_3. Thus, to avoid cycling s_3 must be chosen as the leaving variable for the first iteration. The second and third iterations are carried out by usual minimum ration rule. The whole working is demonstrated in the following combined table showing all iteration to reach at optimality.

Table 6

B V	s_1	s_2	s_3	$x_1 \downarrow$	x_2	Soln
z	0	0	0	-2	-1	0
s_1	1	0	0	4	3	12
s_2	0	1	0	4	1	8
$\leftarrow s_3$	0	0	1	$\boxed{4}$	-1	8
z	0	0	1/2	0	$-3/2 \downarrow$	4
s_1	1	0	-1	0	4	4
$\leftarrow s_2$	0	1	-1	0	$\boxed{2}$	0
x_1	0	0	1/4	1	$-1/4$	2
z	0	3/4	$-1/4 \downarrow$	0	0	4
$\leftarrow s_1$	1	-2	$\boxed{1}$	0	0	4
x_2	0	1/2	$-1/2$	0	1	0
x_1	0	1/8	1/8	1	0	2

Table 6 (Contd.)

B V	s_1	s_2	s_3	x_1	x_2	Soln
z	1/4	1/4	0	0	0	5
s_3	1	−2	1	0	0	4
x_2	1/2	−1/2	0	0	1	2
x_1	−1/8	3/8	0	1	0	3/2

Optimal solution: $x_1 = 3/2$, $x_2 = 2$; max value $= 5$.

Remark. To avoid cycling there is a very simple method called *Bland's rule*. It specifies the choice of both the entering and leaving variable. In this rule variables are first ordered in sequence, then

1. Among all nonbasic variables with negative reduced costs (max problem), choose the one with the smallest index to enter the basis.

2. When there is a tie for in the minimum ratio test, choose the basic variable with the smallest index to leave the basis.

Note that according to Bland's rule in the first iteration x_1 enters and s_2 leaves the basis and the above LPP (which has been solved by Charne's perturbation) is solved just in two iterations.

This example pertains to the temporary degeneracy. For cycling degeneracy, see Problem 18. Apply Charne's perturbation method or Bland's rule to deal with cycling process.

Finally, we commend how Bland's rule works. It creates the following monotone property:

If a variable x_q enters the basis, then it can not leave the basis until some other variable with a larger index, which was nonbasic when x_q enters, also enters the basis. This monotone property prevents cycling, because in a cycle any variable that enters the basis must also leave the basis, which implies that there is some largest indexed variable that enters and leave the basis. This certainly contradicts the monotone property.

Problem Set 3

1. Let the LPP be min $z = C^T X$ subject to $AX = b$, $X \geq 0$, $b \geq 0$. If, for any BFS, X_B all $z_j - c_j \leq 0$, show that X_B is the optimal solution of the problem.

2. Find the optimal solutions of the following LPP's by the simplex method

   (a) max $z = x_2 + 3$
       s.t. $4x_1 + x_2 \leq 5$
             $-2x_1 + x_2 \leq 1$
             $x_1, x_2 \geq 0$

   (b) max $z = 3x_1 + 4x_2$
       s.t. $-x_1 + x_2 \leq 5$
             $x_1 + x_2 \leq 4$
             $x_1 \geq 2, x_2 \geq 1$

   (c) max $z = -3x_1 + x_2$
       s.t. $\dfrac{x_1}{2} + x_2 \leq 1$
             $-x_1 + x_2 \leq 1$
             $x_1, x_2 \geq 0$

   (d) max $z = -(x_1 + 0.6x_2)$
                $+ 0.4(5x_1 + 3x_2)$
       s.t. $\dfrac{x_1}{2} + x_2 \leq 30$
             $2x_1 + x_2 \leq 40$
             $x_1, x_2 \geq 0$

3. The optimal table of a LPP is given as

B V	x_1	x_2	s_1	s_2	Soln
z	2	0	0	1	1
s_1	1	0	1	3	4
x_2	-2	1	0	1	1

   Given the staring BFS as $(s_1, s_2)^T$. By doing one simplex iteration back in the above table write the original LPP.

4. For a maximization problem, the starting table is given as

B V	x_1	x_2	s_1	s_2	Soln
z	-6	2	0	0	0
s_1	2	-1	1	0	2
s_2	1	0	0	1	4

Perform one simplex iteration to find basic solution that is not feasible.

Suggestion. Violate the leaving variable rule.

5. Solve the example of Section 3.4, if the objective function is taken as $-2x_1 + x_2$ and constraints remain the same.

6. Solve the following LPP using simplex method

$$\begin{aligned}
\max \quad & z = 2x_1 - 3x_2 \\
\text{s.t.} \quad & -x_1 + x_2 \leq 2 \\
& 2x_1 - x_2 \leq 2 \\
& -x_1 - x_2 \leq 2 \\
& x_1 \geq 0, \ x_2 \text{ unrestricted}
\end{aligned}$$

Remark. Below nonbasic variable y_2, $z_2 - c_2 = 0$ in the optimal table, but alternative optimal BFS does not exist. Thus, if in optimal table below nonbasic variable $z_j - c_j = 0$, then alternative BFS solution is expected but not essential. Here, it happened due to unboundedness of P_F. Any way other alternate optimal solutions will be nonbasic.

7. Let $\max z = C^T X$, subject to $AX = b$, $X \geq 0$, $b \geq 0$. (a) How the optimal solution is affected when the cost vector C is replaced by λC, $\lambda > 0$; (b) Interpret the assertion (a) when C is replaced by $C + \lambda$, $\lambda \in \mathbb{R}$.

Suggesting. Let X^* be an optimal solution. Then $C^T X^* \geq C^T X$ for all X. Replacing C by λC, $\lambda > 0$, we have $(\lambda C)^T X^* = \lambda C^T X^* \geq (\lambda C^T)X = \lambda C^T X$. Thus, X^* remains optimal solution of the changed problem.

For part (b) $(C + \lambda)^T X^* = C^T X^* + \lambda X^{*^T}$. So, for $\lambda > 0$ no change occurs in the optimal solution, while for $\lambda < 0$, $(C + \lambda)^T X^* \leq (C + \lambda)^T X$ ensures a change in the optimal value.

8. The following is an optimal table of a LPP with $\leq$ constraints

and nonnegative right-hand side vector.

B V	x_1	x_2	x_3	x_4	s_1	s_2	s_3	Soln
z	0	4	0	1	4	2	0	40
	0	1/2	1	0	1/2	0	0	5/2
	1	−1/2	0	1/3	−1/6	1/3	0	5/2
	0	0	0	2/3	−1/3	−1/3	1	3

(i) What variables are in the basis?

(ii) What is the original LPP?

Suggestion. Use $(c_3, c_1, 0)B^{-1} = (4, 2, 0)^T$. Use this C_B^T to find c_2, c_4. Finally, compute A_j's and (b_1, b_2, b_3).

9. Consider the following LPP

$$\max \quad x_0 = 3x_1 + 2x_2 + 5x_3$$
$$\text{s.t.} \quad x_1 + 2x_2 + x_3 \leq 430$$
$$- 3x_1 - 2x_3 \geq -460$$
$$x_1 + 4x_2 \leq 420$$
$$x_1, x_2, x_3 \geq 0$$

One of the simplex iteration table of this LPP is

B V	x_1	x_2	x_3	s_1	s_2	s_3	Soln
x_0	4						
x_2					−1/4		
x_3					1/2		
s_3					1		

Without performing simplex iteration, find the following missing entries in the above table. Justify your answer by writing the formula used and doing all calculations.

(i) x_1-column;

(ii) Entry below s_2 in x_0-row;

(iii) The BFS corresponding to the above table;

(iv) The value of the objective function corresponding to the above table.

10. Solve the following LPP using big M-method and two phase method, separately.

$$\text{max} \quad z = 3x_1 - 3x_2 + x_3$$
$$\text{s.t.} \quad x_1 + 2x_2 - x_3 \geq 5$$
$$- 3x_1 - x_2 + x_3 \leq 4$$
$$x_1, x_2, x_3 \geq 0.$$

11. Consider the system

$$x_1 + x_2 \geq 1$$
$$- 2x_1 + x_2 \geq 2$$
$$2x_1 + 3x_2 \leq 7$$
$$x_1, x_2 \geq 0.$$

   (a) Using simplex algorithm find BFS of the system;
   (b) Find a BFS so that both decision variable function as basic variables.

   Suggestion. Use only the first phase of two phase method.

12. Detect the redundant constraint in the following system

$$x_1 + x_2 + x_3 \geq 7$$
$$2x_1 + 3x_2 + 3x_3 \geq 1$$
$$3x_1 + 4x_2 + 2x_3 \geq 5$$
$$x_1, x_2, x_3 \geq 0$$

13. Avoiding big M-method or two phase method find the optimal solution of the following LPP by simplex method

$$\text{max} \quad z = x_1 + x_2 + x_3 + x_4 + x_5$$
$$\text{s.t.} \quad 3x_1 + 2x_2 + x_3 = 1$$
$$5x_1 + x_2 + x_3 + x_4 = 2$$
$$2x_1 + 5x_2 + x_3 + x_5 = 4$$
$$x_1, x_2, x_3, x_4, x_5 \geq 0$$

   Suggestion. See Remark (4) in Section 3.4.

14. During simplex iterations if the minimum ratio rule fails, then the LPP has an unbounded solution. Establish the validity of this statement mathematically.

Suggestion. Let the LPP be a max problem and $(x_{B_1}, x_{B_2}, \ldots, x_{B_m})$ is any BFS and $B = [B_1, B_2, \ldots, B_m]$ is the corresponding basis matrix. Suppose a nonbasic variable x_k (A_k) wants to enter the basis. Then

$$\sum_{i=1}^{m} x_{B_i} B_i = b \text{ or } \sum_{i=1}^{m} x_{B_i} B_i - \theta A_k + \theta A_K = b.$$

Since $\alpha^k = B^{-1} A_k$, it follows that

$$\sum_{i=1}^{m} \left(x_{B_i} - \theta \alpha_i^k \right) B_i + \theta A_k = b.$$

Fix $\theta > 0$. Then the above equation gives a feasible solution, but not basic as $m + 1$ variables are nonzero. The value of the objective function for this solution is

$$\hat{z} = \sum_{i=1}^{m} c_{B_i} \left(x_{B_i} - \theta \alpha_i^k \right) + \theta c_k \text{ or } \hat{z} = z - \theta(z_k - c_k).$$

Since θ is arbitrary but fixed, $\hat{z}$ can be made arbitrarily large with θ.

15. Consider the LPP

$$\begin{aligned}
\max \quad & z = 20x_1 + 10x_2 + x_3 \\
\text{s.t.} \quad & 3x_1 - 3x_2 + 5x_3 \le 50 \\
& x_1 + x_3 \le 10 \\
& x_1 - x_2 + 4x_2 \le 20 \\
& x_1, x_2, x_3 \ge 0
\end{aligned}$$

By inspection determine the direction (x_1, x_2 or x_3) in which the solution space is unbounded.

16. Consider the LPP

$$\begin{aligned}
\max \quad & x_0 = 3x_1 - x_2 - x_3 \\
\text{s.t.} \quad & x_1 + 2x_2 + 4x_3 \le 5 \\
& x_1 - x_2 - x_3 \le 8 \\
& x_1, x_2, x_3 \text{ unrestricted in sign}
\end{aligned}$$

Find the optimal solution of the above LPP using four restricted decision variables.

Suggestion. Use Proposition of Section 1.2.

17. Find the optimal solution of the following LPP using simplex method

$$\begin{aligned}
\max \quad & z = -x_1 + 2x_2 - x_3 \\
\text{s.t.} \quad & 3x_1 - x_2 + 2x_2 \leq 7 \\
& -2x_2 + 4x_2 \leq 8 \\
& -4x_1 + 3x_2 + 8x_3 \leq 1 \\
& x_1, x_2, x_3 \geq 0
\end{aligned}$$

Does there exist alternative optimal solution? if yes, find this solution.

18. In a max (min) LPP, there is a tie between x_1 and x_2 to enter the basis, provided we look at relative cost of these variables. Given that $\theta_1 \neq \theta_2$. Which variable should be preferred to enter the basis so that larger increase (decrease) in objective function is ensured?

19. Show that the LPP:

$$\begin{aligned}
\max \quad & z = 3x_1 + 3x_2 \\
\text{s.t.} \quad & 3x_1 + 2x_2 \leq 18 \\
& x_1 \leq 4 \\
& 2x_2 \leq 12 \\
& x_1, x_2 \geq 0
\end{aligned}$$

verifies the statement in Problem 18.

Suggestion. If x_1 is permitted to enter the basis then it takes three iterations to reach at optimal solution, while for selecting x_2 as entering variable it takes only two iterations to reach at optimal solution.

20. Let $z_j - c_j > 0$ for all nonbasic variables in optimal table of a max LPP. Then show that the LPP has a unique solution.

Suggestion. This is equivalent to prove that if $z_j - c_j = 0$ for at least one nonbasic variable then the LPP has alternate optimal solution.

21. Consider the LPP

$$\min \quad z = -\frac{3}{4}x_1 + 20x_2 - \frac{1}{2}x_3 + 6x_4$$

$$\text{s.t.} \quad \frac{1}{4}x_1 - 8x_2 - x_3 + 9x_4 \leq 0$$

$$\frac{1}{2}x_1 - 12x_2 - \frac{1}{2}x_3 + 3x_4 \leq 0$$

$$x_3 \leq 1$$

$$x_1, x_2, x_3, x_4 \geq 0$$

Show that forever cycling occurs when the problem is put on simplex iterations. Use Charnes perturbation technique to find the optimal solution.

Suggestion. By applying simplex method with usual rules, it can be verified that if $\{x_1, x_2, x_3\}$ is the starting basis, then the successive bases are $\{x_4, x_2, x_3\}$, $\{x_4, x_5, x_3\}$, $\{x_6, x_5, x_3\}$, $\{x_6, x_7, x_3\}$, $\{x_1, x_7, x_3\}$ and return to $\{x_1, x_2, x_3\}$. If the same sequence of pivots is repeated again and again, the simplex method will cycle forever among these bases without giving optimal solution. That's why Charne's perturbation method or Bland's rule is recommended to find its optimal solution. This LPP was proposed by E. M. L. Beale in 1955.

Chapter 4

Duality Theory

This chapter is devoted to the dual linear programming problem associated with a linear programming problem. Before defining the dual of a LPP we introduce with its canonical forms. The economic interpretation of duality is also explained in Section 7.3. In the end the dual simplex method is included.

4.1 Dual Linear Program

The notion of duality is one of the most important concept in linear programming. To each linear program defined by the matrix A, right hand side vector b and the cost vector C, there corresponds another linear program with the same set of data A, b and C. To begin with this chapter we introduce duality for LPP with greater than equal to constraints.

Canonical form. A LPP is said to be in canonical form if it can be written as

$$\begin{aligned} \min \quad & x_0 = C^T X \\ \text{s.t.} \quad & AX \geq b \\ & X \geq 0, \end{aligned} \qquad (4.1)$$

as usual A is matrix of order $m \times n$ and

$$C = (c_1, c_2, \cdots, c_n)^T, \ X = (x_1, x_2, \cdots, x_n)^T, b = (b_1, b_2, \cdots, b_m)^T.$$

Note that in canonical form, there is no restriction on the entries b_i as these may be any real numbers.

Definition. The dual LPP of above canonical form is defined as

$$\begin{aligned} \max \quad & y_0 = b^T Y \\ \text{s.t.} \quad & A^T Y \le C \\ & Y \ge 0, \end{aligned} \qquad (4.2)$$

where $Y = (y_1, y_2, \cdots, y_m)^T$, and y_i is the dual variable associated with the ith constraint.

System (4.1) is termed as primal, while (4.2) as its dual.

Example 1. Consider the LPP

$$\begin{aligned} \max \quad & x_0 = 3x_1 + 4x_2 \\ \text{s.t.} \quad & x_1 + 7x_2 \ge 35 \\ & 2x_1 + 5x_2 \le 60 \\ & x_1, x_2 \ge 0 \end{aligned}$$

To write the dual of this LPP, first convert the problem into canonical form

$$\begin{aligned} \min \quad & x_0' = -3x_1 - 4x_2 \\ \text{s.t.} \quad & x_1 + 7x_2 \ge 35 && y_1 \\ & -2x_1 - 5x_2 \ge -60 && y_2 \\ & x_1, x_2 \ge 0, \end{aligned}$$

where y_1 and y_2 are the dual variables associated with the first and second constraints of the primal.

By definition, its dual is written as

$$\max \quad y_0 = (35, -60) \begin{bmatrix} y_1 \\ y_2 \end{bmatrix}$$

$$\text{s.t.} \quad \begin{bmatrix} 1 & -2 \\ 7 & -5 \end{bmatrix} \begin{bmatrix} y_1 \\ y_2 \end{bmatrix} \le \begin{bmatrix} -3 \\ -4 \end{bmatrix}$$

$$y_1, y_2 \ge 0$$

or, in simplified form

$$\begin{aligned} \max \quad & y_0 = 35y_1 - 60y_2 \\ \text{s.t.} \quad & y_1 - 2y_2 \le -3 \\ & 7y_1 - 5y_2 \le -4 \qquad\qquad (4.3) \\ & y_1, y_2 \ge 0 \end{aligned}$$

We detect an interesting phenomenon, if we write further, the dual of (4.3). First, express (4.3) in canonical form

$$\begin{aligned} \min \quad & y_0' = -35y_1 + 60y_2 \\ \text{s.t.} \quad & -y_1 + 2y_2 \ge 3 \\ & -7y_1 + 5y_2 \ge 4 \\ & y_1, y_2 \ge 0, \end{aligned}$$

Now, its dual is

$$\begin{array}{lll} \max \quad z_0' = 3z_1 + 4z_2 & \qquad \text{or} \qquad & \max \quad x_0 = 3x_1 + 4x_2 \\ \text{s.t.} \quad -z_1 - 7z_2 \le -35 & & \text{s.t.} \quad x_1 + 7x_2 \ge 35 \\ \qquad\; 2z_1 + 5z_2 \le 60 & & \qquad\; 2x_1 + 5x_2 \le 60 \\ \qquad\; z_1, z_2 \ge 0 & & \qquad\; x_1, x_2 \ge 0 \end{array}$$

This is nothing but the original LPP from where we started to write the dual and again dual. Thus, we got an idea that dual of dual is primal. We are shortly proving this result theoretically.

So far, we noticed the following points for a minimization problem in canonical form:

(i) Its dual is max problem.

(ii) The objective function of dual is obtained by multiplying the right-hand side entries by the dual variables y_i.

(iii) The constraints of the dual are of the type $\le$.

(iv) The left-hand side of the ith constraint is $A_i^T Y$.

(v) The right-hand side of the dual constraints are respectively the cost coefficients of the primal.

(vi) The dual variables y_i are nonnegative.

Theorem 1. The dual of a dual is primal.

Proof. Let the LPP be given in canonical form (4.1). Then its dual is written as (4.2). Now, we want to write the dual of the dual. To write dual of the system (4.2), we first write this into canonical form as

$$\text{min} \quad y_0' = -b^T Y$$
$$\text{s.t.} \quad -A^T Y \geq -C$$
$$Y \geq 0,$$

The dual of the above LPP is

$$\text{max} \quad z_0' = (-C)^T Z \qquad \text{or} \qquad \text{min} \quad z_0 = C^T Z$$
$$\text{s.t.} \quad (-A^T)^T Z \leq (-b^T)^T \qquad\qquad \text{s.t.} \quad AZ \geq b$$
$$Z \geq 0, \qquad\qquad\qquad\qquad\qquad Z \geq 0$$

The last LPP is nothing but the starting problem (4.1).

Theorem 1 concludes that

(a) the primal may be a minimization problem with $\geq$ type constraints and nonnegative restriction on the decision variables.

(b) the primal may be a maximization problem with $\leq$ type constraints and nonnegative restriction on the decision variables;

Thus, dual of (a) type of problem is (b) type; and dual of (b) type of problem is (a) type problem. This suggests that LPP is in canonical form if it is minimization problem with $\geq$ constraints or maximization problem with $\leq$ constraints.

Through examples 3 and 4, we demonstrate two facts:

Fact 1. If the primal contains some equality constraint, then the dual variable corresponding to this constraint will be unrestricted in sign.

Fact 2. If a LPP contains any variable x_k of unrestricted in sign, then the kth dual constraint will be written as equality constraint.

Example 2. Write the dual of the following LPP and verify Fact 1.

$$\begin{aligned} \max \quad & x_0 = 2x_1 - 3x_2 \\ \text{s.t.} \quad & x_1 + 2x_2 \leq 4 \\ & 3x_1 - x_2 = 5 \\ & x_1, x_2 \geq 0 \end{aligned}$$

Write the equality constraints into its equivalent inequality constraints, and associate the dual variables for each constraints as

$$\begin{aligned} \max \quad & x_0 = 2x_1 - 3x_2 & \\ \text{s.t.} \quad & x_1 + 2x_2 \leq 4 & y_1 \\ & 3x_1 - x_2 \leq 5 & y_2^+ \\ & -3x_1 + x_2 \leq -5 & y_2^- \\ & x_1, x_2 \geq 0 & \end{aligned}$$

The dual of the above LPP is

$$\begin{aligned} \min \quad & y_0 = 4y_1 + 5(y_2^+ - y_2^-) \\ \text{s.t.} \quad & y_1 + 3(y_2^+ - y_2^-) \geq 2 \\ & 2y_1 - (y_2^+ - y_2^-) \geq -3 \\ & y_1, y_2^+, y_2^- \geq 0 \end{aligned} \qquad \text{or} \qquad \begin{aligned} \min \quad & y_0 = 4y_1 + 5y_2 \\ \text{s.t.} \quad & y_1 + 3y_2 \geq 2 \\ & 2y_1 - y_2 \geq -3 \\ & y_1 \geq 0,\ y_2 \text{ unrestricted} \end{aligned}$$

Example 3. Find the dual of the following LPP to verify Fact 2.

$$\begin{aligned} \min \quad & x_0 = 3x_1 + x_2 \\ \text{s.t.} \quad & -x_1 + x_2 \leq 1 \\ & 3x_1 + 5x_2 \geq 3 \\ & x_1 \geq 0,\ x_2 \text{ unrestricted} \end{aligned}$$

First, convert the above LPP in canonical form and then write $x_2 = x_2^+ - x_2^-$ to have

$$\begin{aligned} \max \quad & x_0 = 3x_1 + x_2^+ - x_2^- & \\ \text{s.t.} \quad & -x_1 + x_2^+ - x_2^- \geq -1 & y_1 \\ & 3x_1 + 5x_2^+ - 5x_2^- \geq 2 & y_2 \\ & x_1, x_2^+, x_2^- \geq 0 & \end{aligned}$$

The dual of the above LPP is

$$
\begin{array}{ll}
\min & y_0 = -y_1 + 2y_2 \\
\text{s.t.} & -y_1 + 3y_2 \le 3 \\
& y_1 + 5y_2 \le 1 \\
& -y_1 - 5y_2 \le -1 \\
& y_1, y_2 \ge 0
\end{array}
\qquad \text{or} \qquad
\begin{array}{ll}
\min & y_0 = -y_1 + 2y_2 \\
\text{s.t.} & -y_1 + 3y_2 \le 3 \\
& y_1 + 5y_2 = 1 \\
& y_1, y_2 \ge 0
\end{array}
$$

4.2 Duality Theorems

In this section we prove two duality theorems and mention their applications to develop the Duality theory. Our primal problem will be a minimization problem $\min z = C^T X$, $AX \ge b$, $X \ge 0$.

Theorem 2 (Weak Duality Theorem). If X_0 is a primal feasible solution and Y_0 is dual feasible, then

$$C^T X_0 \ge b^T Y_0.$$

Proof. The dual feasibility of Y_0 implies that $A^T Y_0 \le C$, $Y_0 \ge 0$. If X_0 is a primal feasible, then $X_0 \ge 0$, and

$$X_0^T A^T Y_0 \le X_0^T C.$$

Note that $AX_0 \ge b$. Hence

$$C^T X_0 = X_0^T C \ge X_0^T A^T Y_0 = (AX_0)^T Y_0 \ge b^T Y_0.$$

Several corollaries can immediately be deduced from the weak duality theorem.

Corollary 1. If X_0 is primal feasible, Y_0 is dual feasible, and $C^T X_0 = b^T Y_0$, then X_0 and Y_0 are the optimal solutions to the respective problems.

Proof. Theorem 2 indicates that $C^T X \ge b^T Y_0 = C^T X_0$, for each primal feasible solution X. Thus, X_0 is an optimal solution to the primal. A similar argument holds for the dual.

Corollary 2. If the primal is unbounded below, then the dual problem is infeasible.

Proof. Whenever the dual problem has a feasible solution Y_0, the weak duality theorem prevents the primal objective from falling below $b^T Y_0$.

Similarly, we have the following result :

Corollary 3. If the dual problem is unbounded above, then the primal problem is infeasible.

It is to be noted that the converse statement of either of two foregoing corollaries is not true. For example, when the primal problem is infeasible, the dual could be either unbounded above or infeasible.

The next result is a stronger result than Theorem 2 and to introduce this we standardize some notations here. For a given cost vector C (column vector) and the coefficient matrix A, we can always denote these as

$$C = \begin{bmatrix} C_B \\ C_N \end{bmatrix} \quad \text{and} \quad A = \begin{bmatrix} B \mid N \end{bmatrix},$$

where C_B is the cost vector of basic variables, C_N is the cost vector referred to nonbasic variables, B is a $m \times m$ nonsingular matrix (basis matrix), and N is a matrix corresponding to nonbasic variables called as nonbasic matrix with dimensionality $m \times (n - m)$.

Theorem 3 (Strong Duality Theorem).

1. If either the primal or the dual linear program has a finite optimal solution, then so does the other and they achieve the same optimal objective value.

2. If either problem has an unbounded objective value, then the other has no feasible solution.

Proof. For the first part, without loss of generality, we can assume that the primal problem has reached a finite optimum at a basic feasible solution X. If we utilize the simplex algorithm at X and define $Y^T = C_B^T B^{-1}$, then (see Proposition 3, Chapter 3)

$$A^T Y - C = \begin{bmatrix} B^T \\ N^T \end{bmatrix} Y - \begin{bmatrix} C_B \\ C_N \end{bmatrix} = r \leq 0.$$

Therefore, Y is dual feasible. Moreover, since X is a basic feasible solution, we have

$$C^T X = C_B^T X_B = C_B^T B^{-1} b = Y^T b = b^T Y.$$

Due to Corollary 1, we can say Y is an optimal solution to the dual linear problem.

The proof of the second part is a direct consequence of Corollaries 2 and 3.

Remark. As a byproduct of the proof, we have established the fact that if, optimal solution of primal exists then $C_B^T B^{-1}$ gives the optimal solution of the dual. It is to be noted that B^{-1} is available from the optimal table below the starting basis. Since starting basis is usually in terms of slack and surplus variable, to write B^{-1} we multiply by -1 to each entry of the column in optimal table below each surplus variable.

4.3 Complementary Slackness Theorem

This theorem explains the fact how primal and dual are closed related. The interesting relationship between the primal and dual reveals so many facts involving optimal solution of one from the other.

Theorem 4 (Complimentary slackness conditions).

(a) If, in optimal table of primal the decision variable x_k appears as basic variable then the kth dual constraint is satisfied as equality constraint, i.e., slack or surplus variable associated with kth dual constraint assumes zero value.

(b) If, in optimal table of primal the slack or surplus variable s_k appears as basic variable then the dual variable y_k associated with kth primal constraint assumes zero value in the optimal solution of dual.

Proof. (a) Since $z_k - c_k = 0$ for all basic variables, it follows that if x_k is a basic variable then $z_k = c_k$. It means $C_B^T B^{-1} A_k = c_k \Rightarrow Y^T A_k = c_k$. This implies

$$y_1 a_{1k} + y_2 a_{2k} + \cdots + y_m a_{mk} = c_k, \qquad (4.4)$$

i.e., kth dual constraint is satisfied as equality constraint.

(b) If s_k is slack or surplus variable, then $c_k = 0$ and $A_k = (0, 0, \ldots 1, \ldots 0)^T$. Using this data in (4.4), we have

$$y_1 \times 0 + y_2 \times 0 + \cdots + y_k \times 1 + \cdots + y_m \times 1 = 0 \implies y_k = 0.$$

Similarly, for surplus variable s_k, $A_k = -e_k = (0, 0, \ldots -1, \ldots, 0)$. Again $-Y^T e_k = 0$, and this also implies $y_k = 0$.

Any way in both cases $y_k = 0$, as asserted.

Remark. At any simplex iteration of the primal or dual, the direct consequence of complementary slackness theorem is

$$\left(\begin{array}{c} \text{Objective coefficient (relative cost)} \\ \text{of variable } j \text{ in one problem} \end{array} \right) =$$

$$\left(\begin{array}{c} \text{Left-hand side minus right-hand side} \\ \text{of constraint } j \text{ in other problem} \end{array} \right)$$

Remark. This property is very much useful for finding optimal solution of primal or dual when the optimal solution of one is known.

Example 4. Verify complimentary slackness conditions by discussing the optimal solution of the LPP:

$$\begin{aligned} \min \quad & x_0 = x_1 - 2x_2 + x_3 \\ \text{s.t.} \quad & x_1 + 2x_2 - 2x_3 \le 4 \\ & x_1 - x_3 \le 3 \\ & 2x_1 - x_2 + 2x_3 \le 2 \\ & x_1, x_2, x_3 \ge 0 \end{aligned}$$

The dual of the above problem is

$$\begin{aligned} \max \quad & y_0 = -4y_1 - 3y_2 - 2y_3 \\ \text{s.t.} \quad & -y_1 - y_2 - 2y_3 \le 1 \\ & -2y_1 + y_3 \le -2 \\ & 2y_1 + y_2 - 2y_3 \le 1 \\ & \text{all var } \ge 0 \end{aligned}$$

The optimal table of the primal is given by

B V	x_1	x_2	x_3	s_1	s_2	s_3	Soln
x_0	$-9/2$	0	0	$-3/2$	0	-1	-8
x_2	3	1	0	1	0	1	6
s_2	$7/2$	0	0	$1/2$	1	1	7
x_3	$5/2$	0	1	$1/2$	0	1	4

Note the observations:

Optimal solution of the dual is $Y^T = C_B^T B^{-1}$ (simplex multiplier), i.e.,

$$(-y_1, -y_2, -y_3) = (-2, 0, 1) \begin{bmatrix} 1 & 0 & 1 \\ 1/2 & 1 & 1 \\ 1/2 & 0 & 1 \end{bmatrix} = (-3/2, 0, -1)$$

Hence, $y_1 = 3/2$, $y_2 = 0$, $y_3 = 1$, optimal value $= -8$.

To verify (a), note that s_2 appears as basic variable in primal optimal table and hence, $y_2 = 0$ is justified.

For (b), we find that x_2 and x_3 appear as basic variables in primal optimal table. Then

$$\text{IInd constraint of the dual :} \quad -2 \times (3/2) + 1 = -2$$
$$\text{IIIrd constraint of the dual :} \quad 2 \times (3/2) - 2 = 1$$

are satisfied as equality constraints.

Remarks. 1. We have taken $(-y_1, -y_2, -y_3)$ in our calculation as to write the LPP in canonical form we multiplied each constraint by -1.

2. Instead of calculating $C_B^T B^{-1}$, we can also read this from the optimal table. This is available in x_0-row below the starting BFS in optimal table of primal with the precaution that cost vector of the starting BFS must be zero. Visiting the above table, we have $(-3/2, 0, -1) = (-y_1, -y_2, -y_3)$.

Consider the primal in the form

$$\min \quad C^T X$$
$$\text{s.t.} \quad AX \geq b, \ X \geq 0$$

Its dual is given by

$$\max \quad b^T Y$$
$$\text{s.t.} \quad A^T Y \leq C, \ Y \geq 0$$

This is called symmetric pair of primal and dual programs.

Define the primal slackness vector

$$s = AX - b \geq 0$$

and dual slackness vector

$$r = C - A^T Y \geq 0.$$

Theorem 5 (Complimentary slackness theorem). Let X be a primal feasible solution and Y be a dual feasible solution to a symmetric pair of linear programs. Then X and Y become an optimal solution pair if and only if

$$\text{either} \quad r_j = (C - A^T Y)_j = 0$$
$$\text{or} \quad X_j = 0, \quad j = 1, 2, \ldots, n$$

and

$$\text{either} \quad s_i = (AX - b)_i = 0$$
$$\text{or} \quad Y_i = 0, \quad i = 1, 2, \ldots, m$$

are satisfied.

Here X_j and Y_i represent the jth and ith component of the feasible vectors X and Y, respectively.

Proof. For any primal feasible X and dual feasible Y, we have

$$0 \leq r^T X + s^T Y$$
$$= (C^T - Y^T A)X + Y^T(AX - b)$$
$$= C^T X - b^T Y$$

Therefore, the quantity $r^T X + b^T Y$ is equal to the duality gap between the primal feasible solution X and dual feasible solution Y. The duality gap vanishes if and only if

$$r^T X = 0 \quad \text{and} \quad s^T Y = 0 \tag{4.5}$$

In this case X and Y become optimal solution of primal and dual respectively. Note that (4.5) requires that "either $r_j = 0$ or $X_j = 0$ for $j = 1, 2, \ldots, n$" and "either $s_i = 0$ or $Y_i = 0$ for $i = 1, 2, \ldots, m$". This proves the theorem.

Remark. If the primal is given in standard form

$$\min \quad C^T X$$
$$\text{s.t.} \quad AX = b$$
$$X \geq 0$$

Its dual is given by

$$\max \quad b^T Y$$
$$\text{s.t.} \quad A^T Y \leq C$$
$$Y \text{ unrestricted}$$

Since the primal has zero slackness (being tight equalities), the condition $s^T Y = 0$ is automatically met. Thus, complimentary slackness is simplified to $r^T X = 0$. This situation will be seen when we find the optimal solution of transportation problem, see Section 7.2.

4.4   An Economic Interpretation of Duality

Let us consider a LPP,

$$\min \quad x_0 = C^T X$$
$$\text{s.t.} \quad AX \geq b$$
$$X \geq 0 \qquad\qquad (4.6)$$

Its dual problem is

$$\max \quad y_0 = b^T Y$$
$$\text{s.t.} \quad A^T Y \leq C$$
$$Y \geq 0$$

First, we consider the scenario of the primal LPP. Consider a manufacturer who makes n products out of m resources. To make one unit of product $j(j = 1, 2, \ldots, n)$ it takes a_{ij} units of resource i for $i = 1, 2, \ldots, m$. The manufacturer has obtained b_i units of resource $i(i = 1, 2, \ldots, m)$ in hand, and the unit price of product $j(j = 1, 2, \ldots, n)$ is c_j at current market. Therefore, the primal problem leads the manufacturer to find an optimal production plan that maximizes the sales with available resources.

Next, we take the dual scenario. We assume that the manufacturer gets the resources from supplier. The manufacture wants to negotiate the unit purchasing price y_i for resources $i(i = 1, 2, \ldots, m)$ with the supplier. Therefore, the manufacturer's objective is to minimize the total purchasing price $b^T Y$ in obtaining the resources $b_i(i = 1, 2, \ldots, m)$. Since the marketing price c_j and the "product-resource" conversion ratio a_{ij} are open information on market, the manufacturer knows that, at least ideally, a "smart" supplier would like to change him as much as possible, so that

$$a_{1j}y_1 + a_{2j}y_2 + a_{3j}y_3 + \cdots + a_{mj}y_m \geq c_j.$$

In this way, the dual linear program leads the manufacturer to come up with a least-cost plan in which the purchasing prices are acceptable to the "smart" supplier.

The foregoing scenarios not only provide economic interpretations of the primal and dual linear programming problems, but also explain the implications of the complementary slackness conditions. Assume that the manufacturer already has $b_i (i = 1, 2, \ldots, m)$ units of resources at hand. Then,

1. the ith component of the optimal dual vector y_i^* represents the maximum marginal price that the manufacture is willing to pay in order to get an additional unit of resource i from the supplier;

2. when the i resource is not fully utilized, i.e., $a_i X^* < b_i$, where a_i is the ith row of A and X^* is an optimal primal solution, the complementary slackness condition requires that $y_i^* = 0$, which means the manufacturer is not willing to pay a penny to get an additional amount of that resource;

3. when the supplier asks too much, i.e., when $A_j Y^* \geq c_j$,, where A_j is the jth column of A, the complementary slackness condition requires that $x_j^* = 0$, which means that the manufacture is no longer willing to produce any amount of product j.

4.5 The Dual Simplex Method

The dual simplex method developed by Lemke finds immense applications in sensitivity analysis to be discussed in the next chapter. This is applicable, when in the starting simplex table the optimal criteria is satisfied but the feasibility remains disturbed, while identity submatrix is manipulated to exist in A, the coefficient matrix. Hence, observe immediately the objective function. Its name is justified because the rules for leaving and entering variables are derived from the dual problem but are used in the primal problem.

Adopt the following procedure to find the optimal solution.

1. After introducing slack or surplus variable write the problem in the format

$$\text{opt} \quad x_0 = C^T X$$
$$\text{s.t.} \quad AX = b$$
$$X \geq 0,$$

where A contains the identity matrix as submatrix, and at least one of the b_i in the right-hand side vector $b = (b_1, b_2, \ldots, b_m)^T$ is negative.

2. Express the objective function in terms of nonbasic variables.

Algorithm.

Step 1. The leaving variable is decided to be the most negative entry of the solution column, i.e.,

$$x_r = \min_i \{ x_i, \ x_i < 0 \}.$$

Step 2. To decide the entering variable, we look for negative entries in row of leaving variable and find the ratio of these entries with the corresponding $(z_j - c_j)$'s in x_0-row. Fix the entering variable by

$$\min_j \left\{ \left| \frac{z_j - c_j}{\alpha_r^j} \right|, \ \alpha_r^j < 0 \right\}.$$

Step 3. When the entering and leaving variables are decided by Steps 1 and 2, perform the simplex iterations to have the next table. If all the entries in solution column of the resulting table after the iteration assume nonnegative values, then stop otherwise continue iterations through Steps 1 and 2 till all the entries in solution column are nonnegative, i.e., the feasibility is attained. This is our optimal table.

Remarks. 1. Use of the artificial variables should be avoided to produce the identity submatrix, while applying the dual simplex method.

2. Suppose jth variable is qualified to leave the basis, but all the entries $\alpha_j^k, k = 1, 2, \ldots, n$ are positive which means that no variable can enter the basis. In this situation the LPP has no feasible solution.

3. The simplex method discussed in Chapter 3 is called primal simplex method. The basic difference between two methods is that

feasibility is at hand and we restore optimality in simplex method, while the process is reverse in dual simplex method.

Example 5. Solve the following LPP by dual simplex method

$$\text{max} \quad z = -4x_1 - 6x_2 - 18x_3$$
$$\text{s.t.} \quad x_1 + 3x_3 \geq 3$$
$$x_2 + 2x_3 \geq 5$$
$$x_1, x_2, x_3 \geq 0$$

Also, using the optimal table of the above LPP, find the optimal solution of its dual.

Observing the objective function we at once conclude that the optimal criteria will be preserved and hence dual simplex method is advisable. Write the LPP in appropriate format as

$$\text{max} \quad z = -4x_1 - 6x_2 - 18x_3$$
$$\text{s.t.} \quad -x_1 - 3x_3 + s_1 = -3$$
$$-x_2 - 2x_3 + s_2 = -5$$
$$\text{all var} \geq 0$$

Now, we solve the problem step by step. In first iteration, note that most negative entry of the solution column is -5 and this corresponds to s_2, and hence s_2 is the leaving variable. Next look for negative entries in s_2-row and compute

$$\min\left\{\left|\frac{6}{-1}\right|, \left|\frac{18}{2}\right|\right\} = 6.$$

This corresponds to x_2, and hence x_2 is the entering variable.

Thus, s_2 leaves and x_2 enters and -1 is the pivotal element. Now, complete the second part of table exactly like simplex method, i.e., we perform row operations.

Similarly, in second iteration, s_1 leaves and x_3 enters and -3 is the pivotal element. Divide by -3 to the entire s_1-row and write the resulting row as x_3-row in next table. Further, perform row operations to complete this table. The third part of the combined table gives the optimal solution because all all $z_j - c_j$ are nonnegative (it is a max problem). All computations are shown in the combined table given below.

B.V.	x_1	$x_2 \downarrow$	x_3	s_1	s_2	Soln
z	4	6	18	0	0	0
s_1	-1	0	-3	1	0	-3
$\leftarrow s_2$	0	$\boxed{-1}$	-2	0	1	-5
z	4	0	$6 \downarrow$	0	6	-30
$\leftarrow s_1$	-1	0	$\boxed{-3}$	1	0	-3
x_2	0	1	2	0	-1	5
z	2	0	0	2	6	-36
x_3	$1/3$	0	1	$-1/3$	0	1
x_2	$-2/3$	1	0	$2/3$	-1	3

Optimal solution: $x_1 = 0$, $x_2 = 3$, $x_3 = 1$, max value $= -36$.

We have already mentioned that the solution of dual is kept below starting BFS as $C_B^T B^{-1}$. Hence the optimal solution of dual is

$$y_1 = 2, \ y_2 = 6, \ y_0 = -36.$$

Remarks. 1. Sometimes the optimal table of the LPP is not given but relative costs of appropriate number of variables are given. In this situation $C_B B^{-1}$ is not available, we use the property mentioned in remark following Theorem 4.

Suppose it is given that x_3 and x_2 are basic variables in optimal basis. Hence relative costs of these variables are zero. This prompts us that the third and second constraints will be satisfied as equality constrains id dual problem. The dual of the above problem is

$$\begin{aligned}
\min \quad & z' = -3y_1 - 5y_2 \\
\text{subject} \quad & -y_1 \leq -4 \\
& -y_2 \leq -6 \\
& -3y_1 - 2y_2 \leq -18 \\
& y_1, y_2 \geq 0
\end{aligned}$$

As the third and second constraints are satisfied as equality constraints, we have

$$-y_2 = -6$$

$$-3y_1 - 2y_2 = -18$$

Solving these equations, it follows that $y_1 = 2$, $y_2 = 6$, and inserting these values in objective function, we get $z' = -36$.

2. It is also possible to say that relative cost of x_1 is 2 and x_3 is basic variable. The same property helps us to give the solution of dual, see Problem 18(b).

Before winding up this section, we would like to point three facts:

1. Solving an LPP in the standard form by the dual simplex method is equivalent to solving its dual using simplex method.

2. Solving an LPP by the dual simplex method absorbs the same amount of efforts as the simplex method.

3. The dual simplex method is very handy in sensitivity analysis with an additional constraint. This will be discussed in Chapter 6.

KKT conditions. Given a linear program in its standard form, vector X is an optimal solution to the problem if and only if there exist vectors W and r such that

1. $AX = b$, $X \geq 0$ (primal feasibility)
2. $A^T W + r = c$, $r \geq 0$ (dual feasibility)
3. $r^T X = 0$ complimentary slackness

In this case W is an optimal solution of the dual problem.

To justify Fact 1 and KKT conditions, let us solve the dual of the LPP solved in this section by dual simplex method. The dual in standard form is

$$\begin{aligned}
\min \quad & z' = -3y_1 - 5y_2 \\
\text{s.t.} \quad & y_1 + r_1 = 4 \\
& y_2 + s_2 = 6 \\
& 3y_1 + 2y_2 + r_3 = 18 \\
& \text{all var } \geq 0
\end{aligned}$$

The solution by simplex method is given in the following table:

B.V.	y_1	$y_2 \downarrow$	r_1	r_2	r_3	Soln
z'	3	5	0	0	0	0
r_1	1	0	1	0	0	4
$\leftarrow r_2$	0	[1]	0	1	0	6
r_3	3	2	0	0	1	18
z'	$3 \downarrow$	0	0	−5	0	−30
r_1	1	0	1	0	0	4
y_2	0	1	0	1	0	6
r_3	[3]	0	0	−2	1	6
z'	0	0	0	−3	−1	−36
s_1	0	0	1	2/3	−1/3	2
y_2	0	1	0	1	0	6
y_1	1	0	0	−2/3	1/3	2

In both tables the optimal value is same and note that the corresponding optimal basic feasible solutions are

$$x_1 = 0 \quad r_1 = 2$$
$$x_2 = 3 \quad r_2 = 0$$
$$x_3 = 0 \quad r_3 = 0$$
$$s_1 = 0 \quad y_1 = 2$$
$$s_2 = 0 \quad y_2 = 6$$

Obviously, $r^T X = 0$ and $s^T Y = 0$, where X and Y are optimal solutions of primal and dual, respectively.

Problem Set 4

1. Under what conditions a primal and its dual are same?

2. Is it possible to solve the example of Section 4.5 by using simplex method? If yes, what type of modification do you need?

3. Write dual of the following LPP and find the optimal solution of primal by determining the optimal solution of dual.

$$\min \quad x_0 = -2x_1 - 3x_2 - x_3$$
$$\text{s.t.} \quad x_1 + 2x_2 \leq 1$$
$$- x_1 - x_3 \leq 2$$
$$x_1 - x_2 + 2x_3 \leq 2$$
$$\text{all var} \geq 0$$

4. Solve the following LPP using two phase method

$$\min \quad x_0 = -4x_1 + x_2$$
$$\text{s.t.} \quad x_1 + x_2 \geq 1$$
$$- 2x_1 + x_2 = 2$$
$$2x_1 + 3x_2 \leq 7$$
$$\text{all var} \geq 0$$

and find the optimal solution of its dual.

5. Write the dual program of the following problem

$$\min \quad x_0 = 9x_1 + 6x_2 - 4x_3 + 100$$
$$\text{s.t.} \quad 3x_1 + 8x_2 - 5x_3 \geq 14$$
$$5x_1 - 2x_2 + 6x_3 = 17$$
$$2x_1 + 4x_2 \leq 19$$
$$x_1 \leq 0, x_2 \geq 0, x_3 \text{ unrestricted}$$

6. Construct an example to show that both the primal and dual problem has no feasible solution. This indicates that the infeasibility of one problem does not imply the unboundedness of the other one in a primal-dual pair.

7. Write dual of the following LPP and solve it by dual simplex method. From the optimal table of the dual find the optimal solution of the primal.

$$\max \quad x_0 = 2x_1 + 3x_2$$
$$\text{s.t.} \quad - 2x_1 + x_2 \geq 3$$
$$3x_1 + x_2 \leq 5$$
$$x_1, x_2 \geq 0$$

8. With the help of the following LPP

$$\max \quad x_0 = x_1 + 3x_2$$
$$\text{s.t.} \quad 2x_1 + x_2 \geq 1$$
$$3x_1 + 2x_2 \leq 5$$
$$x_1 \geq 0, x_2 \text{ unrestricted,}$$

verify that the second dual constraint will be an equality constraint.

9. Let the primal be

$$\max \quad x_0 = C^T X$$
$$\text{s.t.} \quad AX = b$$
$$X \geq 0$$

Suppose Y is an optimal solution of the dual. Then verify the queries: (a) If the kth constraint of the primal is multiplied by $\lambda \neq 0$, how the solution of its dual is affected; (b) If ith constraint of the primal is changed by adding $\lambda \neq 0$ times the kth constraint, then what will be the optimal solution of the dual?

10. Let the optimal solution of the primal be degenerate. Then show that the dual problem has alternative optimal solution. Is converse also true? if yes, prove it.

11. Construct a primal in three variables which is self dual.

12. Suppose a LPP in standard form has two columns that are proportional with a positive constant of proportionality. Construct an equivalent LP with one fewer column.

13. Suppose a standard LPP has a unique optimal solution. Does it follow that the dual has a nondegenerate optimal solution? Does the converse hold?.

14. The relative cost of a nonbasic variable x_k in optimal table of the primal is the difference of the left- and right hand side of kth constraint in dual. Prove this statement and verify for the example in Section 4.3.

15. Consider the data of Problem 8, Problem set 3. (i) Write the dual of the original; (ii) Find the optimal solution of the dual.

16. Find the optimal solution of the following LPP by solving its dual. The dual must be maximization problem and must be solved by two phase method

$$
\begin{aligned}
\min \quad & z = x_1 + x_2 \\
\text{s.t.} \quad & x_1 + 2x_2 = 4 \\
& x_2 \leq 4 \\
& 3x_1 - 2x_2 \leq 0 \\
& x_2 \geq 0, \ x_1 \text{ unrestricted}
\end{aligned}
$$

17. Consider the following linear program

$$
\begin{aligned}
\max \quad & z = x_1 + 5x_2 + 3x_3 \\
\text{s.t.} \quad & x_1 + 2x_2 + x_3 = 3 \\
& 2x_1 - x_2 = 4 \\
& x_1, x_2, x_3 \geq 0
\end{aligned}
$$

If x_1 and x_3 are the basic variables in optimal solution of the primal, then find the optimal solution of its dual.

18. Consider the linear program

$$
\begin{aligned}
\max \quad & z = 2x_1 + x_2 + 2x_3 - 3x_4 \\
\text{s.t.} \quad & 3x_1 + x_2 + x_3 = 7 \\
& 2x_1 - x_2 + x_4 = 5 \\
& x_1, x_2, x_3, x_4 \geq 0
\end{aligned}
$$

(a) Verify that the basic feasible solution (x_2, x_4), i.e., $(0, 7, 0, 26)$ is not optimal solution of the LPP;

(b) Determine the optimal solution of the associated dual linear program when it is known that the primal has optimal solution $x_1 = 7/3$, $x_4 = 1/3$, $z = 11/3$.

19. Find the optimal solution of the following problem by dual simplex method

$$
\begin{aligned}
\max \quad & z = 2x_1 - x_2 + x_3 \\
\text{s.t.} \quad & 2x_1 + 3x_2 - 5x_2 \geq 4 \\
& -x_1 + 9x_2 - x_3 \geq 3 \\
& 4x_1 + 6x_2 + 3x_3 \geq 8 \\
& x_1, x_2, x_3 \geq 0
\end{aligned}
$$

This is the problem where optimality and feasibility are disturbed simultaneously in the simplex table. The dual simplex method is not applicable directly. Remove infeasibility first by applying a version of the dual simplex feasibility condition that selects s_1 as the entering variable. To determine the entering variable, we need a nonbasic variable whose constraint coefficient in the s_1-row is strictly negative irrespective of maintaining optimality because nonexistent at this stage. The procedure is repeated as necessary until the feasibility is satisfied. The next step is to pay attention to restore optimality by the simplex method. This procedure is usually referred to as generalized simplex algorithm.

The other method is to add the constraint $x_1 + x_3 \leq M$ ($M > 0$ large enough) so that no point of the original feasible region is eliminated. Use the new constraint as pivot row and take x_1 as the entering variable. The next table will be amenable to solution by dual simplex method.

20. Use the technique of the preceding problem to solve the following LPP.

$$\begin{aligned}
\min \quad & z = -x_1 + x_2 \\
\text{s.t.} \quad & x_1 - 4x_2 \geq 5 \\
& x_1 - 3x_2 \leq 1 \\
& 2x_1 - 5x_2 \geq 1 \\
& x_1, x_2 \geq 0
\end{aligned}$$

Suggestion. See the effect of generalized simplex algorithm. To apply the second method add the constraint $x_1 \leq M$ and take x_1 as the entering variable.

21. Solve the following LPP by simplex method (avoid big-M or two phase method) and from the optimal table find the optimal solution of its dual.

$$\begin{aligned}
\min \quad & z = x_1 + x_2 + x_3 - 3x_4 + 6x_5 + 4x_6 \\
\text{s.t.} \quad & x_1 + x_2 + 3x_4 - x_5 + 2x_6 = 6 \\
& x_2 + x_3 - x_4 + 4x_5 + x_6 = 3 \\
& x_1 + x_3 - 2x_4 + x_5 + 5x_6 = 5 \\
& x_1, x_2, x_3, x_4, x_5, x_6 \geq 0
\end{aligned}$$

Suggestion. Use the property given in remark following complimentary slackness theorem.

22. Find the dual problem of the following linear programming problem

$$\min \quad z = C^T X$$
$$\text{s.t.} \quad AX = b$$
$$\ell \le X \le u,$$

where ℓ and u are lower and upper bounding vectors.

23. Write the dual of the famous linear programming problem by Karmarkar

$$\min \quad z = C^T X$$
$$\text{s.t.} \quad AX = 0$$
$$e^T X = 1$$
$$X \ge 0$$

where $e = (1, 1, \ldots, 1)^T$ is a n-dimensional vector.

24. The system
$$AX = b, \quad X \ge 0$$

has no solution if and only if the system

$$A^T Y \le 0, \quad b^T Y > 0$$

has a solution.

Suggestion. This is referred to as the *Farka's lemma*.

Chapter 5

Advanced Linear Programming

This chapter includes some advanced techniques such as the revised simplex method, bounded variable technique and decomposition principle. These all are the modified extensions of the simplex algorithm to solve different types of linear programming problems. In the end, the interior point algorithm due to Karmarkar is introduced.

5.1 The Revised Simplex Algorithm

The simplex method was revisited for the purpose of increasing its computational efficiency. During simplex iterations a computer has to store a lot of data which is not required at later stage computations. The revised simplex method follows exactly the same path used in simplex method. The only difference is that next iteration is computed by row operations in simplex method, while in the revised simplex method next iteration is computed by inversion of the basis matrix. We do not find inverse directly but use product form to be explained shortly.

In simplex method, the successive basis B and B_{next} differ only in one column, resulting from interchanging the entering and leaving vectors. Besides formulas to compute various entries of any simplex table we need some more tools for the revised simplex method. Let the LPP be in standard form

$$\max \quad z = C^T X$$
$$\text{s.t.} \quad AX = b$$
$$X \geq 0,$$

where $X = (x_1, x_2, \ldots, x_n)^T$, $b = (b_1, b_2, \ldots, b_m)^T$, $C = (c_1, c_2, \ldots, c_n)^T$ and the coefficient matrix $A = (A_1, A_2, \ldots, A_n) = (a_{ij})_{m \times n}$.

We calculate the inverse of basis matrix at each iteration using the inverse of basis matrix of the previous iteration. Remaining entries are computed with the help of this inverse matrix. That's why the revised simplex method is sometimes referred as the inverse matrix method.

Suppose A_k enters and B_r leaves the basis $B = (B_1, B_2, \ldots, B_r, \ldots, B_m)$, where B_i, $i = 1$ to m is some A_j, $j = 1, 2, \ldots, n$. Then, the new basis matrix is

$$B_{\text{next}} = (B_1, B_2, \ldots, B_{r-1}, A_k, B_{r+1} \ldots, B_m),$$

where

$$A_k = \alpha_1^k B_1 + \alpha_2^k B_2 + \cdots + \alpha_r^k B_r + \cdots + \alpha_m^k B_m, \quad \alpha_r^k \neq 0. \quad (5.1)$$

Note that entering and leaving rules of the simplex method ensure that B_{next} is nonsingular, and hence is a basis. From (5.1)

$$B_r = -\left(\frac{\alpha_1^k}{\alpha_r^k}\right) B_1 - \left(\frac{\alpha_2^k}{\alpha_r^k}\right) B_2 - \cdots + \left(\frac{1}{\alpha_r^k}\right) A_k - \cdots - \left(\frac{\alpha_m^k}{\alpha_m^k}\right) B_m$$

$$= B_{\text{next}} \xi, \quad (5.2)$$

where

$$\xi = \left(-\frac{\alpha_1^k}{\alpha_r^k}, -\frac{\alpha_2^k}{\alpha_r^k}, \frac{1}{\alpha_r^k}, -\frac{\alpha_m^k}{\alpha_r^k}\right)^T.$$

Let E be the $m \times m$ elementary matrix defined by

$$E = (e_1, e_2, \ldots, e_{r-1}, \xi, e_{r+1}, \ldots, e_m)$$

where e_i is the standard unit vector in $\mathbb{R}^m$. E differs with identity matrix in the rth column only. Thus, in view of (5.2), we have

$$B_{\text{next}} E = (B_{\text{next}} e_1, \ldots, B_{\text{next}} e_{r-1}, B_{\text{next}} \xi, B_{\text{next}} e_{r+1}, \ldots, B_{\text{next}} e_m) = B.$$

This implies $B^{-1} = E^{-1}B_{\text{next}}^{-1}$, and hence,

$$B_{\text{next}}^{-1} = EB^{-1}, \tag{5.3}$$

where B^{-1} is the inverse of previous basis matrix, designed as current basis matrix.

Now, we discuss the computational procedure.

Standard form-I. If a LPP in standard form is of the type that the coefficient matrix A after introducing slack and surplus variables contains identity submatrix, then we classify this as standard form-I, see Section 3.3.

First, we workout an example for standard form-I. Before all this, we recollect how different entries of a simplex table are computed using various formulas.

Simplex Table

Basis	x_1	x_2	$\cdots$	x_j	$\cdots$	x_n	Solution
z	$z_1 - c_1$	$z_2 - c_2$	$\cdots$	$z_j - c_j$	$\cdots$	$z_n - c_n$	$C_B^T X_B$
X_B	$B^{-1}A_1$	$B^{-1}A_2$	$\cdots$	$B^{-1}A_j$	$\cdots$	$B^{-1}A_n$	$B^{-1}b$

In the above table X_B is the basic vector and B is the corresponding basis matrix. Also, in view of Proposition 1 (Section 3.2) and relation (5.3), one has

$$z_j - c_j = C_B^T B^{-1} A_j - c_j, \quad j = 1, 2, \ldots, n.$$

Note. We have noticed that simplex method (see Section 3.2) requires the computing of a whole new table at each iteration. But much of the information, particularly in the body matrix is not required at all. This saves space in computer memory and time of computation is reduced. What we need are the following items.

1. Relative costs $z_j - c_j$ of all nonbasic variables (this helps in deciding the entering variable)

2. Column vectors α^j for most positive or most negative nonbasic (max or min problem) variable (this helps in deciding the leaving variable)

3. All entries of the solution column, i.e., $B^{-1}b$, and the objective function value $f(X_B)$.

At each iteration to find the above three items, one needs B^{-1}. Direct methods of finding B^{-1} are not used. A simple way of doing is to use formula (5.3).

Example 1. Solve the following LPP by revised simplex method.

$$
\begin{aligned}
\max \quad & z = 3x_1 - x_2 + x_3 \\
\text{s.t.} \quad & x_1 + x_2 + x_3 \le 10 \\
& -x_2 + 2x_3 \le 2 \\
& 3x_1 - 2x_2 + 2x_3 \le 0 \\
& x_1, x_2, x_3 \ge 0
\end{aligned}
$$

Write the LPP in standard form-I as

$$
\begin{aligned}
\max \quad & z = 3x_1 - x_2 + x_3 \\
\text{s.t.} \quad & x_1 + x_2 + x_3 + s_1 = 10 \\
& -x_2 + 2x_3 + s_2 = 2 \\
& 3x_1 - 2x_2 + 2x_3 + s_3 = 0 \\
& \text{all var} \ge 0
\end{aligned}
$$

The coefficient matrix of the above system is

$$
A = \begin{bmatrix}
1 & 1 & 1 & 1 & 0 & 0 \\
0 & -1 & 2 & 0 & 1 & 0 \\
3 & -2 & 2 & 0 & 0 & 1
\end{bmatrix}
$$

The starting table is just the simplex format, see Table 1.

Table 1

B V	$x_1 \downarrow$	x_2	x_3	s_1	s_2	s_3	Soln
z	-3	1	-1	0	0	0	0
s_1	1			1	0	0	10
s_2	0			0	1	0	2
$\leftarrow s_3$	$\boxed{3}$			0	0	1	0

eration 1. Here x_1 (having most negative relative cost), i.e., A_1 enters and s_3, i.e., A_6 leaves the basis B and hence 3 is the pivot element. From (5.2),

$$\xi = \left(-\frac{1}{3},\ 0,\ \frac{1}{3}\right)^T.$$

The easy way of getting ξ is to replace pivot by $1/$pivot and then multiply remaining entries of the pivot column (which contains pivot element) by $-1/$pivot. The third basic variable is leaving and hence, by the definition

$$E = \begin{bmatrix} 1 & 0 & -1/3 \\ 0 & 1 & 0 \\ 0 & 0 & 1/3 \end{bmatrix}$$

Since $B = I$, causes $B^{-1} = I$, and now using the formula $B^{-1}_{\text{next}} = EB^{-1}$, we get

$$B^{-1}_{\text{next}} = EB^{-1} = E.$$

The new basis is s_1, s_2, x_1 and nonbasic variables are x_2, x_3, s_3. Now, perform the operations in the following sequence.

(i) Enter B^{-1}_{next} below the starting BFS.

(ii) Calculate $z_j - c_j$ for each nonbasic variable and write in the z-row. For this purpose, it is convenient to compute the quantity $C_B^T B^{-1}_{\text{next}}$, called the simplex multiplier and is denoted by Π. Thus,

$$\Pi = (0,0,3) \begin{bmatrix} 1 & 0 & -1/3 \\ 0 & 1 & 0 \\ 0 & 0 & 1/3 \end{bmatrix} = (0,0,1)$$

The simplex multiplier computed at each iteration is inserted in z-row below the starting BFS.

The relative cost of the nonbasic variables are

the relative cost of $x_2 = (0,0,1)(1,-1,-2)^T + 1 = -1$;

the relative cost of $x_3 = (0,0,1)(1,2,2)^T - 1 = 1$.

(iii) Since the relative cost of x_3 is positive, we do not compute its coordinate vector α^3 to be mentioned in the body matrix. The relative cost of s_3 is also positive (see simplex multiplier) but its coordinate vector will automatically be computed every time as it is placed at third position in the starting BFS. The only thing we do calculate is the coordinate vector for x_2 as

$$\alpha^2 = B_{\text{next}}^{-1} A_2 = \begin{bmatrix} 1 & 0 & -1/3 \\ 0 & 1 & 0 \\ 0 & 0 & 1/3 \end{bmatrix} \begin{bmatrix} 1 \\ -1 \\ -2 \end{bmatrix} = \begin{bmatrix} 5/3 \\ -1 \\ -2/3 \end{bmatrix}$$

(iv) The solution column is

$$B_{\text{next}}^{-1} b = \begin{bmatrix} 1 & 0 & -1/3 \\ 0 & 1 & 0 \\ 0 & 0 & 1/3 \end{bmatrix} \begin{bmatrix} 10 \\ 2 \\ 0 \end{bmatrix} = \begin{bmatrix} 10 \\ 2 \\ 0 \end{bmatrix}$$

(v) The objective value is $C_B^T X_B = C_B^T B^{-1} b = (0,0,1)(10,2,0)^T = 0$.

Incorporate these entries to complete Table 2 as follows

Table 2

B V	x_1	$x_2 \downarrow$	x_3	s_1	s_2	s_3	Soln
z	0	-1	1	0	0	1	0
$\leftarrow s_1$		$\boxed{5/3}$	1	0	$-1/3$		10
s_2		-1	0	1	0		2
x_1		$-2/3$	0	0	1/3		0

Iteration 2. Now, x_2 (most negative) enters and s_1 leaves. The pivot is $5/3$. Hence

$$\xi = \left(\frac{3}{5}, \frac{3}{5}, \frac{2}{5} \right)^T .$$

In this iteration, the first basic variable in the sequence s_1, s_2, x_1 is leaving, and hence the elementary matrix

$$E = \begin{bmatrix} 3/5 & 0 & 0 \\ 3/5 & 1 & 0 \\ 2/5 & 0 & 1 \end{bmatrix}$$

Again, use $B_{\text{next}}^{-1} = E B_{\text{current}}^{-1}$, see (5.3), to have

$$B_{\text{next}}^{-1} = \begin{bmatrix} 3/5 & 0 & 0 \\ 3/5 & 1 & 0 \\ 2/5 & 0 & 1 \end{bmatrix} \begin{bmatrix} 1 & 0 & -1/3 \\ 0 & 1 & 0 \\ 0 & 0 & 1/3 \end{bmatrix} = \begin{bmatrix} 3/5 & 0 & -1/5 \\ 3/5 & 1 & -1/5 \\ 2/5 & 0 & 1/5 \end{bmatrix}$$

Insert this B_{next}^{-1} below the starting BFS, and all other entries needed are calculated on the pattern of completing Table 2 as

$$\Pi = (-1, 0, 3) \begin{bmatrix} 3/5 & 0 & -1/5 \\ 3/5 & 1 & -1/5 \\ 2/5 & 0 & 1/5 \end{bmatrix} = (3/5, 0, 4/5),$$

$$z_3 - c_3 = (3/5, 0, 4/5)^T (1, 2, 2) - 1 = 6/5$$

Inserting these entries, we have Table 3.

Table 3

B V	x_1	x_2	x_3	s_1	s_2	s_3	Soln
z	0	0	6/5	3/5	0	4/5	6
x_2				3/5	0	1/5	6
s_2				3/5	1	1/5	8
x_1				2/5	0	7/15	4

Since all $z_j - c_j$ entries in z-row are ≥ 0, and hence the optimality has reached. Hence

optimal solution: $x_1 = 4$, $x_2 = 6$, $x_3 = 0$;   max value $z = 6$.

Standard form-II. If a LPP in standard form is of the type that inclusion of artificial is necessary to get the identity submatrix of coefficient matrix A after introducing slack and surplus variables then we classify this as standard form-II, see Section 3.3.

Let us solve a problem concerning standard form-II.

Example 2. Solve the following LPP by revised simplex method.

$$\max \quad x_0 = x_1 + 2x_2 + 3x_3$$
$$\text{s.t.} \quad x_1 + x_2 + 2x_3 \le 16$$
$$2x_1 + x_2 + 4x_3 \ge 24$$
$$x_1 + x_2 + x_3 \ge 10$$
$$x_1, x_2, x_3 \ge 0$$

It is evident that the problem in standard form will not contain identity submatrix. First, we write the standard form-II as

$$\max \quad x_0 = x_1 + 2x_2 + 3x_3$$
$$\text{s.t.} \quad x_1 + x_2 + 2x_3 + s_1 = 16$$
$$2x_1 + x_2 + 4x_3 - s_2 + R_2 = 24$$
$$x_1 + x_2 + x_3 - s_3 + R_3 = 10$$
$$\text{all var} \ge 0$$

We use the revised simplex method in reference to phase-two method, and for Phase-I solve the auxiliary LPP to find the basic feasible solution of the system

$$\min \quad r = R_2 + R_3$$
$$\text{s.t.} \quad x_1 + x_2 + 2x_3 + s_1 = 16$$
$$2x_1 + x_2 + 4x_3 - s_2 + R_2 = 24$$
$$x_1 + x_2 + x_3 - s_3 + R_3 = 10$$
$$\text{all var} \ge 0$$

Construct the starting table (in simplex format) as

B V	x_1	x_2	$x_3 \downarrow$	s_2	s_3	s_1	R_2	R_3	Soln
r	3	2	5	-1	-1	0	0	0	34
	0	0	0	0	0	0	-1	-1	0
s_1			2			1	0	0	16
$\leftarrow R_2$			4			0	1	0	24
R_3			1			0	0	1	10

eration 1. x_3 (A_3) enters and R_2 (A_7) leaves. Hence $\xi = (-1/2, 1/4, -1/4)^T$ and

$$E = \begin{bmatrix} 1 & -1/2 & 0 \\ 0 & 1/4 & 0 \\ 0 & -1/4 & 1 \end{bmatrix}$$

Note that $B_{\text{next}}^{-1} = EB^{-1} = E$, because starting $B^{-1} = I$. Insert B_{next}^{-1} below the starting BFS, and compute the relative cost of all nonbasic variables using $C_B^T B_{\text{next}}^{-1} A_j - c_j$. Here, we do not compute simplex multiplier to be inserted in r-row below the starting basis in this case, because the objective function is not expressed in terms of nonbasic variables (see the first iteration of the auxiliary LPP). Next, calculate $B_{\text{next}}^{-1} A_j$ only for nonbasic with most positive relative cost. This gives Table 4.

Table 4

B V	x_1	$x_2 \downarrow$	x_3	s_2	s_3	s_1	R_2	R_3	Soln
r	1/2	3/4	0	1/4	-1	0	$-5/4$	0	4
s_1		1/2				1	$-1/2$	0	4
x_3		1/4				0	1/4	0	6
$\leftarrow R_3$		3/4				0	$-1/4$	1	4

Iteration 2. In next iteration, x_2 enters and R_3 leaves. As usual,
$\xi = (-2/3, -1/3, 4/3)^T$, and

$$
E = \begin{bmatrix} 1 & 0 & -2/3 \\ 0 & 1 & -1/3 \\ 0 & 0 & 4/3 \end{bmatrix}
$$

Hence

$$
B_{\text{next}}^{-1} = E B_{\text{current}}^{-1} = \begin{bmatrix} 1 & 0 & -2/3 \\ 0 & 1 & -1/3 \\ 0 & 0 & 4/3 \end{bmatrix} \begin{bmatrix} 1 & -1/2 & 0 \\ 0 & 1/4 & 0 \\ 0 & -1/4 & 1 \end{bmatrix}
$$

$$
= \begin{bmatrix} 1 & -1/3 & -2/3 \\ 0 & 1/3 & -1/3 \\ 0 & -1/3 & 4/3 \end{bmatrix}
$$

Insert latest B_{next}^{-1} below the starting BFS, and compute the remaining entries which are needed to write Table 5 as

Table 5

B V	x_1	x_2	x_3	s_2	s_3	s_1	R_2	R_3	Soln
r	0	0	0	0	0	0	-1	-1	0
s_1	$-1/3$					1	$-1/3$	$-2/3$	4/3
x_3	1/3					0	1/3	$-1/3$	14/3
x_2	2/3					0	$-1/3$	4/3	16/3

Table 3 is the optimal table of Phase-I. For Phase-II, we do not need R_1 and R_2. The coordinate vectors of s_2 and s_3 are just negative of R_2 and R_3, respectively.

The initial table of Phase-II (Table 6) is obtained as usual but the positions of R_2 and R_3 are occupied by s_2 and s_3, respectively, i.e., the starting basis variables are taken in the order s_1, s_2, s_3.

Table 6

B V	x_1	x_2	x_3	s_1	s_2	$s_3 \downarrow$	Soln
x_0	4/3	0	0	0	-1/3	-5/3	74/3
	-1	-2	-3	0	0	0	0
$\leftarrow s_1$	-1/3	0	0	1	1/3	$\boxed{2/3}$	4/3
x_3	1/3	0	1	0	-1/3	1/3	14/3
x_2	2/3	1	0	0	1/3	-4/3	16/3

eration 3. Here, s_3 enters and s_1 leaves. $\xi = (3/2, -1/2, 2)^T$ and

$$E = \begin{bmatrix} 3/2 & 0 & 0 \\ -1/2 & 1 & 0 \\ 2 & 0 & 1 \end{bmatrix}$$

Hence

$$B_{\text{next}}^{-1} = EB_{\text{current}}^{-1} = \begin{bmatrix} 3/2 & 0 & 0 \\ -1/2 & 1 & 0 \\ 2 & 0 & 1 \end{bmatrix} \begin{bmatrix} 1 & -1/3 & -2/3 \\ 0 & 1/3 & -1/3 \\ 0 & -1/3 & 4/3 \end{bmatrix}$$

$$= \begin{bmatrix} 3/2 & -1/2 & -1 \\ -1/2 & 1/2 & 0 \\ 2 & -1 & 0 \end{bmatrix}$$

Note that in phase-II, the starting BFS is (s_1, x_3, x_2), and inverse of basis matrix corresponding to this BFS is contained below s_1, R_2, R_3 in body matrix of Table 3. The column vectors corresponding to s_2 and s_3 are just negative of the column vectors for R_2 and R_3, respectively. Insert latest B_{next}^{-1} below s_1, s_2, s_3. However, the columns below s_2 and s_3 will be negative of the second and third column of B_{next}^{-1}.

The solution column is

$$
B_{next}^{-1}b = \begin{bmatrix} 3/2 & -1/2 & -1 \\ -1/2 & 1/2 & 0 \\ 2 & 1 & 0 \end{bmatrix} \begin{bmatrix} 16 \\ 24 \\ 10 \end{bmatrix} = \begin{bmatrix} 2 \\ 4 \\ 8 \end{bmatrix}
$$

and the objective function value is $C_B^T X_B = C_B^T B^{-1}b = (0,3,2)(2,4,8)^T = 28$.

Also, compute $z_j - c_j$ for each variable. Insert all these computations to have Table 7.

Table 7

B V	x_1	x_2	x_3	s_1	s_2	s_3	Soln
x_0	3/2	0	0	5/2	1/2	0	28
s_3				3/2	1/2	1	2
x_3				-1/2	-1/2	0	4
x_2				2	1	0	8

Since all $z_j - c_j \geq 0$ in above table, this ensures that the optimality has reached and the optimal solution is

$$x_1 = 0, \ x_2 = 8, \ x_3 = 4, \ z = 28.$$

Remarks. 1. In every simplex table or revised simplex table, if any variable x_i is in the BFS, then entry at the intersection of x_i-row and x_i-column is unity and all the remaining entries of the column vector are zero (including x_0-row entry).

2. Like revised simplex method, there is revised dual simplex method, see Problem 4, Problem set 4. The revised dual simplex method proceeds like dual simplex method. The leaving and entering variables rules are same for both the methods. The only difference is that in revised dual simplex method, we compute B^{-1} and relative cost of only nonbasic variables at each iteration till optimal table is received. Note that here too the latest B^{-1} is always inserted below the starting basis.

5.2   Complexity of The Simplex Algorithm

The computational complexity of the simplex algorithm depends upon the total number iterations and the number of elementary operations required at each iteration. Different implementation procedures result in different complexity. Variants of the simplex method were designed to achieve better computational performance.

Following the computational procedure in Chapter 3, it is not difficult to estimate that Dantzig's original simplex method requires about $m(n-m)+n+1$ multiplications and $m(n-1+1)$ additions at each iterations. While the revised simplex method requires $m(n-m)+(m+1)^2$ multiplications and $m(n+1)$ additions at each iteration. The conclusion is that both of them are of order $O(mn)$.

How many iterations are required? Each iteration of the simplex algorithm or revised simplex algorithm stems from one extreme point to an adjacent extreme point. For a linear program in its standard form the feasible region contains at most $C(n,m)$ extreme points that an algorithm could possibly visit. Since

$$C(n,m) = \frac{n!}{m!(n-m)!} \geq \left(\frac{n}{m}\right)^m \geq 2^m \text{ whenever } n \geq 2m.$$

it is quite plausible to require an exponential order of iterations. This fear of exponential order was confirmed by some worst-case examples specifically designed for the simplex method and its variants.

The first such example is given by V. Klee and G. Minty in 1971 to show that Dantzig's simplex method traverse all $(2^n - 1)$ extreme points to reach at the optimal solution. For $0 < \delta < 1/2$, this LPP is

$$\begin{aligned}
\max \quad & z = x_n \\
\text{s.t.} \quad & 0 \leq x_1 \leq 1 \\
& \delta x_{i-1} \leq x_i \leq 1 - \delta x_{i-1}, \quad i = 2,3,\ldots n \\
& x_i \geq 0, \quad i = 1,2,\ldots,n
\end{aligned}$$

Obviously, the origin point is a basic feasible solution. If we start with the origin and apply the minimum ratio rule to the entering nonbasic variable, the simplex method takes $2^n - 1$ iterations to visit every extreme point of the feasible region. For the case $n = 2$, Fig. 5.1 illustrates the fact.

Variants of the simplex method may change the entering and leaving rules to avoid traversing every extreme point. But different bad

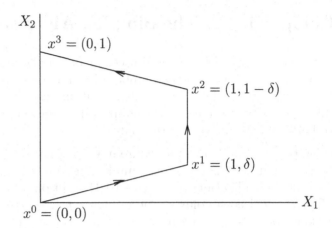

Figure 5.1

examples were reported for different variants. Any way we have to believe that the simplex method and its variants have *exponential complexity*.

However, bad examples rarely happen in real-world problems. It has been observed that more or less all real-life problems require the simplex method to take $4m$ to $6m$ iterations in completing two phases. This explains well the efficiency of the simplex method in practice, although it is of exponential complexity in theory.

5.3   Bounded Variable Technique

Consider the LPP in which the variables x_j are bounded by their lower bounds ℓ_j and upper bound u_j.

$$\max \quad z = C^T X$$
$$\text{s.t.} \quad a_{i1}x_1 + a_{i2}x_2 + \cdots + a_{in}x_n \geq, =, \leq b_i$$
$$\ell_i \leq x_j \leq u_j, \quad i = 1, 2, \ldots, m, \ j = 1, 2, \ldots, n$$

Since $x_j \geq \ell_j \implies y_j = x_j - \ell_j \geq 0$, it follows that the lower bounds may be converted to 0 just replacing x_j by $y_j + \ell_j$. Note that $0 \leq y_j \leq u_j - \ell_j$ and the simplex method is not applicable directly. Such type of LPP in which some or all the variables are having lower bound 0 and upper bound a finite number are solved using bounded variable technique described as follows:

Assume that we have BFS of the above system which satisfies the upper bounds. Note that no upper bound is mentioned for slack or surplus or artificial variable, and is taken at infinity. Suppose that a BFS

$$X_B = \{x_{B_1}, x_{B_2}, \ldots, x_{B_m}\}$$

is available by introducing slack, surplus and artificial variables (if necessary). Suppose x_j is the nonbasic variable with most negative relative cost $z_j - c_j$. Then x_j enters. To decide the leaving variable, the following conditions are desired.

(i) The next solution must be BFS;

(ii) All basic variables in next BFS must satisfy the upper bound limits.

B V	x_1 or A_1	x_j or A_j	Soln
z	$z_1 - c_1$	$z_j - c_j$	$f(X_B)$
x_1 or A_1	α_1^1	α_1^j	x_{B_1}
$\vdots$	$\vdots$	$\vdots$	$\vdots$
x_r or A_r	α_r^1	α_r^j	x_{B_r}
$\vdots$	$\vdots$	$\vdots$	$\vdots$
x_m or A_m	α_m^1	α_m^j	x_{B_m}

The basis matrix corresponding to X_{B_i} is given by (see above table)

$$B = (A_1, \ldots, A_r, \ldots, A_m),$$

where A_i are the columns of the coefficient matrix A and α_i^j is the coordinate vector of the column A_j of A with respect to B, see Chapter 3.

$$A_j = \alpha_1^j A_1 + \cdots + \alpha_r^j A_r + \cdots + \alpha_m^j A_m.$$

We have assumed that x_j enters (column A_j) the basis and x_r (column A_r) leaves. The new basis is

$$B^* = (A_1, \ldots, A_{r-1}, A_j, A_{r+1}, \ldots, A_m).$$

So far, we are performing simplex iterations. The value of the basic variables are given by

$$x_{B_i}^* = x_{B_i} - \alpha_i^j x_j = x_{B_i} - \alpha_i^j \frac{x_{B_r}}{\alpha_r^j}. \tag{5.4}$$

In the above, the leaving variables x_{B_r} should be determined so that the desired conditions (i) and (ii) are satisfied.

From the theory of the simplex method, we know that the next solution will be a BFS if the leaving variable is decided by the minimum ratio rule, i.e.,

$$\theta_1 = \frac{x_{B_r}}{\alpha_r^j} = \min\left\{\frac{x_{B_i}}{\alpha_i^j}, \ \alpha_i^j > 0\right\} = x_{B_r}^* = x_j. \tag{5.5}$$

The variable enters at level θ_1. This satisfies (i), but we have the additional condition that no variable must exceed its upper limit. This is achieved by (5.4) as

$$x_{B_i}^* = x_{B_i} - \alpha_i^j x_j \le u_i. \tag{5.6}$$

If $\alpha_i^j \ge 0$, then condition (ii) is met since $x_{B_i} \le u_i$. In case $\alpha_i^j < 0$, then the bound may exceed. The relation (5.6) holds true in this situation, provided

$$x_j \le \frac{u_i - x_{B_i}}{-\alpha_i^j}, \text{ for all those } i \text{ such that } \alpha_i^j < 0.$$

This means that

$$x_j \le \min_i\left\{\frac{u_i - x_{B_i}}{-\alpha_i^j}, \ \alpha_i^j < 0\right\} = \theta_2. \tag{5.7}$$

Thus, x_j should not exceed its upper limit, i.e.,

$$x_j \le u_j. \tag{5.8}$$

From (5.6), (5.7) and (5.8), it follows that the largest value of x_j which meets (i) and (ii) is

$$\theta = \min\{\theta_1, \theta_2, u_j\}. \tag{5.9}$$

Now, we discuss all three possibilities:

Situation $\theta = \theta_1$, since minimum is θ_1 and x_j enters at θ_1 and $\theta_1 \le \theta_2$ and $\theta \le u_j$, the conditions (i) and (ii) will be met. Thus, iteration after deciding leaving and entering variable is nothing but exactly simplex iteration.

Situation $\theta = \theta_2$ ensures that minimum in (5.9) is θ_2. Hence, $\theta_1 \geq \theta_2$ and x_j enters at θ_2 level such that

$$\frac{u_r - x_{B_r}}{-\alpha_r^j}, \ \alpha_r^j < 0.$$

Since, $\theta_1 \geq \theta_2$, the solution may not be a basic solution. The next solution will be a basic solution provided x_j enters at θ_1 level. It can be made basic by the substitution

$$x_r = u_r - x_r^1, \ 0 \leq x_r^1 \leq u_r. \tag{5.10}$$

The substitution (5.10) means: Replace x_r by x_r^1 and the column of x_r^1 will be just negative of the column of x_r. Transfer u_r times x_r column to the solution column. We are not going into details as this will give BFS.

Situation $\theta = u_j$ implies that x_j enters at its upper bound. To make x_j at its upper bound, make the substitution

$$x_j = u_j - x_j^1, \ 0 \leq x_j^1 \leq u_j.$$

In fact x_j does not enter the basis but remains nonbasic at its upper limit. Since, $\theta = u_j < \theta_1$, the new solution will not be basic. If it enters at θ_1, then the new solution will be BFS.

What we have done theoretically so far will be more clear by solving a numerical problem.

Example 3. Solve the following LPP

$$\begin{aligned} \max \quad & z = 4x_1 + 2x_2 + 6x_3 \\ \text{s.t.} \quad & 4x_1 - x_2 - 3x_3 \leq 9 \\ & -x_1 + x_2 + 2x_3 \geq 8 \\ & -3x_1 + x_2 + 4x_3 \leq 11 \\ & 1 \leq x_1 \leq 3, \ 0 \leq x_2 \leq 5, \ 0 \leq x_3 \leq 2 \end{aligned}$$

Since the variables are bounded, we shall find the optimal solution by bounded variable technique. First, we make lower bounds at 0 level. Replace x_1 by $y_1 + 1$, and hence $0 \leq y_1 \leq 2$. By inspection, we see that two phase method is convenient to apply.

Phase-I. To find initial BFS, solve the auxiliary problem

$$\min \quad r = R_2$$

$$\text{s.t.} \quad 4y_1 - x_2 - 3x_3 + s_1 = 5$$
$$- y_1 + x_2 + 2x_3 - s_2 + R_2 = 9$$
$$- 3y_1 + x_2 + 4x_3 + s_3 = 14$$
$$0 \le y_1 \le 2, \ 0 \le x_2 \le 5, \ 0 \le x_3 \le 2$$
$$s_1, s_2, s_3, R_2 \ge 0$$

The initial table is

Table 8

B V	y_1	x_2	x_3	s_2	s_1	R_2	s_3	Soln
z	-1	1	2	-1	0	0	0	9
s_1	4	-1	-3	0	1	0	0	5
R_2	-1	1	2	-1	0	1	0	9
s_3	-3	1	4	0	0	0	1	14

In Table 8, x_3 enters. To decide the leaving variable compute

$$\theta_1 = \min\left\{\tfrac{9}{2}, \tfrac{14}{4}\right\} = \tfrac{7}{2}$$

$$\theta_2 = \min\left\{\tfrac{\infty - 5}{3}\right\} = \infty, \text{ upper bound on } s_1$$

$$u_3 = 2, \text{ upper bound on entering variable } x_3$$

Thus, $\theta = \min\{\theta_1, \theta_2, u_3\} = u_3 = 2$. This is the case $\theta = u_j$. Thus, x_3 enters at its upper bound as nonbasic variable. Use the relation

$$x_3 = u_3 - x_3^1, \ 0 \le x_3^1 \le 2.$$

In Table 8, change x_3 to x_3^1 with column of x_3^1 is just negative of the column of x_3. In fact x_3 does not enter into the basis but it remains as nonbasic at its upper bound, see Table 9.

Table 9

B V	y_1	x_2	x_3^1	s_2	s_1	R_2	s_3	Soln
z	-1	1	-2	-1	0	0	0	5
s_1	4	-1	3	0	1	0	0	11
R_2	-1	1	-2	-1	0	1	0	5
s_3	-3	1	-4	0	0	0	1	6

Note that all entries of the solution column have been changed. For illustration, see z-row

$$z - y_1 + x_2 + 2x_3 - s_2 - 0s_1 - 0R_2 - 0s_3 = 9$$

If we substitute $2 - x_3^1$ for x_3, we get the z-row of Table 2. Similarly, by writing constraints of the constraint matrix, we compute other entries of the solution column.

In Table 9, x_2 enters. To decide leaving variable, compute

$$\theta_1 = \min\left\{\frac{5}{1}, \frac{6}{1}\right\} = 5$$

$$\theta_2 = \min\left\{\frac{\infty - 11}{-(-1)}\right\} = \infty, \text{ upper bound on } s_1$$

$$u_2 = 5 \text{ upper bound on } x_2$$

$$\theta = \min\{\theta_1, \theta_2, u_2\} = 5$$

The minimum θ is for θ_1 or u_2. Obviously, we prefer θ_1, and R_2 leaves. Make simplex iteration. This is the end of Phase-I and we discontinue with R_2-column, see Table 10.

Table 10

B V	y_1	x_2	x_3^1	s_2	s_1	s_3	Soln
z							
s_1	3	0	1	-1	1	0	16
x_2	-1	1	-2	-1	0	0	5
s_3	-2	0	-2	1	0	1	1

Phase-II. Rewrite the fresh objective function as

$$\max z = 4(y_1 + 1) + 2x_2 + 6(2 - x_3^1) = 4y_1 + 2x_2 - 6x_3^1 + 16.$$

Insert this objective function in Table 10 to get the next table

Table 11

B V	y_1	x_2	x_3^1	s_2	s_1	s_3	Soln
z	−6	0	2	−2	0	0	26
s_1	3	0	1	−1	1	0	16
x_2	−1	1	−2	−1	0	0	5
s_3	−2	0	−2	1	0	1	1

In Table 11, y_1 enters. To decide leaving variable, we compute

$$\theta_1 = \min\left\{\frac{16}{3}\right\} = \frac{16}{3}$$

$$\theta_2 = \min\left\{\frac{5-5}{1}, \frac{\infty-2}{2}\right\} = 0$$

$$u_1 = 2$$

The minimum 0 corresponds to the variable x_2. Thus, y_1 enters and x_2 leaves. This is the case $\theta = \theta_2$. To achieve this we just make simplex iteration and get Table 12.

Table 12

B V	y_1	x_2	x_3^1	s_2	s_1	s_3	Soln
z	0	−6	14	4	0	0	−4
s_1	0	3	−5	−4	1	0	31
y_1	1	−1	2	1	0	0	−5
s_3	0	−2	2	3	0	1	−9

Make x_2 at its upper bound by the substitution $x_2 = u_2 - x_2^1$, $0 \leq x_2^1 \leq 5$. For example y_1-row gives

$$y_1 - x_2 + 2x_3^1 + s_2 + 0s_1 + 0s_3 = -5$$

Since $u_2 = 5$, make the substitution $5 - x_2^1$ for x_2 in above constraint equation to have

Table 13

B V	y_1	x_2^1	x_3^1	s_2	s_1	s_3	Soln
z	0	6	14	4	0	0	26
s_1	0	3	-5	-4	1	0	16
y_1	1	1	2	1	0	0	0
s_3	0	-2	2	3	0	1	1

Table 13 is the optimal table. The optimal solution is

$$y_1 = 0 \implies x_1 - 1 = 0 \text{ that is } x_1 = 1$$
$$x_2^1 = 0 \implies 5 - x_2 = 0 \text{ that is } x_2 = 5$$
$$x_3^1 = 0 \implies 2 - x_3 = 0 \text{ that is } x_3 = 2$$
$$z = 26$$

Remark. If there is no negative entry in entering column, then $\theta_2 = \infty$.

5.4 Decomposition Principle

For solving problems of large size, it is not advisable to use simplex or revised simplex algorithm. For such problems, Dantzig and Eolfe proposed decomposition algorithm.

If the LPP has the following special structure it is possible to obtain the optimal solution by applying the Decomposition principle.

$$\text{opt} \quad f(X) = C_1^T X_1 + C_2^T X_2 + \cdots + C_r^T X_r \qquad (5.11a)$$
$$\text{s.t.} \quad A_1 X_1 + A_2 X_2 + \cdots + A_r X_r = D_0 \qquad (5.11b)$$
$$B_1 X_1 = D_1$$
$$B_2 X_2 = D_2$$
$$\cdots = \cdots \qquad (5.11c)$$
$$B_r X_r = D_r$$
$$X_1, X_2, \ldots, X_r \geq 0,$$

where $A_j = m_0 \times n_j$, $B_j = m_j \times n_j$, $D_j =$ column vector having m_j components $j = 0, 1, \ldots, r$, C_j and X_j are column vectors having n_j components, $j = 1, 2, \ldots, r$.

Algorithm.

Step 1. Consider the r subsidiary constraint sets

$$B_j X_j = D_j$$
$$X_j \geq 0, \; j = 1, 2, \ldots, r \tag{5.12}$$

Let S_{F_j} be the sets of feasible solutions of the system (5.12), which are bounded convex sets. Let h_j be the number of vertices of S_{F_j}, denoted by $X_1^j, X_2^j, \ldots, X_{h_j}^j$. Then, any point $X_j \in S_{F_j}$ can be written as

$$X_j = \mu_1^j X_1^j + \mu_2^j X_2^j + \cdots + \mu_{h_j}^j X_{h_j}^j \tag{5.13a}$$
$$\mu_1^j + \mu_2^j + \cdots + \mu_{h_j}^j = 1 \tag{5.13b}$$
$$\mu_k^j \geq 0, k = 1, 2, \ldots, h_j; \;\; j = 1, 2, \ldots, r \tag{5.13c}$$

The vertices $X_k^j, k = 1, 2, \ldots, h_j$ for each S_{F_j} can be obtained by solving (5.12) for $j = 1, 2, \ldots, r$.

Step 2. Substitute (5.13a) to (5.13c) in (5.11b) and (5.11c) and eliminate the subsidiary constraint sets from the given problem to have the following equivalent form.

$$\text{opt} \quad f(X) = C_1^T \sum_{k=1}^{h_1} \mu_k^1 X_k^1 + C_2^T \sum_{k=1}^{h_2} \mu_k^2 X_k^2 + \cdots + C_r^T \sum_{k=1}^{h_r} \mu_k^r X_k^r \tag{5.14a}$$

$$\text{s.t.} \quad A_1 \sum_{k=1}^{h_1} \mu_k^1 X_k^1 + A_2 \sum_{k=1}^{h_2} \mu_k^2 X_k^2 + \cdots + A_r \sum_{k=1}^{h_r} \mu_k^r X_k^r = D_0 \tag{5.14b}$$

$$\sum_{k=1}^{h_1} \mu_k^1 = 1, \quad \sum_{k=1}^{h_2} \mu_k^2 = 1, \quad \cdots, \quad \sum_{k=1}^{h_r} \mu_k^r = 1 \tag{5.14c}$$

$$\mu_k^j \geq 0, \;\; j = 1, 2, \ldots, r, \; k = 1, 2, \ldots, h_j \tag{5.14d}$$

The vertices $X_1^j, X_2^j, \cdots, X_{h_j}^j$ are known from the solution of $B_j X_j = D_j, j = 1, 2, \ldots r$, $C_1, C_2, \ldots, C_r$ and $A_1, A_2, \ldots, A_r$ are known as problem data. The unknowns in (5.14b) to (5.14d) are $\alpha_k^{(j)}$. Hence μ_k^j are the new decision variables of the modified problem (5.14).

Step 3. Solve the LPP in (5.14) by the standard techniques available in the literature and find the values μ_k^j.

Let the optimum solution be given by μ_k^{*j}. Substitute μ_k^{*j}'s value in (5.13a) and obtain the optimum solution X_j^*, $j = 1, 2, \ldots, r$ of the original problem. The optimum solution of the problem is

$$X^* = (X_1^*, X_2^*, \ldots, X_r^*)^T.$$

Example 4. (a) Write the following problem amenable to Decomposition principle; (b) Also, find its optimal solution?

$$\begin{aligned}
\max \quad & f = x_1 + 2x_2 + 2x_3 + 3x_4 \\
\text{s.t.} \quad & x_1 + x_2 + x_3 + x_4 \leq 100 \\
& x_1 + x_3 \leq 50 \\
& x_1 + x_2 \leq 60 \\
& x_1 - 2x_2 \leq 0 \\
& - 2x_3 + x_4 \leq 0 \\
& x_i \geq 0, \quad i = 1, 2, 3, 4.
\end{aligned}$$

The given problem can be written as

$$\begin{aligned}
\max \quad & f(X) = C_1^T X_1 + C_2^T X_2 \\
\text{s.t.} \quad & A_1 X_1 + A_2 X_2 = D_0 \\
& B_1 X_1 = D_1 \\
& B_2 X_2 = D_2 \\
& X_1, X_2 \geq 0,
\end{aligned}$$

where

$$A_1 = \begin{bmatrix} 1 & 1 \\ 1 & 0 \end{bmatrix}; \quad A_2 = \begin{bmatrix} 1 & 1 \\ 1 & 0 \end{bmatrix}; \quad B_1 = \begin{bmatrix} 1 & 1 \\ 1 & -2 \end{bmatrix}; \quad B_2 = \begin{bmatrix} -2 & 1 \end{bmatrix};$$

$$C_1 = \begin{bmatrix} 1 \\ 2 \end{bmatrix}; \quad C_2 = \begin{bmatrix} 2 \\ 3 \end{bmatrix}; \quad D_0 = \begin{bmatrix} 100 \\ 50 \end{bmatrix}; \quad D_1 = \begin{bmatrix} 60 \\ 0 \end{bmatrix};$$

$$D_2 = \begin{bmatrix} 0 \end{bmatrix}$$

$$X_1 = \begin{bmatrix} x_1 \\ x_2 \end{bmatrix}; \quad X_2 = \begin{bmatrix} x_3 \\ x_4 \end{bmatrix}$$

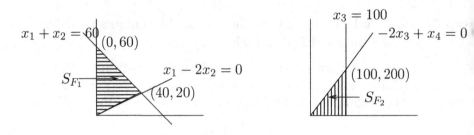

Iteration 1. Consider the subsidiary constraints sets $B_j X_j = D_j$, $j = 1, 2$

The vertices of

$$S_{F_1} \text{ are } X_1^{(1)} = \begin{bmatrix} 0 \\ 0 \end{bmatrix}; \quad X_2^{(1)} = \begin{bmatrix} 0 \\ 60 \end{bmatrix}; \quad X_3^{(1)} = \begin{bmatrix} 40 \\ 20 \end{bmatrix};$$

$$S_{F_2} \text{ are } X_1^{(2)} = \begin{bmatrix} 0 \\ 0 \end{bmatrix}; \quad X_2^{(2)} = \begin{bmatrix} 100 \\ 0 \end{bmatrix}; \quad X_3^{(2)} = \begin{bmatrix} 100 \\ 200 \end{bmatrix}$$

Hence, any point $X_1 \in S_{F_1}$ and $X_2 \in S_{F_2}$ can be given as

$$X_1 = \mu_1^1 \begin{bmatrix} 0 \\ 0 \end{bmatrix} + \mu_2^1 \begin{bmatrix} 0 \\ 60 \end{bmatrix} + \mu_3^1 \begin{bmatrix} 40 \\ 20 \end{bmatrix}$$

$$X_2 = \mu_1^2 \begin{bmatrix} 0 \\ 0 \end{bmatrix} + \mu_2^2 \begin{bmatrix} 100 \\ 200 \end{bmatrix} + \mu_3^2 \begin{bmatrix} 100 \\ 0 \end{bmatrix}$$

$$\mu_1^1 + \mu_2^1 + \mu_3^1 = 1$$
$$\mu_1^2 + \mu_2^2 + \mu_3^2 = 1$$
$$\mu_k^1, \; \mu_k^2 \geq 0, \quad k = 1, 2, 3$$

Iteration 2. Putting the values of X_1 and X_2 in the objective function and the first constraint of the problem (amenable to decomposition principle) and including restrictions on μ_k^j, we have

$$\max \quad f = (1,2) \begin{bmatrix} 40\mu_3^1 \\ 60\mu_2^1 + 20\mu_3^1 \end{bmatrix} + (2,3) \begin{bmatrix} 100\mu_2^2 + 100\mu_3^2 \\ 200\mu_2^2 \end{bmatrix}$$

$$\text{s.t.} \quad \begin{bmatrix} 1 & 1 \\ 1 & 0 \end{bmatrix} \begin{bmatrix} 40\mu_3^1 \\ 60\mu_2^1 + 20\mu_3^1 \end{bmatrix}$$

$$+ \begin{bmatrix} 1 & 1 \\ 1 & 0 \end{bmatrix} \begin{bmatrix} 100\mu_2^2 + 100\mu_3^2 \\ 200\mu_2^2 \end{bmatrix} \leq \begin{bmatrix} 100 \\ 50 \end{bmatrix}$$

$$\mu_1^1 + \mu_2^1 + \mu_3^1 = 1$$
$$\mu_1^2 + \mu_2^2 + \mu_3^2 = 1$$
all var ≥ 0

or, in simplified form

$$\max \quad f = 120\mu_2^1 + 80\mu_3^1 + 800\mu_2^2 + 200\mu_3^2$$
$$\text{s.t.} \quad 60\mu_2^1 + 60\mu_3^1 + 300\mu_2^2 + 100\mu_3^2 \leq 100$$
$$40\mu_3^1 + 100\mu_2^2 + 100\mu_3^2 \leq 50$$
$$\mu_1^1 + \mu_2^1 + \mu_3^1 = 1$$
$$\mu_1^2 + \mu_2^2 + \mu_3^2 = 1$$
$$\mu_k^1, \mu_k^1, \quad k = 1, 2, 3$$

The optimal solution of the above problem is

$$\mu_1^1 = 1, \mu_2^1 = 0, \mu_3^1 = 0, \mu_1^2 = 2/3, \mu_2^2 = 1/3, \mu_3^2 = 0; \ f = 800/3.$$

Inserting these values for the expressions of X_1 and X_2, we have

$$X_1 = \begin{bmatrix} x_1 \\ x_2 \end{bmatrix} = \begin{bmatrix} 0 \\ 0 \end{bmatrix} \implies x_1 = 0, x_2 = 0;$$

$$X_2 = \begin{bmatrix} x_3 \\ x_4 \end{bmatrix} = \frac{2}{3} \begin{bmatrix} 100 \\ 200 \end{bmatrix} = \begin{bmatrix} 100/3 \\ 200/3 \end{bmatrix} \implies x_3 = 100/3, x_4 = 200/3.$$

Thus, the optimal solution is

$$x_1 = 0, \ x_2 = 0, \ x_1 = 100/3, \ x_4 = 200/3, \ f = 800/3.$$

5.5   Karmarkar Interior Point Algorithm

With an increase in the number of variables or constraints causes an increase in multiplications and additions required for any iteration. The complexity of the simplex method is exponential, see Section 5.2. In 1984, Karmarkar proposed an algorithm named as interior point algorithm to solve large-scale linear programming problems efficiently. The beauty of the approach is that it gives polynomial time complexity for the solution. This is remarkably an excellent improvement over the simplex method. However, the analysis is not simple and requires projective geometry. In the simplex method, we move from a vertex to another vertex to find the vertex where the optimal solution lies. For the large LP problems the number of vertices will be quite large and this makes the simplex method very expensive in terms of computational time. It has been reported that the Karmakar's algorithm solved problems involving 150,000 variables and 12,000 constraints in one hour while the simplex method requires 4 hours for solving a smaller problem with 36,000 variables and 10,000 constraints. In fact it was found that the Karmakar's algorithm is 50 times faster than the simplex method.

The algorithm is based on the two observations:

1. If the current solution is near the center of the polytope, we can move along the direction of steepest descent to reduce value of f by maximum amount. From Fig. 5.2, it is clear that the current solution can be improved substantially by moving along the steepest direction if it is near the center (point 2) but not near the other points (point 1) and (point 3).

2. The solution can always be transformed without changing the nature of the problem so that the current solution lies near the center of the polytope.

The Karmarkar algorithm requires the LPP in a specific format:

$$\min \quad f = C^T X \qquad (5.15a)$$
$$\text{s.t.} \quad AX = 0 \qquad (5.15b)$$
$$e^T X = 1 \qquad (5.15c)$$
$$X \geq 0 \qquad (5.15d)$$

where $X = (x_1, x_2, \ldots, x_n)^T$, $C = (c_1, c_2, \ldots, c_n)^T$, $e = (1, 1, \ldots, 1)^T$ and A is a $m \times n$ matrix.

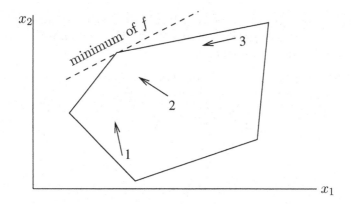

Figure 5.2

A feasible solution vector X of the above problem is defined to be an interior solution if every variable $x_i > 0$. Here the feasible domain is bounded, and hence a polytope. A consistent problem in the Karmakar's standard form certainly has a finite infimum. Karmarkar made two assumptions for his algorithm:

(A1) $Ae = 0$, so that $X_0 = (1/n, 1/n, \dots, 1/n)^T$ is an initial interior solution.

(A2) The optimal objective value of the this problem is zero.

Conversion of a LPP in required form. Let the LPP be given in standard form:

$$\text{min} \quad f = C^T X \qquad (5.16a)$$
$$\text{s.t.} \quad AX = b \qquad (5.16b)$$
$$X \geq 0 \qquad (5.16c)$$

Our objective is to convert this problem into the standard form (5.15) required by Karmarkar, while satisfying the assumptions (A1) and (A2).

The key feature of the Karmakar's standard form is the simplex structure, which of course results in a bounded feasible region. Thus, we have to regularize problem (5.16) by adding a bounding constraint

$$e^T X = x_1 + x_2 + \cdots + x_n \leq Q$$

for some positive integer Q derived from the feasibility and optimality considerations. In the worst case, we can choose $Q = 2^L$, where L is the problem size (number of variables). If this constraint is binding at optimality with objective value $-2^{O(L)}$, then we can show that the LPP has unbounded solution.

By introducing a slack variable x_{n+1}, we have a new linear program

$$\min \quad f = C^T X \tag{5.17a}$$
$$\text{s.t.} \quad AX = b \tag{5.17b}$$
$$e^T X + x_{n+1} = Q \tag{5.17c}$$
$$X \geq 0, \; x_{n+1} \geq 0 \tag{5.17d}$$

In order to keep the matrix structure of A undisturbed for sparsity manipulation, we introduce a new variable $x_{n+2} = 1$ and rewrite the constraints of (5.17) as

$$AX - bx_{n+2} = 0 \tag{5.18a}$$
$$e^T X + x_{n+1} + Qx_{n+2} = 0 \tag{5.18b}$$
$$e^T X + x_{n+1} + x_{n+2} = Q + 1 \tag{5.18c}$$
$$X \geq 0, \; x_{n+1}, \; x_{n+2} \geq 0 \tag{5.18d}$$

Note that the constraint $x_{n+2} = 1$ is direct consequence of (5.18b) and (5.18c). To normalize (5.18c) for the required simplex structure, we apply the transformation

$$x_j = (Q+1)y_j, \quad j = 1, 2, \ldots, n+2.$$

In this way, we have an equivalent linear program

$$\min \quad f = (Q+1)C^T Y \tag{5.19a}$$
$$\text{s.t.} \quad AY - by_{n+2} = 0 \tag{5.19b}$$
$$e^T Y + y_{n+1} - Qy_{n+2} = 0 \tag{5.19c}$$
$$e^T Y + y_{n+1} + y_{n+2} = 1 \tag{5.19d}$$
$$Y \geq 0, \; y_{n+1}, \; y_{n+2} \geq 0 \tag{5.19e}$$

The problem (5.19) is now in the standard form required by the Karmarkar algorithm. In order to satisfy the assumption (A1), we may introduce an artificial variable y_{n+3} with a large cost coefficient M as designed in big-M method and consider the following problem

$$\min \quad f = (Q+1)C^T Y + M y_{n+3} \tag{5.20a}$$

$$\text{s.t.} \quad AY - b y_{n+2} - [Ae - b] y_{n+3} = 0 \tag{5.20b}$$

$$e^T Y + y_{n+1} - Q y_{n+2} - (n+1-Q) y_{n+3} = 0 \tag{5.20c}$$

$$e^T Y + y_{n+1} + y_{n+2} + y_{n+3} = 1 \tag{5.20d}$$

$$Y \geq 0, \ y_{n+1}, \ y_{n+2} \geq 0, \ y_{n+3} \geq 0 \tag{5.20e}$$

Observe that this form satisfies assumption (A1) as $(1/n+3, 1/n+3, \ldots, 1/n+3)$ is the interior point solution. Its minimum value is zero (assumption (A2)) which we shall not prove here.

Example 5. Transform the following LPP into a from required by Karmarkar algorithm.

$$\min \quad x_0 = 2x_1 + 3x_2$$
$$\text{s.t.} \quad 3x_1 + x_2 - 2x_3 = 3$$
$$5x_1 - 2x_2 = 2$$
$$x_1, x_2, x_3 \geq 0$$

In the given LPP, the following data is available

$$A = \begin{bmatrix} 3 & 1 & -2 \\ 5 & -2 & 0 \end{bmatrix}; \quad b = \begin{bmatrix} 3 \\ 2 \end{bmatrix}; \quad Q = 2^3 = 8; \quad n = 3.$$

Fill up this data in (5.20) to get the Karmarkar specific form as

$$\min \quad y_0 = (8+1)(2,3,0)(y_1, y_2, y_3)^T + M y_6$$

$$\text{s.t.} \quad \begin{bmatrix} 3 & 1 & -2 \\ 5 & -2 & 0 \end{bmatrix} \begin{bmatrix} y_1 \\ y_2 \\ y_3 \end{bmatrix} - y_5 \begin{bmatrix} 3 \\ 2 \end{bmatrix} - y_6 \left(\begin{bmatrix} 2 \\ 3 \end{bmatrix} - \begin{bmatrix} 3 \\ 2 \end{bmatrix} \right) = \begin{bmatrix} 0 \\ 0 \end{bmatrix}$$

$$y_1 + y_2 + y_3 + y_4 - 8y_5 + 4y_6 = 0$$
$$y_1 + y_2 + y_3 + y_4 + y_5 + y_6 = 1$$
all var ≥ 0

On simplification, we have

$$\min \quad 7y_0 = 18y_1 + 27y_2 + My_6$$
$$\text{s.t.} \quad 3y_1 + y_2 - 2y_3 - 3y_5 + y_6 = 0$$
$$5y_1 - 2y_2 - 2y_5 - y_6 = 0$$
$$y_1 + y_2 + y_3 + y_4 - 8y_5 + 4y_6 = 0$$
$$y_1 + y_2 + y_3 + y_4 + y_5 + y_6 = 1$$
$$\text{all var } \geq 0$$

Obviously $(1/6, 1/6, \ldots, 1/6)$ is an interior point solution and objective function value is zero.

Algorithm. The Karmarkar algorithm proceeds as

Step 1. Set $k = 0$, $x^o = (1/n, 1/n, \ldots, 1/n)^T$.

Step 2. If the desired accuracy $\varepsilon > 0$ such that $C^T x^k \leq \varepsilon$ is achieved, then stop with x^k is an approximation to the optimal solution. Otherwise, go to Step 3.

Step 3. Here we find a better solution.

$$D_k = \text{diagonal matrix formed with diagonal elements as}$$
$$\text{the components of } x^k$$

$$B_k = \begin{bmatrix} AD_k \\ 1, 1, \ldots, 1 \end{bmatrix}$$

$$d_k = -\left[\mathbf{I} - B_k^T \left(B_k B_k^T \right)^{-1} B_k \right] D_k C$$

$$y^k = \left(\frac{1}{n}, \frac{1}{n}, \ldots, \frac{1}{n} \right)^T + \frac{\alpha}{n} \left(\frac{d_k}{\|d^k\|} \right) \quad \text{for some } 0 < \alpha \leq 1$$

$$x^{k+1} = \frac{D_k y^k}{e^T D_k y^k}$$

Set $k = k + 1$ and go to Step 2.

Note that in this computational procedure x^k is always an interior feasible solution; D_k is an n-dimensional matrix with ith element of vector x^k as its ith diagonal element. B_k is the constraint matrix of the LPP in the Karmarkar's standard form; d_k is the feasible direction of the projected negative gradient; y^k is a new interior feasible solution in the transformed space; and x^{k+1} is a new interior feasible solution in original space.

To make the algorithm more accessible, let us workout a problem.

Example 6. Find the solution of the following LPP using the Karmarkar's method.

$$\begin{aligned}
\min \quad & f = 2x_1 + x_2 - x_3 \\
\text{s.t.} \quad & x_2 - 3_3 = 0 \\
& x_1 + x_2 + x_3 = 1 \\
& x_1, x_2, x_3 \geq 0
\end{aligned}$$

Use tolerance 0.075 for testing the convergence of the procedure.

First we see that the LPP is in the Karmarkar's standard form, which satisfies both assumptions A(1) and A(2). Hence, we start with

$$x^o = \left(\frac{1}{3}, \frac{1}{3}, \frac{1}{3}\right).$$

Note that $A = [0, 1, -1]$ and $C = (2, 1 - 1)^T$.

Now we check Step 2. $f(x^o) = 2/3 => 0.05$, go to Step 3.

For Step 3, we define

$$D_0 = \begin{bmatrix} 1/3 & 0 & \\ 0 & 1/3 & 0 \\ 0 & 0 & 1/3 \end{bmatrix}$$

Then $AD_0 = [0, 1/3, -1/3]$ and

$$B_0 = \begin{bmatrix} 0 & 1/3 & -1/3 \\ 1 & 1 & 1 \end{bmatrix}$$

$$B_0 B_0^T = \begin{bmatrix} 0 & 1/3 & -1/3 \\ 1 & 1 & 1 \end{bmatrix} \begin{bmatrix} 0 & 1 \\ 1/3 & 1 \\ -1/3 & 1 \end{bmatrix} = \begin{bmatrix} 2/9 & 0 \\ 0 & 3 \end{bmatrix}$$

$$\left(B_0 B_0^T\right)^{-1} = \begin{bmatrix} 9/2 & 0 \\ 0 & 1/3 \end{bmatrix}$$

$$D_0 C = \begin{bmatrix} 1/3 & 0 & \\ 0 & 1/3 & 0 \\ 0 & 0 & 1/3 \end{bmatrix} \begin{bmatrix} 2 \\ 1 \\ -1 \end{bmatrix} = \begin{bmatrix} 2/3 \\ 1/3 \\ -1/3 \end{bmatrix}$$

$$\mathbf{I} - B_0^T \left(B_0 B_0^T\right)^T B_0 = \begin{bmatrix} 1 & 0 & 0 \\ 0 & 1 & 0 \\ 0 & 0 & 1 \end{bmatrix} - \begin{bmatrix} 0 & 1 \\ 1/3 & 1 \\ -1/3 & 1 \end{bmatrix} \begin{bmatrix} 9/2 & 0 \\ 0 & 1/3 \end{bmatrix}$$

$$\times \begin{bmatrix} 0 & 1/3 & -1/3 \\ 1 & 1 & 1 \end{bmatrix}$$

or,

$$\mathbf{I} - B_0^T \left(B_0 B_0^T\right)^T B_0 = \begin{bmatrix} 1 & 0 & 0 \\ 0 & 1 & 0 \\ 0 & 0 & 1 \end{bmatrix} - \begin{bmatrix} 1/3 & 1/3 & 1/3 \\ 1/3 & 5/6 & -1/6 \\ 1/3 & -1/6 & 5/6 \end{bmatrix}$$

$$= \begin{bmatrix} 2/3 & -1/3 & -1/3 \\ -1/3 & 1/6 & 1/6 \\ -1/3 & 1/6 & 1/6 \end{bmatrix}$$

Moreover, the moving direction is given by

$$d_0 = - \left[\mathbf{I} - B_0 \left(B_0 B_0^T \right)^T B_0 \right] D_0 C = \begin{bmatrix} -2/3 & 1/3 & 1/3 \\ 1/3 & -1/6 & -1/6 \\ 1/3 & 1/6 & 1/6 \end{bmatrix} \begin{bmatrix} 2/3 \\ 1/3 \\ -1/3 \end{bmatrix}$$

$$= \begin{bmatrix} -4/9 \\ 2/9 \\ 2/9 \end{bmatrix}$$

with norm $\|d^o\|$, length of vector d^o defined by

$$\|d_0\| = \sqrt{(-4/9)^2 + (2/9)^2 + (2/9)^2} = 2\sqrt{6}/9.$$

Let us choose $\alpha = 1/\sqrt{6}$ (see $r = 1/\sqrt{n(n-1)}$) to obtain a new solution in transformed space

$$y^o = \begin{bmatrix} 1/3 \\ 1/3 \\ 1/3 \end{bmatrix} + \frac{1}{3} \cdot \frac{1}{\sqrt{6}} \cdot \frac{9}{2\sqrt{6}} \begin{bmatrix} -4/9 \\ 2/9 \\ 2/9 \end{bmatrix} = \begin{bmatrix} 1/3 \\ 1/3 \\ 1/3 \end{bmatrix} + \begin{bmatrix} -1/9 \\ 1/18 \\ 1/18 \end{bmatrix} = \begin{bmatrix} 2/9 \\ 7/18 \\ 7/18 \end{bmatrix}$$

Hence, the new interior feasible solution is given by

$$x^1 = \frac{D_0 y^o}{e^T D_0 y^o} = \frac{(2/27, 7/54, 7/54)}{1/3} = (2/9, 7/18, 7/18)^T$$

Note that
$$f(x^1) = 4/9 > 0.075$$

Certainly, there is improvement over the first starting estimation, but tolerance limit is not satisfied. Thus, we have to repeat the algorithm until required tolerance limit is satisfied.

However, if we choose $\alpha = 2/\sqrt{6}$, then

$$y^o = \begin{bmatrix} 1/3 \\ 1/3 \\ 1/3 \end{bmatrix} + \frac{1}{3} \cdot \frac{2}{\sqrt{6}} \cdot \frac{9}{2\sqrt{6}} \begin{bmatrix} -4/9 \\ 2/9 \\ 2/9 \end{bmatrix} = \begin{bmatrix} 1/9 \\ 4/9 \\ 4/9 \end{bmatrix}$$

This time $e^T D_0 y^o = 1$ and hence

$$x^1 = (1/27, 4/27, 4/27)^T$$

Now $f(x^1) = 2/27 < 0.075$. Thus, $x^1 = (1/27, 4/27, 4/27)^T$ is taken as an approximation to the optimal solution.

Problem Set 5

1. Solve the following LPP by revised simplex method

$$\min \quad x_0 = x_1 + x_2 + x_3$$
$$\text{s.t.} \quad x_1 - x_4 - 2x_6 = 5$$
$$x_2 + 2x_4 - 3x_5 + x_6 = 3$$
$$x_3 + 2x_4 - 5x_5 + 6x_6 = 5$$
$$x_i \geq 0, \quad i = 1 \text{ to } 6$$

Suggestion. Take x_1, x_2, x_3 as the starting basic variables.

2. Use any method which seems to be convenient to find the optimal solution of the LPP

$$\min \quad x_0 = x_1 + 2x_2 + 3x_3 - x_4$$
$$\text{s.t.} \quad x_1 + 2x_2 + 3x_3 = 15$$
$$2x_1 + x_2 + 5x_3 = 20$$
$$x_1 + 2x_2 + 5x_3 + x_4 = 10$$
$$x_1, x_2, x_3, x_4 \geq 0$$

Suggestion. Follow Remark 4, Section 3.4 or use the revised simplex method.

3. Solve Example 2 using the revised simplex method in reference to big-M method.

4. Use revised dual simplex method to solve the LPP

$$\max \quad z = -2x_1 - x_3$$
$$\text{s.t.} \quad x_1 + x_2 - x_3 \geq 5$$
$$x_1 - 2x_2 + 4x_3 \geq 8$$
$$x_1, x_2, x_3 \geq 0$$

5. Find the optimal solution of the following using bounded variable simplex method.

$$\max \quad z = 3x_1 + 5x_2 + 3x_3$$
$$\text{s.t.} \quad x_1 + 2x_2 + 2x_3 \leq 14$$
$$2x_1 + 4x_2 + 3x_3 \leq 23$$
$$0 \leq x_1 \leq 4, \ 0 \leq x_2 \leq 5, \ 0 \leq x_3 \leq 3$$

6. Solve the following LPP using bounded variable technique

$$\max \quad z = 4y_1 + 2y_2 + 6y_3$$
$$\text{s.t.} \quad 4y_1 - y_2 \leq 9$$
$$- y_1 + y_2 + 2y_3 \leq 8$$
$$- 3y_1 + y_2 + 4y_3 \leq 12$$
$$1 \leq y_1 \leq 3, \ 0 \leq y_2 \leq 5, \ 0 \leq y_3 \leq 2$$

7. Use the bounded variable technique to solve the linear programming problem formulated in Problem 9, Problem set 1.

8. A company manufactures two type of toys A-type and B-type in a machine centre. The productions times for one toy of types A and B are 10 and 12 minutes, respectively. The total machine time available is 2500 minutes per day. In one day the company can sell between 150 and 200 toys of A-type, but not more than 45 toys of B-type. Overtime may be used to meet the demand at an additional cost of $0.5 per minute.

   (a) Assuming that the unit profits for toys of types A and B are $6 and $7.5, respectively, formulate a model and determine the optimal production level for each type of toys as well as any overtime needed at the machine center.

   (b) If the cost per overtime minute is increased to $1.5, should the company use overtime.

Suggestion. Define $x_1 =$ number of toys of A-type per day, $x_2 =$ number of toys of B-type per day, $x_3^+ =$ unused minutes of the machine per day, $x_3^- =$ overtime of the machine in minutes per day. Note that the LP solution can not yield both x_3^+ and x_3^- at positive level simultaneously. This means that either x_3^+ or x_3^- can be positive, but never both at the same time. This observation ensures the validity of the model.

9. Solve the following LPP by decomposition principle, and verify your answer by solving it using revised simplex method.

$$\begin{aligned}
\max \quad & z = 8x_1 + 3x_2 + 8x_3 + 6x_4 \\
\text{s.t.} \quad & 4x_1 + 3x_2 + x_3 + 3x_4 \le 16 \\
& 4x_1 - x_2 + x_3 \le 12 \\
& x_1 + 2x_2 \le 8 \\
& 3x_1 + x_2 \le 10 \\
& 2x_3 + 3x_4 \le 9 \\
& 4x_3 + x_4 \le 12 \\
& x_i \ge 0, \quad i = 1 \text{ to } 4
\end{aligned}$$

10. Consider the LPP with bounded constraints

$$\max \quad z = \{C^T X : L \le AX \le U, \; X \ge 0\},$$

where L and U are constant column vectors. Let us define the slack vector $Y \ge 0$ such that $AX + Y = U$. Show that this LPP is equivalent to

$$\max \quad z = \{C^T X : AX + Y = U, \; 0 \le Y \le U - L, \; X \ge 0\}.$$

Using above procedure solve the LPP

$$\begin{aligned}
\max \quad & z = 5x_1 - 4x_2 + 6x_3 \\
\text{s.t.} \quad & 20 \le x_1 + 7x_2 + 3x_3 \le 50 \\
& 10 \le 3x_1 - x_2 + x_3 \le 20 \\
& 15 \le 2x_1 + 3x_2 - x_3 \le 40 \\
& x_1, x_2, x_2 \ge 0
\end{aligned}$$

This is referred to as bounded interval programming.

11. Convert the LPP of Problem 19, Problem set 4 into Karmarkar's standard form.

12. Apply Karmarkar's algorithm to solve the following linear pro-

gramming problems

(a) min $z = -x_1 + 1$

s.t. $x_2 - x_3 = 0$

$x_1 + x_2 + x_3 = 1$

$x_1, x_2, x_3 \geq 0$

(b) min $z = -x_1 - 2x_2 + 4x_5$

s.t. $x_2 - x_3 = 0$

$2x_1 - 2x_2 + 4x_3 - 4x_5 = 0$

$x_1 + 2x_2 + x_4 - 5x_5 = 0$

$x_1 + x_2 + x_3 + x_4 + x_5 = 1$

$x_1, x_2, x_3, x_4, x_5 \geq 0$

Chapter 6

Sensitivity Analysis

Sensitivity analysis, sometimes referred to as post optimal analysis is an essential part of the optimization techniques. This chapter is devoted to sensitivity analysis, a process applied to the optimal table of any linear programming problem when some changes are proposed in the original problem. The last section gives an introduction to systematic sensitivity analysis or parametric programming.

6.1   Introduction

Given a LPP in standard form, the problem is completely specified by the constraint matrix A, the right hand side vector (availabilities) b, and the cost vector C. We assume that the LPP has an optimal solution with the data set (A, b, C). In many cases, we find the data set (A, b, C) needs to be changed within a range after we obtained the optimal solution, and we are interested to find the new optimal solution.

Thus, the possible changes are

(i) Change in the cost vector;

(ii) Change in the right hand side vector (availability);

(iii) Change in the constraint matrix.

With the help of sensitivity analysis a relatively small amount of work is applied to the optimal table of the LPP to find the optimal solution of the changed problem which saves time, labour and space.

Remarks. 1. There are cases when rigorous changes are made in the original linear programming problem, there is no alternative but to go back to the beginning and re-solve the problem.

2. In this chapter we shall be interested in discrete changes in the data. The effect of the continuous (parametric) change in data on the optimal solution will be discussed in Chapter 18.

When we incorporate any of the above changes in the original problem, either the optimal table remains unchanged or the optimality criteria is disturbed or the feasibility is disturbed or both are disturbed. To restore optimality we shall use the simplex method, while the disturbed feasibility is restored by using the dual simplex method.

The combined disturbance in optimal criteria and the feasibility in the optimal table will be treated as a comprehensive problem in the problem set.

Now, we discuss the above possible changes in a sequential manner. Before we do any thing, let us recall the structure of the simplex table.

Simplex Table

Basis	x_1	$\cdots$	x_m	Starting BFS	Solution
z	$z_1 - c_1$	$\cdots$	$z_m - c_m$	$\cdots$	$C_B^T X_B = C_B^T B^{-1} b$
X_B	$B^{-1}A_1$	$\cdots$	$B^{-1}A_m$	B^{-1}	$B^{-1}b$

Here $z_j - c_j = C_B^T B^{-1} A_j - c_j$, $j = 1, 2, \ldots, n$, A_j is the jth column of the coefficient matrix A, when the LPP is written in standard form.

6.2   Change in the Cost Vector

The change in cost of variables has a direct impact on optimal criteria (z-row) which has its entries as $z_j - c_j = C_B^T B^{-1} A_j - c_j$. If the optimal criteria is disturbed due to cost change, then use simplex method to restore optimality which results in a new solution.

The two types of changes are possible:

(i) *Change in cost of a nonbasic variable.* With the change in cost of a nonbasic variables the relative cost of this variable is changed.

Obviously, there is no change in relative cost of any other variable. If sign of the relative cost is changed then bring this variable into the basis to get the new optimal solution.

Let the LPP be max $z = C^T X$, subject to $AX = b$, $X \geq 0$. Suppose x_k is a nonbasic variable and its cost c_k is changed to $c_k \Delta c_k$, where $k \in \tilde{N}$, the index set of nonbasic variables. The new relative cost of x_k turns up

$$C_B^T B^{-1} A_k - (c_k + \Delta c_k), \quad k \in \tilde{N},$$

Since C_B^T is fixed and cost of all remaining variables are kept fixed, there will be no change relative cost of any other variable. The optimal solution remains same if

$$C_B^T B^{-1} A_k - (c_k + \Delta c_k) \geq 0,$$

otherwise the optimality is disturbed which can be restored by simplex method to find new optimal solution.

(ii) *Change in cost of a basic variable.* With change in the cost of a basic variable all $z_j - c_j$ will change except for the basic variables. Note that $C_B^T B^{-1}$ can not be taken from the optimal table as C_B^T has changed. There will also be a change in the objective function value.

Let c_j, the cost of jth basic variable is shifted to $c_j + \Delta c_j$, where $j \in \tilde{B}$, the index set of basic variables. Then relative cost of each nonbasic variable is changed as

$$(C_B^T + \Delta c_j e_j) \alpha^k - c_k, \quad j \in \tilde{B}, \ k \in \tilde{N},$$

where α^k and c_k are the coordinate vector and cost of kth nonbasic variable, respectively.

To stay optimal solution as it is, we must have

$$C_B^T \alpha^k - c_k + \Delta c_j \alpha_j^k \geq 0$$

This implies

$$\Delta c_j \geq \frac{z_k - c_k}{-\alpha_j^k}, \quad \text{for } k\text{th nonbasic variable}$$

Hence, we can define

$$\Delta \underline{c}_j = \max_k \left\{ \frac{z_k - c_k}{-\alpha_j^k}, \ \alpha_j^k > 0, \ k \in \tilde{N} \right\}. \tag{6.1}$$

$$\Delta \bar{c}_j = \min_k \left\{ \frac{z_k - c_k}{-\alpha_j^k}, \ \alpha_j^k < 0, \ k \in \tilde{N} \right\}. \tag{6.2}$$

which ensures the variation limits in cost of jth basic variable

$$\Delta \underline{c}_j \leq \Delta c_j \leq \Delta \bar{c}_j.$$

If Δc_j goes out of these limits for at least one nonbasic variable, this implies optimality is disturbed. Calculate fresh objective function value using

$$(c_1, c_2, \ldots, c_j + \Delta c_j, \ldots, c_m) B^{-1} b,$$

and apply simplex method to restore optimality which results in new optimal solution.

Example 1. Suppose that we are given the LPP

$$\begin{aligned}
\max \quad & x_0 = 2x_1 + 3x_2 + 4x_3 \\
\text{s.t.} \quad & x_1 + 2x_2 + 3x_3 \leq 11 \\
& 2x_1 + 3x_2 + 2x_3 \leq 10 \\
& x_1, x_2, x_3 \geq 0
\end{aligned}$$

with its optimal table as

Table 1

B V	x_1	x_2	x_3	s_1	s_2	Soln
x_0	0	1/2	0	1	1/2	16
x_3	0	1/4	1	1/2	−1/4	3
x_1	1	5/4	0	−1/2	3/4	2

(a) Within what range the cost of x_1 varies so that the optimality remains unaffected.

(b) Within what range the cost of x_2 varies so that the optimal solution remains unaffected.

(c) Discuss the effect of changing the costs $2, 3, 4$ of the decision variables x_1, x_2, x_3 to $1, 2, 2$.

For part (a), there will be a change in relative cost of x_1. Note that x_1 is a basic variable, and hence, the cost of basis C_B^T has changed, and $C_B^T B^{-1}$ (simplex multiplier) can not be taken from the optimal table. Let cost of x_1 be c_1 for which we have to determine the variation. The cost of basis is $(4, c_1)^T$. For no change in the optimal solution, we must have $z_j - c_j = C_B^T B^{-1} A_j - c_j \geq 0$ for all nonbasic variables. Calculate

$$z_2 - c_2 = (4, c_1) \begin{bmatrix} 1/4 \\ 5/4 \end{bmatrix} - 3 \geq 0 \implies c_1 \geq 8/5$$

$$z_4 - c_4 = (4, c_1) \begin{bmatrix} 1/2 \\ -1/2 \end{bmatrix} \geq 0 \implies c_1 \leq 4$$

$$z_5 - c_5 = (4, c_1) \begin{bmatrix} -1/4 \\ 3/4 \end{bmatrix} \geq 0 \implies c_1 \geq 4/3$$

The common value of $c_1 \in [8/5, 4]$ satisfies all the above three inequalities, and this is the range in which c_1 may vary without affecting the optimality.

For part (b), note that x_2 is a nonbasic variable. There is no change in C_B^T. Hence, take $C_B^T B^{-1}$ from the table which is available below starting BFS. Hence,

$$z_2 - c_2 = (1, 1/2) \begin{bmatrix} 2 \\ 3 \end{bmatrix} - c_2 \geq 0 \implies c_2 \leq 7/2.$$

Thus, for $c_2 \in (-\infty, 7/2]$, there is no change in the optimal solution.

For part (c), we observe that cost of basic and nonbasic variables have changed. We shall not calculate $z_j - c_j$ for all basic variables as these are bound to be zero in every simplex table. With the change of the objective coefficients from $2, 3, 4$ to $1, 2, 2$, the new cost of the basis is $C_B^T = (2, 1)^T$. Calculate new $z_j - c_j$ for all nonbasic variables as

$$\text{New } z_2 - c_2 = (2, 1) \begin{bmatrix} 1/4 \\ 5/4 \end{bmatrix} - 2 = -1/4$$

$$\text{New } z_4 - c_4 = (2, 1) \begin{bmatrix} 1/2 \\ -1/2 \end{bmatrix} - 0 = 1/2$$

$$\text{New } z_5 - c_5 = (2, 1) \begin{bmatrix} -1/4 \\ 3/4 \end{bmatrix} - 0 = 1/4$$

Thus the optimality is disturbed as the new relative cost of x_2 has turned to be negative. Before proceeding further, calculate

$$\text{New } f(X_B) = (2, 1) \begin{bmatrix} 3 \\ 2 \end{bmatrix} = 8.$$

After incorporating these changes in Table 1, we have

Table 2

B V	x_1	$x_2 \downarrow$	x_3	s_1	s_2	Soln
x_0	0	$-1/4$	0	$1/2$	$1/4$	8
x_3	0	$1/4$	1	$1/2$	$-1/4$	3
$\leftarrow x_1$	1	$\boxed{5/4}$	0	$-1/2$	$3/4$	2

The variable x_2 enters (having most negative relative cost) and x_1 leaves the basis (minimum ratio rule), see Table 2. Table 3 gives the optimal solution of the changed problem.

Table 3

B V	x_1	x_2	x_3	s_1	s_2	Soln
x_0	$1/5$	0	0	$2/5$	$2/5$	$42/5$
x_3	$-1/5$	0	1	$3/5$	$-2/5$	$13/5$
x_2	$4/5$	1	0	$-2/5$	$3/5$	$8/5$

The optimal solution: $x_1=0$, $x_2=8/5$, $x_3=13/5$,   optimal value $x_0=42/5$.

Remarks. 1. For parts (a), this is sufficient to mention the optimal table only, if cost of x_3 is specified.

However, if formulas (6.1) and (6.2) are used then only optimal table will serve our purpose. Note that $\tilde{B} = \{3, 1\}$ and $\tilde{N} = \{2, 4, 5\}$. Hence,

$$\Delta \underline{c}_1 = \max\{-2/5, -2/3\} = -2/5$$

$$\Delta \bar{c}_1 = \min\{2\} = 2.$$

Thus,

$$-2/5 \leq \Delta c_1 \leq 2 \implies 8/5 \leq c_1 \leq 4.$$

2. For part (b), mentioning the optimal table is sufficient provided we add the information that s_1, s_2, are slack variables and $A_2 = (2, 3)^T$. Extract $C_B^T B^{-1}$ from the given optimal table.

3. Case (c) is related to simultaneous change in the costs of basic and nonbasic variables.

6.3   Changes in the Right-hand Side Vector

In a LPP, if the change in right hand side of the constraints is made then the solution column, $B^{-1}b$, and the objective function value, $f(X_B) = C_B^T B^{-1} b$ are affected. This change corresponds to two cases.

(i) If all entries of the new solution column turn out to be non-negative, then the existing table remains optimal with the new solution and new optimal value.

Let the LPP be max $z = C^T X$, subject to $AX = b$, $X \geq 0$. Suppose right hand side entry b_k of the vector $b = (b_1, b_2, \ldots, b_m)^T$ is shifted to $(b_1 + b_2 + \ldots + b_k + \Delta b_k, \ldots, b_m)^T$. Then the new solution column is

$$X'_B = B^{-1}(b + \Delta b_k e_k),$$

where $e_k = (0, 0, \ldots, 1, \ldots, 0)^T$ has 1 at kth position.

Let $B^{-1} = (\beta_{ik})_{m \times m}$. The optimal basis remains the same if

$$X'_B = B^{-1}b + \Delta b_k B^{-1} e_k \geq 0.$$

or

$$X_B + \Delta b_k \begin{bmatrix} \beta_{1k} \\ \beta_{2k} \\ \vdots \\ \beta_{mk} \end{bmatrix} \geq \begin{bmatrix} 0 \\ 0 \\ \vdots \\ 0 \end{bmatrix}$$

If $\tilde{B}$ denotes the index set of basic variables, then we have

$$X_q + \Delta b_k \beta_{ik} \geq 0, \quad q \in \tilde{B}, \quad i = 1, 2, \ldots, m \qquad (6.3)$$

Further, (6.3) gives

$$\Delta b_k \geq \frac{X_q}{-\beta_{ik}},$$

Obviously, we define

$$\Delta \underline{b}_k = \max_i \left\{ \frac{X_q}{-\beta_{ik}}, \ \beta_{ik} > 0 \right\}. \qquad (6.4)$$

$$\Delta \overline{b}_k = \min_i \left\{ \frac{X_q}{-\beta_{ik}}, \ \beta_{ik} < 0 \right\}. \qquad (6.5)$$

Thus, if Δb_k satisfies

$$\Delta \underline{b}_k \leq \Delta b_k \leq \Delta \overline{b}_k,$$

the same basis remains intact.

In this way the same set of basic variable with changed values according to formula (6.3) produce the new solution and new objective value is $C_B^T X_B'$.

(ii) The second possibility is that Δb_k is assigned beyond the above limits. Then at least one of the entry in new $B^{-1}b$ is negative, i.e., feasibility is disturbed. Restore the feasibility using the dual simplex method to get the optimal solution of the revised problem.

Example 2. Consider the LPP

$$\begin{aligned}
\max \quad & x_0 = 3x_1 + 2x_2 + 5x_3 \\
\text{s.t.} \quad & x_1 + 2x_2 + x_3 \leq 43 \\
& 3x_1 + 2x_3 \leq 46 \\
& x_1 + 4x_2 \leq 42 \\
& x_1, x_2, x_3 \geq 0
\end{aligned}$$

The optimal table of this LPP is

Table 4

B V	x_1	x_2	x_3	s_1	s_2	s_3	Soln
x_0	4	0	0	1	2	0	135
x_2	$-1/4$	1	0	$1/2$	$-1/4$	0	10
x_3	$3/2$	0	1	0	$1/2$	0	23
s_3	2	0	0	-2	1	1	2

Find the optimal solution when following modifications are proposed in the above LPP.

(a) $b = (43, 46, 42)^T$ is changed to $b' = (60, 64, 59)^T$.

(b) $b = (43, 46, 42)^T$ is changed to $b' = (45, 46, 40)^T$.

(c) Within what range the first component of right hand side vector varies so that the optimal basis remains unaltered.

For part (a), new solution column is $B^{-1}b'$, i.e.,

$$\begin{bmatrix} 1/2 & -1/4 & 0 \\ 0 & 1/2 & 0 \\ -2 & 1 & 1 \end{bmatrix} \begin{bmatrix} 60 \\ 64 \\ 59 \end{bmatrix} = \begin{bmatrix} 14 \\ 32 \\ 3 \end{bmatrix}$$

Since $B^{-1}b' \geq 0$, it follows that feasibility remains unaltered, and only the value of the basic variables are changed. New objective function value is $C_B^T X_B = C_B^T B^{-1} b'$, i.e.,

$$(1, 2, 0)(60, 64, 59)^T = 188.$$

The same basis remains optimal, and hence, the optimal solution of changed problem is

$$x_1 = 0, \ x_2 = 14, \ x_3 = 32, \ \text{max value } x_0 = 188.$$

For part (b), note that

$$\begin{bmatrix} 1/2 & -1/4 & 0 \\ 0 & 1/2 & 0 \\ -2 & 1 & 1 \end{bmatrix} \begin{bmatrix} 45 \\ 46 \\ 40 \end{bmatrix} = \begin{bmatrix} 11 \\ 23 \\ -4 \end{bmatrix}$$

Thus, the feasibility is disturbed. Calculate new $f(X_B)$ and then restore the feasibility by the dual simplex method. The new $f(X_B) = C_B^T B^{-1} b'$ is

$$(1, 2, 0)(45, 46, 40)^T = 137.$$

Incorporate the new values of the solution column and $f(X_B)$ in Table 4 to write next table.

Table 5

B V	x_1	x_2	x_3	$s_1 \downarrow$	s_2	s_3	Soln
x_0	4	0	0	1	2	0	137
x_2	$-1/4$	1	0	1/2	$-1/4$	0	11
x_3	3/2	0	1	0	1/2	0	23
$\leftarrow s_3$	2	0	0	$\boxed{-2}$	1	1	-4

From the dual simplex method, s_3 leaves and s_1 enters, see Table 5, which yields the following optimal table.

B V	x_1	x_2	x_3	s_1	s_2	s_3	Soln
x_0	5	0	0	0	5/2	1/2	135
x_2	1/4	1	0	0	0	1/4	10
x_3	3/2	0	1	0	1/2	0	23
s_1	-1	0	0	1	$-1/2$	$-1/2$	2

The revised optimal solution remains unchanged, except slack is shifted to the first constraint.

(c) For this part of the example, we use the formulas (6.4) and (6.5). Note that $\tilde{B} = \{2, 3, 6\}$ and $i = 1, 2, 3$ and $k = 1$. Thus,

$$\Delta \underline{b}_1 = \max\{-20\} = -20,$$

$$\Delta \overline{b}_1 = \min\{1\} = 1,$$

This ensures that for same optimal basis to stay

$$-20 \le \Delta b_1 \le 1 \implies 23 \le b_1 \le 44.$$

Remarks. 1. For parts (a) and (b) there is no need of mentioning the original LPP, provided we are given that the constraints are of the type $\le$ with nonnegative right hand side vector.

2. In the above analysis, only one iteration has restored the feasibility. Usually, it may take more iterations to restore feasibility.

3. Note that in part (c), we have considered the variation of b_1 only, while b_2 and b_3 are fixed. We have not derived any result which accounts for simultaneous change in all components, a complicated situation due to appearance of a system of inequalities.

6.4 Change in the Constraint Matrix

So far, we have dealt with the changes in cost vector and right hand side vector. Further, we proceed to analyze the situation with the changes in constraint matrix. In general, the changes made in constraint matrix may result in different optimal basis and optimal solutions. We discuss five possible changes, viz., adding a constraint or a variable, removing a constraint or variable, and change in some column of the constraint matrix A.

Let us consider the LPP

$$\begin{aligned}
\min \quad & z = x_1 - 2x_2 + x_3 \\
\text{s.t.} \quad & x_1 + 2x_2 - 2x_3 \leq 4 \\
& x_1 - x_3 \leq 3 \\
& 2x_1 - x_2 + 2x_3 \leq 2 \\
& \text{all var } x_i \geq 0
\end{aligned}$$

The optimal table of the above LPP is

Table 6

B.V.	x_1	x_2	x_3	s_1	s_2	s_3	Soln
z	$-9/2$	0	0	$-3/2$	0	-1	-8
x_2	3	1	0	1	0	1	6
s_2	$7/2$	0	0	$1/2$	1	1	7
x_3	$5/2$	0	1	$1/2$	0	1	4

1. Addition of a constraint. If a new constraint is added to a LPP, then we have to make two observations:

(i) Whether the constraint to be added is satisfied by the given optimal solution, then there will be no effect on adding this constraint;

(ii) If this constraint is not satisfied, then addition will affect the optimal solution. Addition of such a constraint first will disturb the simplex format. When the simplex format is restored the feasibility will get disturbed. Restore the feasibility by the dual simplex method to find the new optimal solution.

Remarks. 1. It is worth mentioning to point out that addition of a constraint (provided it affects the optimal solution) always worsens the current optimal value of the objective function as the set of basic feasible solutions has shrunk.

2. For equality constraint to be added, we split this into two inequality constraints. Certainly one of the inequality constraint will be satisfied by the optimal solution and other one is considered for addition in the LPP.

Example 3. Consider the LPP whose optimal solution is given in Table 6.

(a) Add the constraint $x_1 + x_2 \leq 4$ to the LPP, and find the solution of the new problem.

(b) Add the constraint $x_1 + x_2 \geq 7$ to the LPP, and find the solution of the changed problem.

For part (a), Note that the constraint $x_1 + x_2 \leq 4$ is not satisfied by the given optimal solution, and hence, its addition will impact the optimum solution. Since this is the fourth constraint, we add

$$x_1 + x_2 + s_4 = 4$$

in the last row of Table 6.

Make zero in z-row below basic variables that will bring the table into simplex format. This operation causes the disturbance in feasibility. We use the dual simplex method to restore the feasibility. All iterations are shown in Table 7.

Table 7

B V	x_1	x_2	x_3	s_1	s_2	$s_3 \downarrow$	s_4	Soln
z	$-9/2$	0	0	$-3/2$	0	-1	0	-8
x_2	3	1	0	1	0	1	0	6
s_2	$7/2$	0	0	$1/2$	1	1	0	7
x_3	$5/2$	0	1	$1/2$	0	1	0	4
$\leftarrow s_4$	-2	0	0	-1	0	$\boxed{-1}$	1	-2
	1	1	0	0	0	0	1	4
z	$-5/2$	0	0	$-1/2$	0	0	0	-6
x_2	1	1	0	1	0	0	0	4
s_2	$3/2$	0	0	$1/2$	1	0	0	5
x_3	$1/2$	0	1	$1/2$	0	0	0	2
s_3	2	0	0	1	0	1	-1	2

New optimal solution: $x_1 = 0$, $x_2 = 4$, $x_3 = 2$, min value $z = -6$.

For part (b), the addition of $x_1 + x_2 \geq 7$ means we add $x_1 + x_2 - s_4 = 7$. Multiply the inserted s_4-row by -1 and add to x_2-row to bring the table into simplex format, see Table 8.

Table 8

B V	x_1	x_2	x_3	s_1	s_2	s_3	s_4	Soln
z	$-9/2$	0	0	$-3/2$	0	-1	0	-8
x_2	3	1	0	1	0	1	0	6
s_2	$7/2$	0	0	$1/2$	1	1	0	7
x_3	$5/2$	0	1	$1/2$	0	1	0	4
$\leftarrow s_4$	2	0	0	1	0	1	0	-1
	1	1	0	0	0	0	-1	7

The variable s_4 desires to leave the basis but there is no entering variable and hence, no feasible solution exists.

2. Addition of a variable. Addition of a new variable causes addition of some column in the optimal table which may affect the optimal criteria. Suppose that a new variable, say x_{n+1} is identified after we obtained the optimal solution X^* of the original LPP. Assume that c_{n+1} is the cost coefficient associated with x_{n+1}, and A_{n+1} is the associated column in the new constraint matrix. Our aim is to find an optimal solution of the new linear program

$$\begin{aligned}
\max \quad & z = C^T X + c_{n+1} x_{n+1} \\
\text{s.t.} \quad & AX + A_{n+1} x_{n+1} = b \\
& X \geq 0, x_{n+1} \geq 0
\end{aligned}$$

Observe that we can set $x_{n+1} = 0$, then $(X, 0)^T$ becomes a basic feasible solution to the new LPP. Hence, the simplex algorithm can be started right away. Also, note that X^* is an optimal solution to the original problem, the relative costs $z_j - c_j$, $j = 1, 2, \ldots, n+1$ must be nonnegative. There fore, we have to check additional relative cost

$$z_{n+1} - c_{n+1} = C_B^T B^{-1} A_{n+1}.$$

If $z_{n+1} - c_{n+1} \geq 0$, then the current solution X^* with $x_{n+1} = 0$ is the optimal solution to the new problem. and we have not to do anything. On the other hand if $z_{n+1} - c_{n+1} < 0$, then x_{n+1} should be included in the basis. Continue simplex iterations till an optimal solution to the new LPP is available.

Consider the LPP

$$\begin{aligned}
\min \quad & z = 2x_1 + x_2 + MR_1 + MR_2 \\
\text{s.t.} \quad & 3x_1 + x_2 + R_1 = 3 \\
& 4x_1 + 3x_2 - s_2 + R_2 = 6 \\
& x_1 + 2x_2 + s_3 = 3
\end{aligned}$$

The optimal table of this LPP is

Table 9

B V	x_1	x_2	s_2	R_1	R_2	s_3	Soln
z	0	0	$-1/5$	$2/5 - M$	$1/5 - M$	0	$12/5$
x_1	1	0	$1/5$	$3/5$	$-1/5$	0	$3/5$
x_2	0	1	$-3/5$	$-4/5$	$3/5$	0	$6/5$
s_3	0	0	1	1	-1	1	0

Example 4. If in the above LPP, a variable x_3 with cost $1/2$ and column $A_3 = (0.5, 2)^T$ in constraint matrix is added to the LPP, then reprocess the optimal table of the given problem to get the optimal solution of the new LPP.

The relative cost of x_3 and its coordinate vector are computed. The analysis is given as

$$z_3 - c_3 = C_B^T B^{-1} A_3 - c_3 = (2/5, 1/5, 0)(0, 5, 2)^T - 1/2 = 1/2.$$

Thus, the optimal criteria is disturbed. Calculate the column below x_3, i.e., $\alpha^3 = B^{-1}A_3$ as

$$\alpha^3 = \begin{bmatrix} 3/5 & -1/5 & 0 \\ -4/5 & 3/5 & 0 \\ 1 & -1 & 1 \end{bmatrix} \begin{bmatrix} 0 \\ 5 \\ 2 \end{bmatrix} = \begin{bmatrix} -1 \\ 3 \\ -3 \end{bmatrix}$$

Insert variable x_3, $z_3 - c_3$ and α^3 in Table 9. Further iterations are shown below.

B V	x_1	x_2	$x_3 \downarrow$	s_2	R_1	R_2	s_3	Soln
z	0	0	$1/2$	$-1/5$	$2/5 - M$	$1/5 - M$	0	$12/5$
x_1	1	0	-1	$1/5$	$3/5$	$-1/5$	0	$3/5$
$\leftarrow x_2$	0	1	$\boxed{3}$	$-3/5$	$-4/5$	$3/5$	0	$6/5$
s_3	0	0	-3	1	1	-1	1	0
z	0	$-1/6$	0	$-1/5$	$8/15 - M$	$1/10 - M$	0	$11/5$
x_1	1	$1/3$	0	0	$1/3$	0	0	1
x_3	0	$1/3$	1	$-1/5$	$-4/5$	$1/5$	0	$2/5$
s_3	0	1	0	$2/5$	$1/5$	$-2/5$	1	$6/5$

Optimal solution: $x_1 = 1$, $x_2 = 0$, $x_3 = 2/5$, $z = 11/5$.

Remark. In case a new variable x_3 having cost 3 and column $(1, 2, 3)^T$ is added to a LPP, then for the changed problem

$$\text{new } z_3 - c_3 = C_B^T B^{-1} A_3 - c_3 = (2/5, 1/5, 0)(1, 2, 3)^T - 3 = -11/5.$$

We need not to go further in this case as the optimal solution remains unchanged.

3. Deletion of a constraint. While deleting a constraint we observe two situations:

(i) If any constraint is satisfied on the boundary, i.e., slack or surplus variable corresponding to this constraint is at zero level then deletion of such a constraint may cause change in the optimal solution.

(ii) If any constraint is satisfied in interior of P_F, i.e., slack or surplus variable corresponding to this constraint are positive, then deletion of such a constraint will not affect the optimal solution.

In other words situation (i) is a binding on the optimal solution, while situation (ii) is a nonbinding on the optimal solution.

Remark. We shall observe that deletion of a constraint amounts to addition of a variable.

Example 5. Consider the LPP

$$\max \quad z = 6x_1 + 2x_2 + 10x_3$$
$$\text{s.t.} \quad 3x_1 - x_2 + x_3 \leq 10$$
$$x_1 + x_3 \leq 8$$
$$x_2 + 2x_3 \leq 5$$
$$x_1, x_2, x_3 \geq 0$$

The optimal table of this LPP is

B V	x_1	x_2	x_3	s_1	s_2	s_3	
z	0	0	0	2	0	4	40
x_1	1	$-1/2$	0	$1/3$	0	$1/6$	$5/2$
s_2	0	0	0	$-1/3$	1	$-1/3$	3
x_3	0	$1/2$	1	0	0	$1/2$	$5/2$

Show that (a) deletion of Ist and IIIrd constraints affects on the optimal solution; (b) deletion of IInd constraint does not change the optimal solution.

Let us work out the deletion of Ist constraint. This constraint is of the type

$$\cdots + s_1 = \cdots$$

and is satisfied on the boundary of the feasible region because $s_1 = 0$. Thus its deletion will impact the optimal solution.

Add a variable $-s_1'(s_1' \geq 0)$ to this constraint to have

$$\cdots + s_1 - s_1' = \cdots$$

Now, treat the problem as addition of a variable. Note that, $z_j - c_j$ for $s_1' = -$relative cost of $s_1 = -2$, and

$$\text{Column below } s_1' = -\text{column below } s_1 = \begin{bmatrix} -1/3 \\ 1/3 \\ 0 \end{bmatrix}$$

B V	x_1	x_2	x_3	$s \downarrow$	s_1	s_2	s_3	Soln
z	0	0	0	-2	2	0	4	40
x_1	1	$-1/2$	0	$-1/3$	$1/3$	0	$-1/6$	$5/2$
$\leftarrow s_2$	0	0	0	$\boxed{1/3}$	$-1/3$	1	$-1/3$	3
x_3	0	$1/2$	1	0	0	0	$1/2$	$5/2$
z	0	0	0	0	0	6	2	58
x_1	1	$-1/2$	0	0	0	1	$1/2$	$11/2$
s	0	0	0	1	-1	3	-1	9
x_3	0	$1/2$	1	0	0	0	$1/2$	$5/2$

Optimal solution: $x_1 = 11/2$, $x_2 = 0$, $x_3 = 5/2$, $z = 58$.

Similarly, the deletion of third constraint can be carried out. However, it requires three iterations to reach at the conclusion that deletion of third constraint results in unbounded solution.

For case (b), the second constraint is satisfied as a strict inequality, because $s_2 = 3$, i.e., inside the feasible region P_F, and hence, its deletion causes no changes in the optimal solution.

Note that in part (a) we may also get alternate optimal solution as relative cost of nonbasic variable x_2 is zero.

Remarks 1. For deleting a constraint having surplus variable, we add $s_i'(\geq 0)$ as
$$\cdots - s_i' + s_i' = \cdots .$$
To apply deletion, bring s_i' into the basis.

2. For equality constraint, add $s_i - s_1'$ and bring any one of these variables into the basis. Calculate freshly $z_j - c_j$ and α^j for s. These entries for one will be negative of the other.

3. Sometimes it is more profitable (provided a LPP is given) to workout all iterations freshly after removing the constraint physically from the LPP.

4. Deletion of a variable. We observe for the two situations:

(i) If we delete nonbasic variable or a basic variable (at zero level) in the optimal solution, there will be no change in the optimal solution.

(ii0 However, deletion of a positive basic variable will change the optimal solution.

But, deleting a basic variable with positive value is equivalent to convert this into nonbasic. For the purpose, first we remove the entire column from the optimal table associated with the basic variable to be deleted and then multiply the entire row in front of this variable by -1. This will certainly disturb the feasibility. Now, use the dual simplex method to restore feasibility. This is clear from the following example.

Example 6. Delete the variable x_3 from the LPP with the optimal Table 6.

Since x_3 is a basic variable with positive value 4 in the optimal table (see Table 6), its deletion will affect the optimal solution. Cancel the column corresponding to x_3 and multiply by -1 to all the entries in x_3-row. Note that the feasibility has been disturbed due to the occurrence of -4 in the solution column. Now, apply the dual simplex method to restore the feasibility.

B V	x_1	x_2	x_3	s_1	s_2	$s_3 \downarrow$	Soln
z	$-9/2$	0		$-3/2$	0	-1	-8
x_2	3	1		1	0	1	6
s_2	$7/2$	0		$1/2$	1	1	7
$\leftarrow x_3$	$-5/2$	0		$-1/2$	0	$\boxed{-1}$	-4
z	-2	0		-1	0	0	-4
x_2	$1/2$	1		$1/2$	0	0	2
s_2	1	0		0	1	0	3
s_3	$5/2$	0		$1/2$	0	1	4

Optimal solution: $x_1 = 0$, $x_2 = 2$, minimum value $z = -4$.

Remark. In working out the above example, we started the dual simplex iteration, even though the body matrix does not contain identity submatrix. This is possible by adding a nonbasic variable x_3' with cost $-c_3$ and the coefficient column $-A_3$ so that the column corresponding to x_3' comes out to be $(0; 0, 0, -1)^T$. Removing the column associated with x_3 and adding the column associated with x_3' will not affect the existing optimal solution. Now, multiplying by -1 will produce the identity submatrix. Note that x_3' will not enter the basis as $\alpha_3' \geq 0$ (the coordinate vector of x_3').

5. Change in column of the constraint matrix. The change in coefficients associated with a variable may effect the optimal criteria. We consider the LPP max $z = c^T X$ subject to $AX = b$, $X \geq 0$. Two cases arise:

First, we discuss the change in coefficients of constraint matrix associated with a nonbasic variable. This will change the whole column (coordinate vector) below this variable in the optimal table.

Suppose the a_{ik}th entry of the column A_k corresponding to kth nonbasic variable is shifted to $A'_k = A_k + \delta_{ik} e_i$, here $k \in \tilde{N}$, the index set of nonbasic variables and e_i is the column vector with 1 at ith position and zero elsewhere. We decide the limits of variation a_{ik} such that the optimal solution remains same.

When A_k is changed to A'_k, this will affect the relative cost of x_k. The new relative cost of x_k becomes $C_B^T B^{-1} A'_k - c_k$. The optimal table remains same if

$$C_B^T B^{-1} (A_k + \delta_{ik} e_i) - c_k \geq 0,$$

or

$$C_B^T \alpha^k + \delta_{ik} C_B^T B^{-1} e_i - c_k \geq 0$$

Let $B^{-1} = [\beta_1, \beta_2, \ldots, \beta_m]$. Then the above expression is simplified to

$$z_k - c_k + \delta_{ik} C_B^T \beta_i.$$

But β_i is the coordinate vector of ith variable in the starting basis. Hence,

$$\delta_{ik} \geq -\frac{z_k - c_k}{\beta_i}.$$

This gives the variation in element a_{ik} of the column A_k in constraint matrix so that the optimal solution remains same. If it is violated then obviously the optimal criteria is disturbed. Restore the optimal criteria by the simplex method to get the new optimal solution.

Suppose the coefficients of constraint matrix associated with some basic variable in a LPP considered above are changed. Let the column A_k associated with basic variable x_k is changed to A'_k, where $k \in \tilde{B}$. Then add a variable x'_k with same cost as that of x_k and column A'_k in constraint matrix. Compute

$$z'_k - c_k = C_B^T B^{-1} A k' - c_k,$$

if $z'_k - c_k \geq 0$, there is no effect of such change, otherwise the optimality is disturbed and to restore optimality bring x'_k into the basis. Treat x_k

as the artificial variable and force to come out of the basis. In the last optimal table the value of x'_k in the solution is nothing but x_k with the new column.

Example 7. Suppose the column of x_1 in the LPP with the optimal Table 6 is changed to $A_1 + G$, where $G = (-4, 2, 1)^T$. Then, find the optimal solution of the changed problem.

Compute new $z_1 - c_1 = C_B^T B^{-1}(A_1 + G) - c_1 = C_B^T B^{-1} A_1 - c_1 + C_B^T B^{-1} G$, i.e.,

$$-9/2 + (-3/2, 0, -1)(-4, 2, 1)^T = 1/2.$$

New $\alpha^1 = B^{-1}(A_1 + G) = B^{-1} A_1 + B^{-1} G$, i.e.,

$$
\begin{bmatrix} 3 \\ 7/2 \\ 5/2 \end{bmatrix}
+
\begin{bmatrix} 1 & 0 & 1 \\ 1/2 & 1 & 1 \\ 1/2 & 0 & 1 \end{bmatrix}
\begin{bmatrix} -4 \\ 2 \\ 1 \end{bmatrix}
=
\begin{bmatrix} 3 \\ 7/2 \\ 5/2 \end{bmatrix}
+
\begin{bmatrix} -3 \\ 1 \\ -1 \end{bmatrix}
=
\begin{bmatrix} 0 \\ 9/2 \\ 3/2 \end{bmatrix}
$$

Since the optimal criteria is disturbed, we use the regular simplex method to restore optimality and the optimal solution is given in the next table.

B V	$x_1 \downarrow$	x_2	x_3	s_1	s_2	s_3	Soln.
z	1/2	0	0	-3/2	0	-1	-8
x_2	0	1	0	1	0	1	6
← s_2	9/2	0	0	1/2	1	1	7
x_3	3/2	0	1	1/2	0	1	4
z	0	0	0	-14/9	-1/9	-10/9	-79/9
x_2	0						6
x_1	1			1/9	2/9	2/9	14/9
x_3	0						5/3

The new optimal solution: $x_1 = 14/9$, $x_2 = 6$, $x_3 = 5/3$ $z = -79/9$.

Remark. If the LPP is given and $A_1 = (1, 1, 2)$ is changed to $(-1, 4, 3)^T$, then it can be verified that optimality is retained and hence, it is not needed at all to calculate the new α^1.

Example 8. Let the column in the constraint matrix of a LPP associated with x_1 be changed to $(6, 0, -7)^T$ for which the optimal solution is given in Table 9. Find the optimal solution of the new problem.

Add a new variable x_3 with the same cost as that of x_1, i.e., 2 and column $(6, 0, -7)^T$ in A. The relative cost of x_3 is

$$z_3 - c_3 = C_B^T B^{-1} A_3 - 2 = (2/5, 1/5, 0)(6, 0, -7)^T - 2 = 2/5,$$

and

$$\alpha^3 = \begin{bmatrix} 3/5 & -1/5 & 0 \\ -4/5 & 3/5 & 0 \\ 1 & -1 & 1 \end{bmatrix} \begin{bmatrix} 6 \\ 0 \\ -7 \end{bmatrix} = \begin{bmatrix} 18/5 \\ -24/5 \\ -1 \end{bmatrix}$$

B V	x_1	x_2	$x_3 \downarrow$	s_2	R_1	R_2	s_3	Soln
z	0	0	2/5	-1/5	$2/5 - M$	$1/5 - M$	0	12/5
$\leftarrow x_1$	1	0	18/5	1/5	3/5	-1/5	0	3/5
x_2	0	1	-24/5	-3/5	-4/5	3/5	0	6/5
s_3	0	0	-1	1	1	-1	1	0
z	0	-1/9	0	-2/9	$1/3 - M$	$2/5 - M$	0	7/3
x_3	5/18	0	1	0	1/3	0	0	1/6
x_2			0					2
s_3			0					1/6

Observe that x_3 has the same cost as that of x_1 and its column is the proposed column for x_1. Hence, in the revised LPP x_3 plays the role of x_1. Thus,

optimal solution: $x_1 = 1/6$, $x_2 = 2$, min value $z = 7/3$.

6.5 Special Cases

In such type of situations we are encountered simultaneous changes in

(a) cost and availability or

(b) cost and coefficient matrix A or

(c) availability and coefficient matrix.

We discuss only one case and others can be understood by combining the analysis done earlier. We solve one problem pertaining to simultaneous changes in the cost vector and the right hand side vector.

Example 9. Consider the LPP

$$\begin{aligned}
\max \quad & 7x_0 = 5x_1 + 4x_2 \\
\text{s.t.} \quad & 6x_1 + 2x_2 \le 24 \\
& x_1 + 2x_2 \le 6 \\
& -x_1 + x_2 \le 1 \\
& x_2 \le 2 \\
& x_1, x_2 \ge 0
\end{aligned}$$

The optimal table of the above LPP is

B V	x_1	x_2	s_1	s_2	s_3	s_4	Soln
z	0	0	3/4	1/2	0	0	21
x_1	1	0	1/4	−1/2	0	0	3
x_2	0	1	−1/8	3/4	0	0	3/2
s_3	0	0	3/8	−5/4	1	0	3/2
s_4	0	0	1/8	−3/4	0	1	1/2

Two types of changes are proposed in the above LPP:

(a) Suppose that the profit coefficients are assigned the values $1,000 and 4,000;

(b) The right hand side vector is changed to $(28, 8, 1, 2)^T$.

Show that these changes render the LPP table nonoptimal and nonfeasible. Find the optimal solution of the revised LPP.

Let us calculate the right-hand entries

$$\text{New } B^{-1}b = \begin{bmatrix} 1/4 & -1/2 & 0 & 0 \\ -1/8 & 3/4 & 0 & 0 \\ 3/8 & -5/4 & 1 & 0 \\ 1/8 & -3/4 & 0 & 1 \end{bmatrix} \begin{bmatrix} 28 \\ 8 \\ 1 \\ 2 \end{bmatrix} = \begin{bmatrix} 3 \\ 5/2 \\ 3/2 \\ -9/4 \end{bmatrix}$$

This shows that feasibility is disturbed.

Since x_1, x_2, s_3, s_4 are basic variables, it follows that with change in cost vector there will be change in relative cost of nonbasic variables s_1 and s_2.

Calculate the entries:

New $z_3 - c_3 = (1000, 4000, 0, 0)(1/4, -1/8, 3/8, 1/8)^T = -250$

New $z_4 - c_4 = (1000, 4000, 0, 0)(-1/2, 3/4, -5/4, -3/4)^T = 2500$

This implies that optimality is also disturbed.

To deal with this situation, apply the technique suggested in Problem 17, Problem set 4 as follows to have

B V	x_1	x_2	s_1	$s_2 \downarrow$	s_3	s_4	Soln
z	0	0	-250	2500	0	0	13000
x_1	1	0	$1/4$	$-1/2$	0	0	3
x_2	0	1	$-1/8$	$3/4$	0	0	$5/2$
s_3	0	0	$3/8$	$-5/4$	1	0	$3/2$
$\leftarrow s_4$	0	0	$1/8$	$\boxed{-3/4}$	0	1	$-9/4$
z	0	0	$500/3$	0	0	$10000/3$	5500
x_1	1	0		0	0		$9/2$
x_2	0	1		0	0		$1/4$
s_3	0	0		0	1		$21/4$
s_2	0	0	$-1/6$	1	0	$-4/3$	3

The new optimal solution and optimal value: $x_1 = 10/3$, $x_2 = 2$, $z = 34000/3$.

6.6   Parametric Programming

The effect of discrete changes in the model parameters C, b and A have been studied in Sections 6.1-6.5, just concluded. Another common approach to sensitivity analysis is to vary one or more parameters continuously over some interval(s) to see when the optimal solution changes. Here we discuss how continuous changes affect the optimal solution of the problem.

For instance in Section 6.3, rather than beginning by testing the specific changes from $b_2 = 46$ to $b_2 = 64$, might instead set $b_2+\alpha t$. Here α is the constant which decides the relative change of the parameter t in b_i. Suppose b_i is the production capacity of Plant $i = 1, 2, 3$. It is possible to shift some of the current production of a company product from Plant 2 to Plant 3, there by increasing b_2 and by decreasing b_3. If b_3 is decreased twice as fast as b_2 increases, then $b_2 = 46 + t$ and $b_3 = 42 - 2t$, where t measures the amount of production shifted. Then $\alpha_1 = 0$, $\alpha_2 = 1$ and $\alpha_3 = -2$ in this case.

Systematic Change in the c_j Parameters. For this case the objective function of the ordinary linear programming model

$$\max z = \sum_{j=1}^{n} c_j x_j$$

is replaced by

$$\max z(t) = \sum_{j=1}^{n} (c_j + \alpha_j t) x_j,$$

where the α_j are given input constants representing the relative rate at which the coefficients are being changed. Therefore, gradually, increasing t from 0 changes the coefficients at these relative rates. This may also be based on how the coefficients (e.g.,, unit profits) would change together with respect to some factor measured by t. For any given value of t, the optimal solution of the corresponding LPP can be obtained by the simplex method. Observe that for $t = 0$, this is just the original problem.

Our objective is to find the optimal solution of the modified LPP (maximize $z(t)$ with respect to original constraints) as t increases from 0 to any specified positive number. This is portrayed graphically in Figure 6.1. The function $z(t)$ must be piecewise linear and convex, see Problem 14.

The solution procedure is based directly upon the sensitivity analysis procedure for investigating changes in the c_j parameters. The only basic difference that changes are now expressed in terms of t rather than as specific numbers.

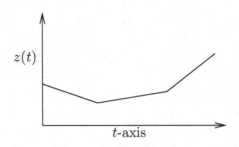

Figure 6.1

Algorithm. The algorithm follows as

Step 1. Solve the problem for $t = 0$, by the simplex method.

Step 2. Remove the z-row and insert the new objective function $z(t)$. Bring the table in simplex format using elementary row operation.

Step 3. Increase $t \geq 0$ until one of the nonbasic variable has its coefficient negative (or until t has been increased just to change the sign).

Step 4. Use this variable as the entering variable for an iteration of the simplex method to find the new optimal solution. Return to Step 3.

Example 10. Consider the LPP

$$\begin{aligned}
\max \quad & z = 3x_1 + 5x_2 \\
\text{s.t.} \quad & x_1 \leq 4 \\
& 2x_2 \leq 12 \\
& 3x_1 + 2x_2 \leq 18 \\
& x_1, x_2 \geq 0
\end{aligned}$$

Suppose the cost of the first item increases two times faster than the decrease of the cost of second item, i.e., $\alpha_1 = 2$ and $\alpha_2 = -1$. Now, the modified problem is

$$\begin{aligned}
\max \quad & z(t) = (3 + 2t)x_1 + (5 - t)x_2 \\
\text{s.t.} \quad & x_1 \leq 4 \\
& 2x_2 \leq 12 \\
& 3x_1 + 2x_2 \leq 18 \\
& x_1, x_2 \geq 0
\end{aligned}$$

Discuss all cases of the parametric changes.

As Step 1, the optimal table for $t = 0$ is

B V	x_1	x_2	s_1	s_2	s_3	Soln
z	0	0	0	$3/2$	1	36
s_1	0	0	1	$1/3$	$-1/3$	2
x_2	0	1	0	$1/2$	0	6
x_1	1	0	0	$-1/3$	$1/3$	2

In Step 2, the z-row is replaced by $z(t)$-row. This disturbs the simplex format in z-row. To return to simplex execute the following operation

$$z(t)\text{-row} + (3 + 2t) \times x_1\text{-row} + (5 - t) \times x_2\text{-row}.$$

The new relative cost of nonbasic variables s_2 and s_3 are

$$-\frac{3 + 2t}{3} + \frac{5 - t}{2} = \frac{9 - 7t}{6}, \quad \frac{3 + 2t}{3} + 0 = \frac{3 + 2t}{3};$$

and the new objective function value is

$$2(3 + 2t) + 6(5 - t) = 36 - 2t.$$

With these calculations the next table is

B V	x_1	x_2	s_1	s_2	s_3	Soln
$z(t)$	0	0	0	$(9 - 7)/t$	$(3 + 2t)/3$	$36 - 2t$
s_1	0	0	1	$1/3$	$-1/3$	2
x_2	0	1	0	$1/2$	0	6
x_1	1	0	0	$-1/3$	$1/3$	2

Range of $t: 0 \le t \le 9/7$.

The stopping rule says that the current basic feasible solution will remain optimal as long as the relative cost of nonbasic variables remain nonnegative.:

$$\frac{3}{2} - \frac{7}{6}t \ge 0, \quad \text{for } 0 \le t \le \frac{9}{7},$$

$$1 + \frac{2}{3}t \geq 0, \quad \text{for all } t \geq 0.$$

Therefore, after increasing t past $t = 9/7$, s_2 will be the entering variable. s_1 leaves the basis and the new optimal solution is

B V	x_1	x_2	s_1	s_2	s_3	Soln
$z(t)$	0	0	$(-9+7t)/2$	0	$(5-t)/2$	$27+5t$
s_2	0	0	3	1	-1	6
x_2	0	1	$-3/2$	0	$1/2$	3
x_1	1	0	1	0	0	4

Range of $t : 9/7 \leq t \leq 9/5$.

Again, for deciding next entering variable, we compute

$$\frac{5-t}{2} \geq 0 \text{ for } t \geq 5$$

After increasing t beyond 5, we shall get new optimal solution, and hence, the range for second optimal table is $9/7 \leq t \leq 5$. To go beyond this, s_3 is the entering variable and x_2 is the leaving variable. The optimal table is

B V	x_1	x_2	s_1	s_2	s_3	Soln
$z(t)$	0	$-5+t$	$3+2t$	0	0	$12+8t$
s_2	0	2	0	1	0	12
s_3	0	2	-3	0	1	6
x_1	1	0	1	0	0	4

Range of $t : 9/7 \leq t \leq 5$.

Systematic Changes in the b_i Parameters. For this case the one modification made in the LPP is that b_i is replaced by $b_i + \alpha_i t$ for $i = 1, 2, \ldots, m$, where α_i are given input constants. Thus, the LPP becomes

$$\max \quad z(t) = \sum_{j=1}^{n} c_j x_j$$

$$\text{s.t.} \quad \sum_{j=1}^{n} a_{ij} x_j \leq b_i + \alpha_i t, \quad i = 1, 2, \ldots, m$$

$$x_j \geq 0, \quad j = 1, 2, \ldots, n$$

The purpose is to identify the optimal solution as a function of t. With this formulation, the corresponding objective function value has the piecewise linear and concave form as shown in Fig. 6.2.

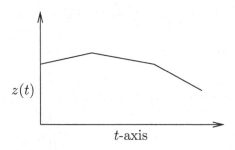

Figure 6.2

Algorithm. The algorithm follows as

Step 1. Solve the problem for $t = 0$, by the simplex method.

Step 2. Use formula $B^{-1}(b + \alpha t)$ to introduce changes in the right-side entries of the optimal table.

Step 3. Increase $t \geq 0$ until one of the basic variable has its value in the right-side column go negative (or until t has been increased just to change the sign).

Step 4. Use this variable as the leaving variable for an iteration of the dual simplex method to find the new optimal solution. Return to Step 3.

Example 11. Let us take the LPP

$$\begin{aligned}
\max \quad & z = 2x_1 + x_2 \\
\text{s.t.} \quad & 4x_1 + x_2 \leq 4 \\
& x_1 + 2x_2 \leq 2 \\
& x_1, x_2 \geq 0
\end{aligned}$$

and modify in the form

$$\begin{aligned}
\max \quad & z(t) = 2x_1 + x_2 \\
\text{s.t.} \quad & 4x_1 + x_2 \leq 4 - t \\
& x_1 + 2x_2 \leq 2 + 2t \\
& x_1, x_2 \geq 0
\end{aligned}$$

Discuss all changes in parameter t giving different optimal solutions in different intervals.

For $t = 0$, the optimal table for the original LPP is

B V	x_1	x_2	s_1	s_2	Soln
z	0	0	3/7	2/7	16/7
x_1	1	0	2/7	$-1/7$	6/7
x_2	0	1	$-1/7$	4/7	4/7

Compute

$$B^{-1} \begin{bmatrix} 4 - t \\ 2 + 2t \end{bmatrix} = \begin{bmatrix} 6/7 \\ 4/7 \end{bmatrix} + t \begin{bmatrix} 2/7 & -1/7 \\ -1/7 & 4/7 \end{bmatrix} \begin{bmatrix} -1 \\ 2 \end{bmatrix} = \begin{bmatrix} (6 - 4t)/7 \\ (4 + 9t)/7 \end{bmatrix}$$

and the objective function value in terms of t as

$$C_B^T B^{-1}(b + \alpha t) = (2, 1) \cdot ((6 - 4t)/7, (4 + 9t)/7) = (16 + t)/7.$$

B V	x_1	x_2	s_1	s_2	Soln
z	0	0	3/7	2/7	$(16 + t)/7$
x_1	1	0	2/7	$-1/7$	$(6 - 4t)/7$
x_2	0	1	$-1/7$	4/7	$(4 + 9t)/7$

Range of $t : 0 \leq t \leq 3/2$.

$$x_1 = \frac{6 - 4t}{7} \geq 0, \text{ for } t \leq \frac{3}{2},$$

$$x_2 = \frac{4 + 9t}{7} \geq 0, \text{ for } t \geq 0.$$

Thus, the optimal basis will remain the same when t varies in the range: $0 \leq t \leq 2/3$. Increasing t past $t = 3/2$ requires making x_1 as the leaving variable for the dual simplex iteration. s_2 enters the basis and the next optimal table is

B V	x_1	x_2	s_1	s_2	Soln
z	2	0	1	0	$4 - t$
s_2	-7	0	-2	1	$-6 + 4t$
x_2	4	1	1	0	$-7t + 30$

Range of $t : 3/2 \leq t \leq 30/7$.

Thus the two basic variables

$$s_2 = -6 + 4t, \quad x_2 = -7t + 30$$

remain nonnegative in the range $3/2 \leq t \leq 30/7$. Increasing t past $t = 30/7$, the x_2 becomes eligible to leave the basis, but no entering variable is found and hence, for $t > 30/7$ the LPP becomes infeasible.

Problem Set 6

1. Let max $z = C^T X$, subject to $AX = b$, $X \geq 0$. Determine how much the components of the cost vector C can be perturbed without affecting the optimal solution.

2. Let max $z = C^T X$, subject to $AX = b$, $X \geq 0$. Discuss the variation in right-hand side vector so that same basis remains optimal.

3. Discuss the effect of changing the column of the coefficient matrix corresponding to nonbasic variable.

4. Consider the LPP and its optimal table given below

$$\min \quad z = 2x_1 + 3x_2$$
$$\text{s.t.} \quad x_1 + 3x_2 \geq 5$$
$$2x_1 + x_2 \geq 6$$
$$x_1, x_2 \geq 0$$

B V	x_1	x_2	s_1	s_2	Soln
x_0	0	0	$-4/5$	$-3/5$	$38/5$
x_2	0	1	$-2/5$	$1/5$	$4/5$
x_1	1	0	$1/5$	$-3/5$	$13/5$

(a) Within what range the cost of x_1 can vary so that the optimal solution remains unchanged.

(b) Within what range the right-hand side entry of the second constraint varies so that the given optimal basis remains unchanged.

5. Find the limit of variation of the column vector A_k so that the optimal solution of the LPP max $z = C^T X$, subject to $AX = b$, $X \geq 0$ stays unchanged when (i) $A_k \in B$; (ii) $A_k \notin B$.

6. Consider the LPP and its optimal table of the preceding problem.

(a) If a constraint $x_1 + x_2 = 4$ is added to the LPP, find the optimal solution of the new LPP;

(b) If a constraint $s_1 \geq 1$ is added, find the optimal solution of the revised problem;

(c) If the column of the constraint matrix associated with x_1 is changed to $(2, 5)^T$, find the optimal solution of the resulting LPP.

Suggestion. For part (b), break $x_1 + x_2 = 4$ in two constraints $x_1 + x_2 \leq 4$ and $x_1 + x_2 \geq 4$. Now, add these two constraints. Addition of $x_1 + x_2 \leq 4$ will cause no effect (as it is satisfied by the given optimal solution). Hence, add $x_1 x_2 \geq 4$ only.

7. Find the optimal solution of the LPP with optimal table 6 in Section 6.4, when

(a) The column corresponding to x_1 is changed to $A_1 + \delta$, $\delta = (-4, 2, 1)^T$;

(b) If the second constraint is replaced by $x_1 + x_2 \geq 4$, then find the optimal solution of changed problem;

(c) If x_1 is replaced by $-x_1$ in the whole LPP, then find the optimal solution of the new LPP.

Suggestion. For part (b), first add $x_1 + x_2 \geq 4$, and then delete $x_1 - x_3 \leq 4$. In case of part (c), multiply by -1 to $z_1 - c_1$ and α^1, and use simplex method to restore optimality.

8. Following is the optimal table of a LPP in which s_1, s_2, $s_3 \geq 0$ are the slack variables when the LPP is written in standard form.

B V	x_1	x_2	s_1	s_2	s_3	Soln
x_0	0	0	7/6	13/6	0	218/3
s_3	0	0	3/2	−25/2	1	5
x_1	1	0	1/3	−2/3	0	16/3
x_2	0	1	−1/6	5/6	0	10/3

Using sensitivity analysis find the optimal solution in each of the cases:

(a) adding a variable x_3 with column vector $(3; 1, 0, 5)^T$;

(b) the variable x_2 is made unrestricted in sign.

Suggestion. For part (b) add a variable x_2' with cost $-c_2$ and column $-A_2$.

9. Consider the following linear program

$$\begin{aligned}
\min \quad & z = 2x_1 + x_2 - x_3 \\
\text{s.t.} \quad & x_1 + 2x_2 + x_3 \leq 8 \\
& -x_1 + x_2 - 2x_3 \leq 4 \\
& x_1, x_2, x_3 \geq 0
\end{aligned}$$

First, use the revised simplex method to find the optimal solution and its optimal dual variables. Then use sensitivity analysis to answer the following questions.

(a) Find a new optimal solution if the cost coefficient of x_1 is changed from 2 to 6.

(b) Find a new optimal solution if the coefficient of x_2 in the first constraint is changed from 2 to 1/4.

(c) Find a new optimal solution if we add one more constraint $x_2 + x_3 = 3$.

(d) If you were to choose between increasing the right-side of the first and second constraints, which one would you prefer? Why? What is the effect of this increase on the optimal solution of the objective function.

(e) Suppose that a new activity x_4 is proposed with a unit cost of 4 and consumption vector $A_4 = (1, 2)^T$. Find the corresponding optimal solution.

10. Given the LPP

$$
\begin{aligned}
\max \quad & z = 14x_1 + 20x_2 + 10x_3 \\
\text{s.t.} \quad & 6x_1 + 10x_2 + 3x_3 \le 100 \\
& 8x_1 + 10x_2 + 6x_3 \le 120 \\
& 4x_1 + 8x_2 + 9x_3 \le 150 \\
& x_1, x_2, x_3 \ge 0
\end{aligned}
$$

(i) Solve by revised simplex method;

(ii) Find the value of c_1 so that x_1 contributes in the optimal solution;

(iii) What is the lower limit for c_2 so that x_2 remains in the optimal solution?

(iv) What will be the impact on optimal solution if b_1 is changed from 100 to 103? from 100 to 200?

11. Solve the LPP

$$
\begin{aligned}
\max \quad & z = 2x_1 + 3x_2 \\
\text{s.t.} \quad & x_1 - x_2 \ge 1 \\
& 3x_1 + 2x_2 \le 5 \\
& x_1, x_2 \ge 0
\end{aligned}
$$

Use the optimal table of the above LPP and then apply the sensitivity analysis to solve

$$
\begin{aligned}
\max \quad & z' = 2x_1 + 3x_2 + x_3 \\
\text{s.t.} \quad & x_1 - x_2 + x_3 \ge 1 \\
& 3x_1 + 2x_2 \le 5 \\
& x_1 + x_2 + x_3 = 3 \\
& x_1, x_2 \ge 0
\end{aligned}
$$

12. Consider the LPP whose optimal solution is given in Table 6. Show that

    (a) deletion of Ist and IIIrd constraints affects on the optimal solution;

    (b) deletion of IInd constraint does not change the optimal solution.

13. Consider the problem

$$\max \quad z = 3x_1 + 4x_2$$
$$\text{s.t.} \quad 2x_1 + 3x_2 \le 120$$
$$2x_1 + x_2 \le 100$$
$$x_2 \le 16$$
$$x_1, x_2 \ge 0$$

Its associated optimal table is

B V	x_1	x_2	s_1	s_2	s_3	Soln
z	0	0	5/4	1/4	0	175
x_1	1	0	$-1/4$	3/4	0	45
s_3	0	0	$-1/2$	1/2	1	6
x_2	0	1	1/2	$-1/2$	0	10

The right hand side vector is changed to $(100, 40, 20)^T$ and the cost coefficients of decision variables x_1, x_2 are assigned the values $4, 1$, respectively. Find the optimal solution of the revised LPP.

Suggestion. New $B^{-1}b = (5, -10, 30)^T$ and new $z_3 - c_3 = -7/2$ and $z_4 - c_4 = -5/2$. Apply generalized simplex method, i.e., x_2 leaves and s_2 enters. This is a problem when the optimality and feasibility are disturbed simultaneously.

14. Consider the function $z(t)$ function shown in Fig. 6.1 for parametric linear programming with systematic changes in the c_j parameters.

    (a) Explain why this function is piecewise linear;

    (b) Show that this function must be convex.

15. Consider the function $z(t)$ function shown in Fig. 6.2 for parametric linear programming with systematic changes in the c_j parameters.

    (a) Explain why this function is piecewise linear?

    (b) Show that this function must be concave.

16. Consider the LPP at the starting of the Section 6.4. Suppose a uniform parametric change is made in each right-side entry of the LPP. Determine the interval in which the parameter may vary so that optimal solution does not change.

Suggestion. New solution column is $B^{-1}(b+\alpha) = B^{-1}b + B^{-1}\alpha$, i.e.,

$$
\begin{bmatrix} 6 \\ 7 \\ 4 \end{bmatrix} + \begin{bmatrix} 1 & 0 & 1 \\ 1/2 & 1 & 1 \\ 1/2 & 0 & 1 \end{bmatrix} \begin{bmatrix} \alpha \\ \alpha \\ \alpha \end{bmatrix} = \begin{bmatrix} 6 \\ 7 \\ 4 \end{bmatrix} + \begin{bmatrix} 2\alpha \\ 5\alpha/2 \\ 3\alpha/2 \end{bmatrix} = \begin{bmatrix} 6 + 2\alpha \\ 7 + 5\alpha/2 \\ 4 + 3\alpha/2 \end{bmatrix}
$$

The same basis remains optimal if $6 + 2\alpha \geq 0 \Rightarrow \alpha \geq -3$, $7 + 5\alpha/2 \geq 0 \Rightarrow \alpha \geq -14/5$, $4 + 3\alpha \geq 0 \Rightarrow \alpha \geq -4/3$. All these inequalities are satisfied for $\alpha \geq -4/3$.

Chapter 7

Transportation Problems

The chapter starts with a description of a transportation problem and different heuristics to find the initial basic feasible solutions. In the next section, the procedure to find an optimal solution is explained. Section 7.3 is on the transportation through transshipment. In the end we introduce the Assignment problem, and the Hungarian algorithm to find its optimal solution.

7.1 Introduction

The transportation problem is a special type of a linear programming problem used for studying the optimal shipping pattern. In this process, any material is to be transported from different sources to different destinations. The transportation model assumes that the shipping cost on a given route is directly proportional to the number of units shipped on that route. In general, the transportation problem can be extended to areas other than the direct transportation of a commodity, including among others, inventory, employment schedule, and personnel assignment.

Let S_i, $i = 1, 2, \ldots, m$ and D_j, $j = 1, 2, \ldots, n$ be m sources and n destinations, respectively. We define the following quantities:

a_i = the quantity of material available at source S_i, $i = 1, 2, \ldots, m$.

b_j = the quantity of material required at destination D_j, $j = 1, 2, \ldots, n$.

c_{ij} = unit cost of transportation from source S_i to destination D_j.

Every transportation problem can be represented by a matrix of order m by n, called the cost matrix or effectiveness matrix. For $m = 3$, $n = 4$, the structure of cost matrix is

Destinations → Sources ↓	D_1	D_2	D_3	D_4	Availabilities or supply ↓
S_1	c_{11} x_{11}	c_{12} x_{12}	c_{13} x_{13}	c_{14} x_{14}	a_1
S_2	c_{21} x_{21}	c_{22} x_{22}	c_{23} x_{23}	c_{24} x_{24}	a_2
S_3	c_{31} x_{31}	c_{32} x_{32}	c_{33} x_{33}	c_{34} x_{34}	a_3
Requirements or demands →	b_1	b_2	b_3	b_4	

Here S_1, S_2, S_3 are the sources, and D_1, D_2, D_3, D_4 are the destinations. The entries a_i, b_j and c_{ij} of the cost matrix are given. The x_{ij} is the number of units of the material to be transported from source S_i to destination D_j. We have to determine x_{ij} such that the product $\sum c_{ij} x_{ij}$ is minimum with the restrictions that sum of ith row is a_i, and that of jth column is b_j. In other words, the objective is to find how much material should be transported from each source S_i to each destination D_j so that the cost of transportation is minimized. Thus, the transportation problem can be mathematically expressed as

$$\min \quad x_0 = \sum_{i=1}^{m} \sum_{j=1}^{n} c_{ij} x_{ij} \qquad (7.1)$$

$$\text{s.t.} \quad \sum_{j=1}^{n} x_{ij} \le a_i, \quad i = 1, 2, \ldots, m \quad \text{(supply)} \qquad (7.2)$$

$$\sum_{i=1}^{m} x_{ij} \ge b_j, \quad j = 1, 2, \ldots, n \quad \text{(demand)} \qquad (7.3)$$

$$x_{ij} \ge 0 \text{ for all } i \text{ and } j.$$

If

$$\sum_{i=1}^{m} a_i = \sum_{j=1}^{n} b_j$$

the LPP is a balanced transportation problem, otherwise it is unbalanced. From (7.2) and (7.3), it is obvious that there are $m + n$ constraints and mn decision variables. However, it can be verified that out of these only $m + n - 1$ constraints are linearly independent, i.e., the rank of coefficient matrix is $m + n - 1$. This is what follows in the next proposition.

Proposition 1. In a balanced transportation problem with m sources and n destinations, only $m+n-1$ constraints are linearly independent.

Proof. The set of constraints of a balanced transportation with m sources and n destinations are given by

$$\sum_{j=1}^{n} x_{ij} = a_i, \quad i = 1, 2, \ldots, m \quad \text{(row constraints)} \tag{7.4}$$

$$\sum_{i=1}^{m} x_{ij} = b_j, \quad j = 1, 2, \ldots, n \quad \text{(column constraints)} \tag{7.5}$$

To show that $m + n - 1$ constraints are linearly independent, it is sufficient to verify that any one of the $m+n$ constraints can be written as linear combination of the others.

Add all the row constraints and $n - 1$ column constraints to have

$$\sum_{i=1}^{m} \sum_{j=1}^{n} x_{ij} = \sum_{i=1}^{m} a_i \tag{7.6}$$

$$\sum_{j=1}^{n-1} \sum_{i=1}^{m} x_{ij} = \sum_{j=1}^{n-1} b_j \tag{7.7}$$

Subtracting (7.7) from (7.6), we get

$$\sum_{i=1}^{m} \sum_{j=1}^{n} x_{ij} - \sum_{j=1}^{n-1} \sum_{i=1}^{m} x_{ij} = \sum_{i=1}^{m} a_i - \sum_{j=1}^{n} b_j + \sum_{i=1}^{m} x_{in}$$

$$= \sum_{i=1}^{m} x_{in} \quad \text{(balanced TP)}$$

Thus, nth column constraint has been written as the difference of the sum of all row constraints and sum of $n - 1$ column constraints, as asserted. This proves the proposition.

Remark. As a consequence of the Proposition 1 the LPP generated by a transportation model with m sources and n destinations will have $m + n - 1$ basic variables and remaining $(m - 1)(n - 1)$ nonbasic variables. Thus, a balanced transportation with m sources and n destinations will have at most

$$\binom{mn}{m+n-1} = \frac{(mn)!}{(m+n-1)!(mn-m-n+1)!}$$

basic feasible solutions.

For many applications, the supply and demand quantities in the model (the a_i and b_j) will have integer values, and the implementation will require that the distribution quantities (the x_{ij}) also be integer. Fortunately, because of the special structure, if such a model has any feasible solution, it always will have an optimal solution with just integer values, and this solution will be found by the solution procedure described in Section 7.2.

Here, we shall be using the simplex type of algorithm to obtain the optimal solution. For this we need a starting basic feasible solution (BFS). There are numerous methods for finding the initial basic feasible solution of a transportation problem. However, we illustrate a few which are commonly used in practice.

(i) North-West rule (N-W rule)

(ii) Least cost method (LCM)

(iii) Vogel approximation method (VAM)

(iv) Russell approximation method (RAM)

North-West rule. In this method, we allocate $\min(a_i, b_j)$, where a_i is the availability at source i and b_j is the requirement at destination j to the north-west corner of the cost matrix. The row or column which is satisfied is ignored for further consideration. Find the north-west corner in the remaining submatrix and allocate in the above manner. Continue the process till all the rows and columns are satisfied.

Consider the transportation given in Table 1.

Table 1

				Supply
5 15	2 15	4	3	30 ~~15~~
6 5	4 35	9	5	~~40~~ 35
2	3	8 5	1 50	~~55~~
~~15~~	20 ~~5~~	~~40~~ ~~5~~	~~50~~	

At the start, the N-W corner is x_{11} position. The maximum possible allocation at this position $= \min\{a_1, b_1\} = \min\{15, 30\} = 15$. Enter 15 at x_{11} position and subtract 15 from a_1 and b_1. The first column is satisfied and hence, ignore this column from Table 1 to decide the next N-W corner. In the remaining table, x_{12} is N-W corner. Now, allocate $min\{15, 20\} = 15$ to this cell and subtract 15 from b_2 and latest a_1. In this way the first row is satisfied. Ignore this row for further allocation. In the remaining table x_{22} is the N-W corner. By the above procedure 5 units are allocated to x_{22} cell. Repeat the procedure till all the rows and columns are satisfied. All calculations are shown in Table 1.

The starting BFS and the cost of transportation are

$x_{11}=15$, $x_{12}=15$, $x_{22} = 5$, $x_{23}=35$, $x_{33}=5$, $x_{34}=50$; $x_0=830$.

Remark. This method is independent of the cost distribution in the transportation table. Due to this reasoning the north-west corner rule is least preferred to other methods to be discussed very shortly for finding initial basic feasible solution of a transportation problem.

Least cost method. Here, we allocate $\min(a_i, b_j)$ to the cell having lowest cost in the cost matrix. Ignore the row or column which is satisfied. Carry on allocations in the same way for the remaining submatrix. In case of tie for the lowest cost cell, choose a cell which can accommodate maximum allocation, otherwise choose arbitrarily. Again, consider the preceding example. The calculations for LCM are shown in Table 2.

The cell with lowest cost is x_{34}. The maximum allocation to this

Table 2

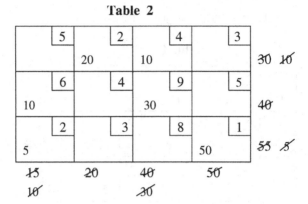

cell is 50. Subtract 50 from a_3 and b_4 which ensures that the fourth column is satisfied. Ignoring this column, the cell with lowest costs in the remaining table are in the cells x_{12} and x_{31}. We arbitrarily choose x_{12} cell and make maximum possible allocation of 20 units. Subtract 20 units from b_2 and a_1. Now, the second column is satisfied. In the remaining table the cell with lowest cost is x_{31}. The maximum possible allocation is in this cell is 5 units. Subtract 5 from b_1 and b_4 (latest) and connote till all the rows and columns are satisfied.

The starting BFS and the cost of transportation are

$$x_{12} = 20, \ x_{13} = 10, \ x_{21} = 10, \ x_{23} = 30, \ x_{31} = 5, \ x_{34} = 50; \ x_0 = 470.$$

Vogel approximation method. In this method, we start with calculating the difference of smallest cost and next higher to the smallest cost for each row and column. These differences are called the row and column penalties, respectively.

Step 1. Take the first row and choose its smallest entry and subtract this from the entry next higher to smallest entry, and write in front of row on the right. This is the penalty for first row. In this way compute penalty for each row. Similarly, calculate column penalties and write them in the bottom of the cost matrix below corresponding columns.

Step 2. Select the highest penalty and observe the row or column for which this corresponds. Then, make allocation $\min(a_i, b_j)$ in the cell having lowest cost in the selected row or column.

Step 3. Ignore for further consideration the row or column which is satisfied. Calculate fresh penalties for the remaining submatrix as in

Step 1 and allocate following the procedure of Step 2. Continue till one row or column remains to be satisfied. This last row or column is made satisfied according to LCM.

Rules for tie. In case of tie for largest penalty choose the lowest cost cell in all tied rows and columns for allocation. Again, if there is a tie for the lowest cost cell, select one for allocation which gives minimum $c_{ij}x_{ij}$.

Remarks. 1. If an allocation is not made in the lowest cost cell of a row or column with largest difference of smallest and next higher to smallest cost of that row or column, then penalty per unit cost will increase for any other choice for allocation. The VAM works on this logic.

2. The penalty for $\{2, 2, 4, 5\} = 0$.

Again consider the same example (for the last time).

Table 3

	5		2		4		3					
				30					3̶0̶	1	-	-
	6		4		9		5					
10		20		10					4̶0̶	1	1	2
	2		3		8		1					
5						50			5̶5̶ 5̶	1	1	1
1̶5̶		2̶0̶		4̶0̶		5̶0̶						
1̶0̶				1̶0̶								
3		1		4		2						
4		1		1		4						
4		1		1		-						

The starting BFS and cost of transportation are

$x_{13}=30$, $x_{21}=10$, $x_{22}=20$, $x_{23}=10$, $x_{31}=5$, $x_{34}=50$; $x_0=410$.

Russell approximation method. For each source i, determine $\bar{u}_i$ which is the largest unit cost c_{ij} in ith row. For each destination j, determine $\bar{v}_j$ which is the largest unit cost of jth column. For each variable x_{ij} calculate $\Delta_{ij} = \bar{u}_i + \bar{v}_j - c_{ij}$. Select the variable x_{ij} with most positive value Δ_{ij} (in case of tie select arbitrarily). This is the cell where allocation is to be made. Ignore the row or column which is satisfied and do the same thing in the remaining matrix. If at any stage only one row or column remains to be unsatisfied, then just fill up entries so that all supply or demand are satisfied. The BFS of the transportation problem based on this method can be seen in Table 4 below.

Table 4

5	2	4	3	
		30		30
6	4	9	5	
	20	10	10	40
2	3	8	1	
15			40	55
15	20	40	50	

The starting BFS and cost of transportation are

$x_{13}=30$, $x_{22}=20$, $x_{23}=10$, $x_{24}=10$, $x_{31}=15$, $x_{34}=40$; $x_0=410$.

Degenerate BFS. While finding the initial basic feasible solution by any method, if a row and column are satisfied simultaneously (except the last allocation), we get a degenerate basic feasible solution. To get the degenerate BFS any one of these is assumed to be satisfied and put '0' for the unsatisfied. The following example (see Table 5) worked out by LCM will make the context clear.

Table 5

2	3	7	1	
20		30		5̶0̶ 3̶0̶
4	1	5	8	
	50	0		5̶0̶ 0̶
3	4	7	1	
		20	60	8̶0̶ 2̶0̶
2̶0̶	5̶0̶	5̶0̶	6̶0̶	

The degenerate BFS and cost of transportation are

$x_{11} = 20$, $x_{13} = 30$, $x_{22} = 50$, $x_{23} = 0$, $x_{33} = 20$, $x_{34} = 60$; $x_0 = 500$.

Remark. We have seen that the N-W corner gives poor starting BFS.

Consequently, more simplex iterations are required to reach the optimality. The LCM gives a reasonable good initial BFS. But the VAM or RAM always give the starting BFS which is nearer to the optimal or sometimes optimal (see Example 3). Hence, the VAM or RAM are usually used to find the initial BFS so that few iterations are required to reach the optimality.

Let us mention two facts which are easy to prove:

Fact 1. The necessary and sufficient condition for a transportation problem with m sources, a_i availability at ith source; and with n destinations, b_j requirement at jth destination to have initial basic feasible solution is that

$$\sum_{i=1}^{m} a_i = \sum_{j=1}^{n} b_j.$$

Fact 2. A balanced transportation problem has always an optimal solution.

7.2 Optimal Solution from BFS

Here, we describe the technique how to reach the optimal solution from a given initial basic feasible solution. The method described below is known as $u-v$ method or modified distribution method (MODI) and goes in the name of Charnes and Cooper.

Take a balanced transportation problem with m sources and n destinations as primal problem

$$\min \ \sum_{i=1}^{m} \sum_{j=1}^{n} c_{ij} x_{ij}$$

$$\text{s.t.} \ \sum_{j=1}^{n} x_{ij} = a_i, \quad i = 1, 2, \ldots, m$$

$$\sum_{i=1}^{m} x_{ij} = b_j, \quad j = 1, 2, \ldots, n$$

$$x_{ij} \geq 0$$

Let u_i be the dual variable associated with ith row constraint $i = 1, 2, \ldots, m$ and v_j be the dual variable associated with jth column constraint. Then the dual of the above problem is

$$\max \quad \sum_{i=1}^{m} a_i u_i + \sum_{j=1}^{n} b_j v_j$$

$$\text{s.t.} \quad u_i + v_j \le c_{ij}$$

$$u_i, \ v_j \text{ unrestricted}$$

We have proved in Proposition 1 that in a basic feasible solution of the primal $m + n - 1$ variables are basic and remaining are nonbasic. Suppose that the dual variables are chosen in such a way that

$$u_i + v_j = c_{ij} \quad \text{in } m + n - 1 \text{ basic cells}$$

$$u_i + v_j \le c_{ij} \quad \text{in remaining cells (nonbasic cells)}$$

Such a choice of the dual variables implies that dual slack variables $r_{ij} = 0$ in basic cells and $r_{ij} \ge 0$ in nonbasic cells. Hence, for basic cells

$$x_{ij} \ge 0, \quad r_{ij} x_{ij} = 0.$$

For the remaining cells

$$x_{ij} = 0, \quad r_{ij} x_{ij} = 0.$$

Observe that this choice of dual variables satisfies the complimentary slackness property (see the remark of Theorem 5, Chapter 4) and hence, it provides an optimal solution of the transportation problem.

The computational procedure may be summarized as

Step 1. Introduce the dual variables u_i and v_j corresponding to each ith row and jth column, respectively. Write u_i in front of each ith row and v_j at the top of each column. Take any one of the u_i or v_j to be zero.

Step 2. For basic cells (which contain allocations), $u_i + v_j = c_{ij}$ (equivalent to $z_j - c_j = 0$ for basic variables in simplex method). This relation assigns values to all u_i and v_j.

Step 3. For nonbasic cells (which have no allocations), calculate $u_i + v_j - c_{ij}$, and write them in S-W corner of the concerned cell.

tep 4. If all S-W entries are ≤ 0, the BFS we are testing is optimal. If, at least one of the S-W entries is positive, then this BFS is not optimal. In this situation, look for the most positive S-W entry in the cost matrix. This decides the entering variable.

tep 5. Assign θ quantity in the cell having the most positive S-W entry, and make a loop as follows.

Rule for making the loop. Start from θ cell and move horizontally and vertically to the nearest basic cell with the restriction that the turn (corner) of the loop must not lie in any nonbasic cell (except θ-cell). In this way, return to the θ-cell to complete the loop.

Add or subtract θ in cornered entries of the loop maintaining feasibility, and value of θ is fixed as the minimum of the entries from which θ has been subtracted. Inserting the fixed value of θ, we get next BFS which improves the initial transportation cost. While inserting the value of θ one cell assumes 0 value. We shall not mention 0 value as this is the leaving variable, i.e., this cell has become nonbasic. Thus, one iteration of the simplex type algorithm is over. This gives the improved value of the objective function. Again, use the latest BFS, and repeat Steps 1 through 5 until every S-W entry turns out to be ≤ 0. This is the optimal solution.

Remarks. 1. A loop may cross itself, and even crossing point may be in nonbasic cell.

2. A loop can skip over some basic cell or nonbasic cell in a row and column. However, the successive lines must be perpendicular.

3. There always exists a unique loop starting from a given nonbasic cell.

Let us workout some problems based on this discussion.

Example 1. Consider the following transportation problem with initial basic feasible solution (obtained by LCM) and find its optimal solution.

For Step 1, introduce the dual variables u_1, u_2, u_3 corresponding to three rows and write them in front of each row. Similarly, introduce v_1, v_2, v_3, v_4 corresponding to the columns and write them at the top of each column. Fix any of the dual variables u_1, u_2, u_3, and v_1, v_2, v_3, v_4 at zero level. However, for convenience, see the row which has maximum basic cells. Assign zero value to the dual variable associated with this row. In case two or more rows have the same maximum number of basic cells, then the tie can be broken arbitrarily.

	1	2	3	4	
	15	15			30
	7	6	2	5	
			25	25	50
	4	3	2	7	
		15		20	35
	15	30	25	45	

In Table 6, each row has two basic cells. Any of the u_1, u_2, u_3 may be assigned zero value. Suppose $u_1=0$.

Table 6

	$v_1 = 1$	$v_2 = 2$	$v_3 = 3$	$v_4 = 6$
$u_1 = 0$	1 15	2 15	3 $\boxed{0}$	4 $\boxed{2}$
$u_2 = -1$	7 $\boxed{-7}$	6 $\boxed{-5}$	2 25−θ	5 25+θ
$u_3 = 1$	4 $\boxed{-2}$	3 15	2 θ $\boxed{2}$	7 20−θ

In Step 2, we calculate the values of other dual variables using the relation: $u_i + v_j = c_{ij}$ for basic cells. The possible equations for basic cells are

$$\begin{aligned}
u_1 + v_1 &= 1 \\
u_1 + v_2 &= 2 \\
u_2 + v_3 &= 2 \\
u_2 + v_4 &= 5 \\
u_3 + v_2 &= 3 \\
u_3 + v_4 &= 7
\end{aligned} \qquad (7.8)$$

With $u_1=0$, the above system of equations gives (in a sequence)

$$v_1 = 1, \quad v_2 = 2, \quad \text{and} \quad u_3 = 1, \quad v_4 = 6, \quad u_2 = -1, \quad v_3 = 3.$$

Thus, we have computed each dual variable. In practice, we do not solve these equations, but determine all the dual variables directly on the table.

In Step 3, we calculate $u_i + v_j - c_{ij}$ for each nonbasic cell and write them in the south-west corner. In the above example for the nonbasic cells x_{13},

$$u_1 + v_3 - c_{ij} = 0 + 3 - 3 = 0$$

and x_{14},

$$u_1 + v_4 - c_{14} = 0 + 4 - 2 = 2.$$

Similarly, compute for other nonbasic cells. We enter these numbers on the south-west corner of these cells. This is done for every nonbasic cell.

To execute Step 4, if all S-W entries are ≤ 0, the basic feasible solution that we are testing is optimal and algorithm stops. Here, all S-W entries are not ≤ 0, see Table 6 and hence, we go to the next step. Look for the most positive entry in S-W corners. This is available at two places, viz., at the intersection of the first row and fourth column, and at the intersection of third row and third column. Hence, there is a tie for the most positive S-W entry, see Table 6. Break the tie arbitrarily to decide the entering variable. Suppose, we consider the nonbasic cell c_{33}, i.e., nonbasic variable x_{33} to enter the basis.

In Step 5, we make a loop according to the rule as explained earlier. Assign θ value to this cell, and subtract and add θ at corners of the loop to maintain feasibility. Now, decide the value of θ by taking minimum of the entries from which θ have been subtracted, i.e.,

$$\theta = \min\{20, 25\} = 20.$$

Table 7

	1	2	3	4	
15	15				30
	7	6	2	5	
			45	5	50
	4	3	2	7	
		15	20		35
15		30	25	45	

Insert this value of θ which causes x_{34} to leave the basis, and the new BFS is given in Table 7. Thus, the first iteration is over.

To carry the second iteration, again introduce freshly the dual variables and find their values with the help of basic cells. Then, compute S-W entries as above. Note that all S-W entries are ≤ 0, and hence, the BFS is optimal one, see Table 8.

Table 8

	$v_1 = 1$	$v_2 = 2$	$v_3 = 1$	$v_4 = 4$
$u_1 = 0$	1 15	2 15	3 $\boxed{-2}$	4 $\boxed{0}$
$u_2 = 1$	7 $\boxed{-5}$	6 $\boxed{-3}$	2 5	5 45
$u_3 = 1$	4 $\boxed{-2}$	3 15	2 20	7 $\boxed{-2}$

Thus, the optimal solution and the optimal value are

$$x_{11} = 15, \ x_{12} = 15, \ x_{23} = 5, \ x_{24} = 45, \ x_{32} = 15, \ x_{33} = 20; \ x_0 = 365.$$

Remarks. 1. It is a point of caution that, if we break a tie by taking c_{14} as θ cell in Table 6, then it requires two iterations to reach the optimality while in the preceding case only one iteration was required to get the optimal table. When we take c_{14} cell as θ cell, the loop will skip a basic cell and two nonbasic cells in the first iteration. This verification is left to the reader.

2. If the size of the problem is large, then it is difficult to draw the loop by using the above technique. A definite way of finding the loop is as follows:

Sketch the flow chart of the order in which dual variables have been determined from (7.8). For the above problem the flow chart is given in Fig. 7.1. Since we have taken θ cell at (u_3, v_3) position, see Table 6. So, join u_3 and v_3, and complete the loop in any direction. One of the direction is

$$(u_3, v_3) \longrightarrow (v_3, u_2) \longrightarrow (u_2, v_4) \longrightarrow (v_4, u_3) \longrightarrow (u_3, v_3)$$

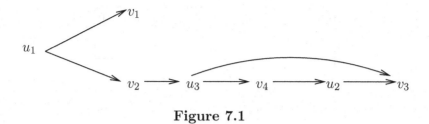

Figure 7.1

Alternative optimal solution. An optimal table of a transportation problem is supposed to have an alternative optimal solution
if at least one of the S-W entry is zero, equivalently relative cost of at
least one nonbasic variable is zero. Convert this nonbasic cell as basic
cell by assigning θ to this cell and make the loop.

Let us consider the optimal solution given in Table 8. Since S-W
entry in c_{14} is zero, decide this cell for assigning θ, i.e., this cell is made
basic. Add and subtract θ to maintain feasibility, and then make a
loop as shown in Table 9. Here, the value of θ is given by

$$\theta = \min\{15, 20, 45\} = 15.$$

Table 9

	$v_1 = 1$	$v_2 = 2$	$v_3 = 1$	$v_4 = 4$
	1	2	3	4
$u_1 = 0$	15	15−θ	−2	θ
		↑		0
	7	6	2	5
$u_2 = 1$			5+θ ←	45−θ
	−5	−3		
	4	3	2	7
$u_3 = 1$		15+θ ←	20−θ	
	−2			−2

Inserting $\theta = 15$ in Table 9, we get the following alternative optimal
table

	1	2	3	4
	15			15
	7	6	2	5
			20	30
	4	3	2	7
		30	5	

The optimal solution and optimal transportation cost are

$$x_{11} = 15, \ x_{14} = 15, \ x_{23} = 20, \ x_{24} = 30, \ x_{32} = 30, \ x_{33} = 5; \ x_0 = 365.$$

Example 2. Take the degenerate BFS of the transportation problem of Table 5, and find the optimal solution.

Introduce the dual variable as in Example 1 and computing the S-W entries, we find that all these entries are ≤ 0, and hence, the degenerate BFS is itself the optimal solution, see Table 10.

Table 10

	$v_1 = 2$	$v_2 = 3$	$v_3 = 7$	$v_4 = 1$
$u_1 = 0$	2 — 20	3 — [0]	7 — 30	1 — [0]
$u_2 = -2$	4 — [-4]	1 — 50	5 — 0	8 — [-9]
$u_3 = 0$	3 — [-1]	4 — [-1]	7 — 20	1 — 60

The optimal solution and optimal transportation cost are

$$x_{11} = 20, \ x_{13} = 30, \ x_{22} = 50, \ x_{23} = 0, \ x_{33} = 20, \ x_{34} = 60; \ x_0 = 500.$$

Note that the problem has an alternate optimal solution, since S-W entry in cells c_{13} and c_{14} are zero. Let us make the nonbasic variable

x_{13} to enter the basis. Write θ here and make a loop. Assigning $\theta = 30$, we get a nondegenerate optimal solution as

$$x_{11} = 20, \ x_{12} = 30, \ x_{22} = 20, \ x_{23} = 30, \ x_{33} = 20, \ x_{34} = 60; \ x_0 = 500.$$

If the c_{14} cell is made basic by assigning θ quantity and making the loop, we get the second alternate alternate optimal solution as

$$x_{11} = 20, \ x_{14} = 30, \ x_{21} = 50, \ x_{13} = 0, \ x_{33} = 50, \ x_{34} = 30; \ x_0 = 500.$$

Transportation problem as a maximization case. Usually, the maximization transportation problem is converted into minimization problem by any one of the procedures:

(i) All c_{ij}'s are multiplied by -1 and the problem is solved as minimization problem. Remember to multiply the minimum value available at the end by -1.

(ii) All c_{ij}'s are subtracted from the largest cost of effectiveness matrix and then problem is solved as a minimization problem. To write the minimum value, allocations are multiplied by the original c_{ij}'s.

7.3 Unbalanced Transportation Problem

To deal with an unbalanced transportation problem $(\sum a_i \neq \sum b_j)$ to obtain its BFS or the optimal solution, we first convert this into a balanced transportation problem as follows:

Create a dummy row (dummy source) if total demand exceeds total supply or a dummy column (dummy destination) if total supply exceeds total demand. The supply or demand at the dummy origin is equal to the symmetric difference of total supply and total demand.

Let us solve an unbalanced transportation problem as our next example.

Example 3. A company has facilities at cities A, B and C which supply warehouses at cities D, E and F. The monthly factory capacities are 50, 150 and 200 units, respectively. The monthly requirements are 100, 130 and 200 units, respectively. The shipping cost per unit are given in the following network.

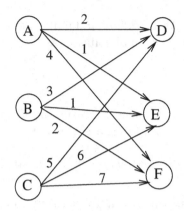

(a) Formulate the LP model for this problem.

(b) Also, write dual of the above problem.

(c) Use the VAM to get the starting BFS.

(d) Find the optimal solution.

(e) Does there exist an alternative optimal solution? If it exists, then find it.

(f) Is the optimal solution degenerate? If it exists, write why? Otherwise give reasons for its absence.

(g) If the penalties per unit cost for the unsatisfied demand are 5, 2, 1 for the cities D, E and F, then find the optimal solution.

The effectiveness matrix of the problem is

	D	E	F	
A	2	1	4	50
B	3	1	2	150
C	5	6	7	200
	100	130	200	

(a) LPP formulation:

$$\min \quad x_0 = 2x_{11} + x_{12} + 4x_{13} + 3x_{21} + x_{22} + 2x_{23} + 5x_{31} + 6x_{32} + 7x_{33}$$

$$\text{s.t.} \quad x_{11} + x_{12} + x_{13} = 50$$
$$x_{21} + x_{22} + x_{23} = 150$$
$$x_{31} + x_{32} + x_{33} = 200$$
$$x_{11} + x_{21} + x_{31} \leq 100$$
$$x_{12} + x_{22} + x_{32} \leq 130$$
$$x_{13} + x_{23} + x_{33} \leq 200$$
$$x_{ij} \geq 0, \ i = 1, 2, 3; \ j = 1, 2, 3.$$

(b) To write its dual, first we write the last three constraints (associated with columns) as $\geq$ constraints to covert this into canonical form, see Chapter 4.

The dual LPP is

$$\max \quad y_0 = 50u_1 + 150u_2 + 200u_3 - 100v_1 - 130v_2 - 200v_3$$

$$\text{s.t.} \quad u_1 - v_1 \leq 2$$
$$u_1 - v_2 \leq 1$$
$$u_1 - v_3 \leq 4$$
$$u_2 - v_1 \leq 3$$
$$u_2 - v_2 \leq 1$$
$$u_2 - v_3 \leq 2$$
$$u_3 - v_1 \leq 5$$
$$u_3 - v_2 \leq 6$$
$$u_3 - v_3 \leq 7$$
$$u_1, u_2, u_3 \text{ unrestricted}, \ v_1, v_2, v_3 \geq 0.$$

Here u_1, u_2, u_3 and v_1, v_2, v_3 are the dual variables corresponding to the row constraints and the column constraints, respectively.

(c) To find its initial basic feasible solution, first we write the problem as balanced balanced transportation problem. Since the demand is more than the supply, we add a fictitious supply, i.e., one additional row is attached with the cost matrix. The supply at the dummy origin is $430 - 400 = 30$ units. The initial BFS (computed using VAM) of the balanced transportation is given in the following table.

The starting BFS and the cost of transportation are

$$x_{12} = 50, \ x_{22} = 80, \ x_{23} = 70, \ x_{31} = 70, \ x_{33} = 130, \ x_{43} = 30; \ x_0 = 1470.$$

	2	1	4	Supply
		50		5̶0̶ 1 1 1
	3	1	2	
			150	1̶5̶0̶ 1 1 −
	5	6	7	
	100	80	20	2̶0̶0̶ 1 1 1
	0	0	0	
			30	3̶0̶ 0 − −

1̶0̶0̶	1̶3̶0̶	2̶0̶0̶
	8̶0̶	1̶7̶0̶
		2̶0̶

2	1	2
1	0	2
3	5	3

(d) Optimal solution. As usual, we apply the $u-v$ method to find its optimal solution. After the first iteration, we observe that all S-W entries ≤ 0, and thus, the BFS obtained in (c) is optimal, see Table 11.

Table 11

	$v_1 = 5$	$v_2 = 6$	$v_3 = 7$
	2	1	4
$u_1 = -5$		50	
	−2		−2
	3	1	2
$u_2 = -5$			150
	−3	0	
	5	6	7
$u_3 = 0$	100	80	20
	0	0	0
$u_4 = -7$			30
	−2	−1	

(e) Since at least one of the S-W entry is '0', and hence, an alternative optimal solution exists. Bring the nonbasic variable x_{22} of this cell into the basis cell. By making a loop we find that $\theta = 80$, and hence, one of the alternative optimal solution is

$$x_{12} = 50, \ x_{22} = 80, \ x_{23} = 70, \ x_{31} = 100, \ x_{33} = 100, \ x_{43} = 30.$$

(f) The optimal solution is not degenerate as all basic variables $4 + 3 - 1 = 6$ are having positive values.

(g) Insert the penalties 5, 2 and 1 in the last row. To find the optimal solution under this restriction, we first find the initial BFS by VAM.

	2	1	4				
		50		~~50~~	1	1	1
	3	1	2				
			150	~~150~~	1	1	−
	5	6	7				
	100	80	20	~~200~~	1	1	1
	5	2	1				
			30	~~30~~	1	−	−
	~~100~~	~~130~~	~~200~~				
		~~80~~	~~170~~				
			~~20~~				
	1	0	1				
	1	0	2				
	3	5	3				

The initial BFS is

$$x_{12} = 50, x_{23} = 150, x_{31} = 100, x_{32} = 80, x_{33} = 20, x_{43} = 30.$$

and the corresponding transportation cost is 1500. It can be easily verified that this BFS is an optimal solution.

Note. The initial basic feasible solution of a transportation problem is usually obtained by North-West rule (NW-rule), Least cost method (LCM) and Vogel approximation method (VAM). In most of the practical problems, the transportation problems are found to be unbalanced. It is but natural to search the initial BFS which is very nearer to the optimal solution so that computational labour and time is saved to reach the optimality. Here, we suggest a simple algorithm which is applicable for all heuristics employed to find the initial BFS.

Algorithm. The algorithm consists of the following steps:

Step 1. Ignore the dummy row or column for calculation, and apply any method to allocate the resources in all cells (except the dummy cells).

Step 2. In Step 1, all row constraints are satisfied (if demand exceeds supply) or all column constraints are satisfied (if supply exceeds demand). Now, allocate the excess demand or supply to dummy cells so that it does not violate supply and demand restrictions.

Let us consider the unbalanced transportation problem

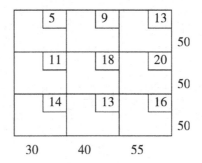

A comparison of initial transportation costs calculated by different approaches is given as

Heuristics	Conventional approach	Suggested
NW Rule	1690	1690
LCM	1670	1570
VAM	1620	1540

The heuristic proposed here most often results in better transportation cost than obtained by the techniques available in the literature. First, this exploits the full potential of any heuristics, and then the distribution is made independent of any priority to the dummy cells.

7.4 Transshipment

When the items to any destination from any source or destination directly or through some other transition node are transported, the process is called the transshipment. Hence, in the transshipment all sources and destinations can function in any direction. Usually, in the absence of the transshipment, the transportation cost goes higher. Hence, the transshipment is very useful to reduce the transportation cost. In the transshipment, we have usually two type of problems, viz., (a) transshipment through sources and destination; (b) transshipment through transition nodes.

(a) **Transshipment through sources and destinations.** As usual the cost matrix is between sources and destinations. The second cost matrix is given among sources, and the third cost is among destinations. Then to solve a transportation problem with transshipment we adopt the following procedure.

(i) Make the fourth cost matrix with sources $S_1, S_2, \ldots, S_m, D_1, D_2, \ldots, D_n$ and destinations $S_1, S_2, \ldots, S_m, D_1, D_2, \ldots, D_n$.

(ii) Add $\sum a_i$ or $\sum b_j$ to each of the availabilities and requirements. In case $\sum a_i$ and $\sum b_j$ are different, then add larger of the $\sum a_i$ and $\sum b_j$, which is called buffer stock. The buffer stock should be sufficiently large so that all the original supply (or demand) units pass through the transshipment nodes.

Let B be the desired buffer stock. Then

$$B = \max\{\text{supply, demand}\}.$$

Example 4. Solve the following transportation problem using the transshipment process.

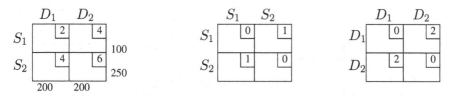

Here S_1 and S_2 are the purely sources and have zero requirement, and to make them destinations, we add B (buffer stock)

$$B = \max\{350, 400\} = 400$$

in the bottom of columns 1 and 2, respectively. Similarly, D_1 and D_2 are purely destinations and as sources these have nothing to supply,

Table 12

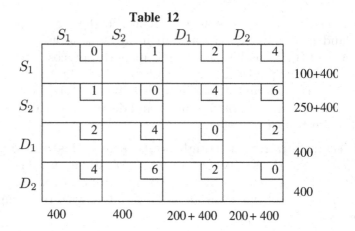

	S_1	S_2	D_1	D_2	
S_1	0	1	2	4	100+400
S_2	1	0	4	6	250+400
D_1	2	4	0	2	400
D_2	4	6	2	0	400
	400	400	200+400	200+400	

and hence, 400 is added in the third and fourth rows. Also, add 400 to the availability in the first and second row, and to the demand at third and fourth column, see Table 12. The resulting transportation model is unbalanced. Hence, add fictitious sources S_F, since demand is higher than supply ($\sum a_i < \sum b_j$). This will make the problem as a balanced transportation problem, see Table 13. First, we find its BFS by VAM.

Table 13

	0	1	2	4	supply	penalties
	150		200	150	~~500~~ ~~300~~ ~~150~~	1 1 1 1 1 1
	250	400			~~650~~ ~~250~~	1 1 1 1 1 1
			400		~~400~~	2 - - - - -
				400	~~400~~	2 2 - - - -
S_F				50	~~50~~	0 0 0 - - -

Costs: row 1: 0, 1, 2, 4; row 2: 1, 0, 4, 6; row 3: 2, 4, 0, 2; row 4: 4, 6, 2, 0; row S_F: 0, 0, 0, 0.

Demand:

demand	~~400~~ 250	~~400~~ 200	~~600~~ 200	~~600~~ 200 150
	0	0	0	0
	0	0	2	2
	0	0	2	4
	1	1	2	2
	1	1	-	2
	1	1	-	-

Now, we check the above BFS for optimality. Introduce the dual variables u_1, u_2, u_3, u_4, u_4, and v_1, v_2, v_3, v_4 corresponding to the rows and columns respectively. Compute their values using the basic cells and the relation $u_i + v_j = c_{ij}$. Then compute the S-W entries of the basic cells using the expression $u_i + v_j - c_{ij}$. This analysis is shown in the following table.

	$v_1 = 0$	$v_2 = -1$	$v_3 = 2$	$v_4 = 4$	
$u_1 = 0$	0 150	1 -2	2 200	4 150	500
$u_2 = 1$	1 250	0 400	4 -1	6 -1	650
$u_3 = -2$	2 -4	4 -7	0 400	2 0	400
$u_4 = -4$	4 -8	6 -11	2 -4	0 400	400
$u_5 = -4$	0 -4	0 -5	0 -2	0 50	50
	400	400	600	600	

All the S-W entries are ≤ 0. This tells us that the basic feasible solution by VAM is optimal. The optimal solution can be read from the flow chart as depicted in Fig. 7.2.

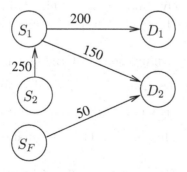

Figure 7.2

D_1 has received 200 from S_2 through S_1

D_2 has received 150 from S_2 through S_1

The optimal cost $= 1250$

Note. This can be verified that the optimal transportation cost without transshipment (solution of the first cost matrix) is 1500.

(b) Transshipment through transition nodes. Let us workout next example of the transshipment in which sources can not function as destinations. However, some destinations are supplied through some transition nodes and within themselves also. The following example will make the context clear.

Example 5. The network in Figure 7.3 gives the shipping routes from source 1 to destinations 5 and 6 through intermediate destinations 3 and 4. The unit cost of transportation is shown on the respective arcs. Formulate the problem as transshipment model and then solve.

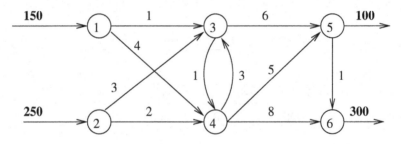

Figure 7.3

First, we formulate the problem. From network, observe that the nodes 1 and 2 are purely supply nodes, while 3, 4 and 5 are transshipment nodes, and node 6 is purely a destination node.

Supply at a purely supply node = Original supply

Supply at transshipment node = Original supply + Buffer

Demand at purely demand node = Original demand

Demand at a transshipment node = Original demand + Buffer

Also, from the network, nodes 1 and 2 (sources) can not supply directly to the destinations nodes 5 and 6, and hence, big M is to be inserted in the cost matrix in corresponding cell. Similarly, nodes 5 can not supply to transshipment nodes 3 and 4, and again big M is inserted in the appropriate cells. The buffer stock B is

$$B = \max\{150 + 250, 300\} = 400.$$

Thus, to make the cost matrix we take nodes 1 to 5 as the sources and the nodes 3 to 6 as the destinations. Add the buffer stock to nodes 3 to 5 as these are transition nodes and have nothing to supply. Also, add the buffer stock to the nodes 3 to 5 when these act as destinations. Note that the node 5 is already having the requirement of 100. Hence, the final requirement at the node 5 will be 500. All these calculations have been soon in the following table:

	3	4	5	6	
1	1	4	M	M	150
	150				
2	3	2	M	M	250
		250			
3	0	1	6	M	400
	250		150		
4	3	0	5	8	400
		150	250		
5	M	M	0	1	400
			100	300	
	400	400	500	300	

In the above table we have shown its BFS. This can be verified to be optimal by the $u-v$ method. The outcome of the transshipment network is shown in Figure 7.4. From Fig. 7.4, it is clear that the destination 6 gets its requirement fulfilled from the destination 5 through the transshipment node 4.

Remark. In Example 5, we have adopted a special technique to find the initial BFS. Ignore all cells with big M. Move satisfying column-wise making allocation at the lowest cost entries. Thus, consider S_1, D_1 and the first column. Then S_2, D_2 and the second column, and similarly, third and fourth columns, respectively.

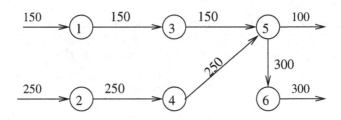

Figure 7.4

7.5   Assignment Problems

An assignment problem is a particular case of the transportation problem. The transportation problem reduces to the assignment problem, if $m = n$ and $a_i = b_i = 1$, $i = 1, 2, \ldots, n$. It means that the variable x_{ii} can take value 0 or 1 only. The LPP formulation of the assignment problem is

$$\min \quad f = \sum_{i=1}^{n} \sum_{j=1}^{n} c_{ij} x_{ij}$$

$$\text{s.t.} \quad \sum_{j=1}^{n} x_{ij} = 1$$

$$\sum_{i=1}^{n} x_{ij} = 1$$

$$x_{ij} = 0 \text{ or } 1, \ 1 \le i, j \le n$$

Since, the assignment problem is a particular case of the transportation problem wherein workers and jobs are treated as sources and destinations, respectively as well as the supply at each source, and demand at each destination are equal to 1. The assignment problem involves assignment of jobs to workers on a one-to-one pattern. The number of jobs is presumed to be equal to the number of workers. However, if this is not the case, either fictitious jobs or workers as required can be created to satisfy this assumption. The time c_{ij} required by the ith worker to complete the jth job is known. The objective here is to assign a job to every worker so that all jobs are completed in the minimum total time.

All the techniques developed in Sections 7.1-7.3 to find the optimal solution of the transportation problem are also applicable here.

However, due to its typical formulation we will discuss special designed technique widely known as the Hungarian method developed by Konig and Egervary.

Theorem 1. If the cost matrix for an assignment problem has non-negative entries and at least n zeroes, then an optimal solution to the problem exists if n of the zeroes lie in the positions of the ones of some $n \times n$ permutation matrix $\mathbf{P}$. The matrix $\mathbf{P}$ represents an optimal assignment.

Proof. In the situation described, the cost can never be smaller than zero, and we have found an assignment for which the cost is zero.

This theorem provides a way to construct an algorithm for assignment problems. We will show that we can modify the cost matrix without changing the optimal solution. The algorithm will then attempt to carry out this modification to reach a situation in which the cost matrix has at least one zero in each row and in each column.

Theorem 2. Suppose that $\mathbf{C} = (c_{ij})$ is the cost matrix for an $n \times n$ assignment problem, and $\overline{X} = [\overline{x_{ij}}]$ is an optimal solution to this problem. Let $\mathbf{C}'$ be the matrix formed by adding α to each entry in the rth row. Then, $\overline{X}$ is an optimal solution to the new assignment problem defined by $\mathbf{C}'$.

Proof. The objective function for the new problem is

$$z' = \sum_{i=1}^{n}\sum_{j=1}^{n} c_{ij}x_{ij}(j \neq r) + \sum_{j=1}^{n}(c_{rj} + \alpha)x_{rj}$$

$$= \sum_{i=1}^{n}\sum_{j=1}^{n} c_{ij}x_{ij} + \alpha\sum_{j=1}^{n} x_{rj}$$

$$= \sum_{i=1}^{n}\sum_{j=1}^{n} c_{ij}x_{ij} + \alpha,$$

since each row sum is 1. Therefore, the smallest value for z' will be obtained when

$$z = \sum_{i=1}^{n}\sum_{j=1}^{n} c_{ij}x_{ij}$$

is smallest; namely, it is obtained when $X = \overline{X}$.

7.6 Hungarian Method

The Hungarian method to solve the assignment problem is outlined with the help of a numerical problem. We assume here that the cost matrix has nonnegative entries. Before begin with the Hungarian method we need a reduced cost matrix defined as

Reduced Matrix. A matrix which contains at least one zero in each row and in each column is called a reduced matrix. A reduced cost matrix is obtained as follows: Subtracting the lowest entry of a row from the remaining entries of this row. Do this for each row. Observe that whether, we have received the reduced cost matrix. If not, then do the same procedure for the columns which do not have at least one zero.

Sometimes it is convenient and the solution converges rapidly, if we reduce the matrix by applying column reduction first and then row reduction. A such type of case has been mentioned in the Problem set of the Chapter 8.

Example 6. Four persons A, B, C, D are assigned to work on four different machines. The following table shows how long it takes for a specific person to finish a job at a specific machine.

		Machines			
		I	II	III	IV
	A	8	26	17	11
Persons	B	13	28	4	26
	C	38	19	18	15
	D	19	26	14	10

Find the optimal allocation, i.e., how the machines should be assigned to A, B, C, D so that the job is completed in minimum time.

Step 1. First, we get the reduced matrix, see Table 14. The reduced cost matrix defines an assignment problem that has the same optimal solutions as that of the original cost matrix.

Step 2. For the first assignment, choose the row having only one zero and box this zero and cross all other zeros of the column and row in which this boxed zero lies. Next, examine other rows containing

Table 14

0	14	9	3
9	20	0	22
23	0	3	0
9	12	14	0

single zero and do the same job. If there are more that one zero in any row, then leave it for the time being. Now, repeat the same procedure for the columns. This gives Table 15.

Table 15

	I	II	III	IV
A	0	14	9	3
B	9	20	0	22
C	23	0	3	0̸
D	9	12	14	0

tep 3. If each zero of the reduced matrix is either boxed or crossed, and each row and column contains exactly one boxed zero, then optimality is reached, and this occurs in Table 15.

The optimal assignment: $A \longrightarrow I$, $B \longrightarrow III$, $C \longrightarrow II$, $D \longrightarrow IV$. The optimal value $= 41$.

If all the zeros are boxed or crossed and even then each row and each column does not contain a boxed zero, it is not possible to get the optimal solution at this stage. We proceed further to deal with this structure. Again, the algorithm is explained by solving a problem.

Example 7. Consider the following assignment problem and find its optimal solution. Does the problem has alternative optimal solutions?

5	5	7	4	8
6	5	8	3	7
6	8	9	5	10
7	6	6	3	6
6	7	10	6	11

Step 1 gives the following reduced matrix, see Table 16. Now, box

Table 16

1	0	0	0	1
3	1	2	0	1
1	2	1	0	2
4	2	0	0	0
0	0	1	0	2

and cross zeros in accordance to Step 2, see Table 17.

Table 17

1	[0]	⨉	⨉	1
3	1	2	[0]	1
1	2	1	⨉	2
4	2	[0]	⨉	⨉
[0]	⨉	1	⨉	2

Every zero is either boxed or crossed, even then each row and column does not contain exactly one boxed zero, i.e., the optimality is not reached.

We proceed by drawing minimum number of horizontal and vertical lines so that all the zeros are covered (each horizontal line must pass through an entire row and each vertical line must pass through an entire column). This can be accomplished easily by first covering zeros in that row or column which has maximum number of zeros. If there exists a tie break it arbitrarily and further search the rows or columns having next lower to highest number of zeros in some column or row and continue this process till all zeros are covered, see Table 18. Note that four lines are needed to cover all zeros this is due to fact that only four assignment could be made previously in Table 17.

Table 18

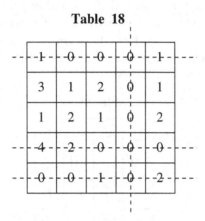

Locate the smallest entry from the uncovered entries, say x. Subtract x from all elements not covered by these lines and add x to all those elements that lie at the intersection of these lines. The entries lying on these lines but not on the intersection must be left unchanged. After these changes are incorporated, the revised cost matrix is given in Table 19.

Table 19

1	0	0	1	1
2	0	1	0	0
0	1	0	0	1
4	2	0	1	0
0	0	1	1	2

Now, apply the Hungarian algorithm freshly to come to the stage so that each row and column contains exactly one boxed zero. In Table 19, there is no single zero in any row or column. We should go for rows with two zeros from above. In the first row cells $(1,2)$ and $(1,3)$ have zeros. We can box any one of them. Let us break the tie arbitrarily, and box zero of cell $(1,2)$. Cross all zeros in first row and second column. Move row-wise and note that now cell $(4,1)$ contains a single zero, and hence, box it and cross all zeros of fourth row and first column. After these steps, no single zero is available in any row. Same thing happens for columns. Again, cells $(2,4)$ and $(2,5)$ contain two zeros each. Let us box zero of cell $(2,4)$ and cross all zeros in second row and fourth column. Now, the third and fourth rows have single zero in cells $(3,3)$ and $(4,5)$. Box these zeros. Thus, we get Table 20 which contains exactly one zero in each row and column, and hence, the optimal assignment is reached.

Table 20

1	[0]	⨯	1	1
2	⨯	1	[0]	⨯
0	1	[0]	⨯	1
4	2	⨯	1	[0]
[0]	⨯	1	1	2

The optimal assignment: $1 \to 2$, $2 \to 4$, $3 \to 3$, $4 \to 5$, $5 \to 1$. The optimal value $= 29$.

The problem has alternative optimal solutions, see just next Remark 1.

Remarks. 1. Sometimes there is no single zero in any row and column of the reduced matrix. In this case arbitrary choice of zero (to be boxed) in the row or column which has minimum number of zeros is advisable, see Table 20 just above. Note that in such situations the assignment problem has alternative solution. For above problem alternate optimal assignments are

The optimal assignment: $1 \to 3$, $2 \to 4$, $3 \to 1$, $4 \to 4$, $5 \to 2$. The optimal value $= 29$.

The optimal assignment: $1 \to 2$, $2 \to 5$, $3 \to 4$, $4 \to 3$, $5 \to 1$. The optimal value $= 29$.

The optimal assignment: $1 \to 3$, $2 \to 2$, $3 \to 4$, $4 \to 5$, $5 \to 1$. The optimal value $= 29$.

2. If the problem is maximization, a simple modifications is done in the cost matrix. First, multiply each entry c_{ij} by -1, then add to each entry the negative of the smallest entry. This results in a new cost matrix with nonnegative entries, and now the Hungarian is applicable.

3. An assignment may be be an unbalanced problem. Suppose five jobs are to assigned to six workers. Then, a dummy job is to be created so that square cost matrix is available and dummy in row or column the costs are taken to be zero.

Finally, we solve a special problem which is solved as an assignment problem.

Example 8. A trip from Chandigarh to Delhi takes six hours by bus. A typical time table of the bus service in both the directions is given below.

Departure Chandigarh	Route number	Arrival Delhi	Departure Delhi	Route number	Arrival Chandigarh
06.00	$a \rightarrow$	12.00	5.30	$1 \rightarrow$	11.30
07.30	$b \rightarrow$	13.30	09.00	$2 \rightarrow$	15.00
11.30	$c \rightarrow$	17.30	15.00	$3 \rightarrow$	21.00
19.00	$d \rightarrow$	01.00	18.30	$4 \rightarrow$	00.30
00.30	$e \rightarrow$	06.00	00.30	$5 \rightarrow$	06.00

The cost of providing this service by the transport depends upon the time spent by the bus crew away from their place in addition to service times. There are five crews and there is a constraint that every crew should be provided at least 4 hours of rest before the return trip again and should not wait for more than 24 hours. The company has residential facilities for the crew at both the cities. Determine the service schedule so that the waiting time is minimum.

As the service time is constant, it does not affect the decision of stationing the crew. If all crew members are asked to reside at Chandigarh (so that they start from and come back to Chandigarh with minimum halt at Delhi), then waiting time at Delhi for different line connections may be calculated as follows:

Since the crew has layover of more than 4 hours between the bus trips, the layover time between the trip a and trip 1 will be from 12.00

(day) to 5.30 AM (next day), i.e., 17.5 hours. Likewise

Bus route no	Layover times
a ⟶ 2	12 day to 9 AM next day = 21
a ⟶ 3	12 day to 15 =24
a ⟶ 4	12 day to 18.30=6.5
a ⟶ 5	12 day to 0.00=12
b ⟶ 1	13.30 to 05.30=16

and so on.

When the crew is assumed to reside at Chandigarh, the cost matrix is constructed as

Layover times when the crew reside at Delhi

	1	2	3	4	5
a	17.5	21	3	6.5	12
b	16	19.5	1.5	5	10.5
c	12	15.5	21.5	1	6.5
d	4.5	8	14	17.5	23
e	23	2.5	8.5	12	17.5

Similarly, if crew is is assumed to reside at Delhi (so that they start from and come back to Delhi with minimum halt at Chandigarh), then minimum time waiting time at Chandigarh for different bus routes is given in the following table.

Layover times when the crew reside at Chandigarh

	1	2	3	4	5
a	18.5	15	9	5.5	0
b	20	16.5	10.5	7	1.5
c	0	20.5	14.5	11	5.5
d	7.5	4	22	18.5	13
e	13	9.5	3.5	0	18.5

Construct the final cost matrix by taking smaller layout times between bus trips with the help of above two tables, provided the value is more than 4 hours. The final cost matrix is

	1	2	3	4	5
a	17.5	15	9	5.5	12
b	16	16.5	10.5	5	10.5
c	12	15.5	14.5	11	5.5
d	4.5	8	14	17.5	13
e	13	9.5	8.5	12	17.5

The Hungarian method can now be applied for finding the optimal connections which give minimum overall layout times, and hence, the minimum cost of bus service operations. This method is well explained well in Examples 6 and 7, therefore we directly write the out come of the method of assignments. The reduced cost matrix is shown in Table 21. It is obvious that the four assignments can not be made at this

Table 21

	1	2	3	4	5
a	12	9.5	3.5	0	6.5
b	11	11.5	5.5	0	5.5
c	6.5	10	9	5.5	0
d	0	3.5	9.5	13	8.5
e	4.5	1	0	3.5	9

stage. Hence, only four horizontal and vertical lines are needed to cover all zeros, see again Table 21.

Now, subtract 3.5 (minimum of all uncovered zeros) from each uncovered entries and add this quantity to the entries being at the intersection of horizontal and vertical lines. The remaining entries are to be kept unchanged, see Table 22. Now, by the usual method, we get the following optimal assignment table.

From Table 22, we get the following assignments:

Table 22

	1	2	3	4	5
a	8.5	5	⊡0	⊠	3
b	7.5	7	2	⊡0	2
c	6.5	9	9	9	⊡0
d	⊡0	2.5	9.5	16.5	8.5
e	4.5	⊡0	⊠	7	9

Crew	Residence at	Service number	Waiting hours
1	Chandigarh	d1	4.5
2	Delhi	2e	9.5
3	Delhi	3a	9.0
4	Chandigarh	b4	5.0
5	Delhi	5c	5.5

Total minimum waiting time $= 4.5 + 9.5 + 9 + 5 + 5.5 = 33.5 = 33$ hours and 30 minutes.

Problem Set 7

1. Plant superintendent of an industry has to assign four different categories of machines to five type of tasks. The number of machines available in the four categories are 25, 30, 20 and 30. The number of jobs available in four categories are 20, 20, 30, 15, and 20. Machine category 4 can not be assigned to task 4. For the unit costs given below, determine the optimal assignment of machines to tasks.

Task type → Machine category ↓	A	B	C	D	E
1	10	2	3	15	9
2	5	10	15	2	4
3	15	5	14	7	15
4	20	15	13	11	8

2. Suddenly, due to change in the government policy on octroie and entry tax the unit transportation from factory A to city F is decreased to 1 in example of Section 7.3, then explain how the optimal solution is affected (without finding the initial BFS).

3. Solve the following transportation problem under the restriction that the requirement at Destination 2 must be shipped only from Source 3.

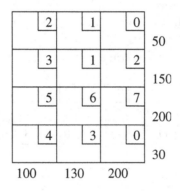

Suggestion. Cross the cells with c_{12}, c_{22}, c_{42}. Allocate 130 in cell c_{32} and find optimal solution for the remaining two columns.

4. If demand at destination F in the example of Section 7.3 must be exactly satisfied, then find the optimal solution under this restriction.

Suggestion. Add the fictitious source to make the problem balanced, and cross the cell with c_{43}. This will not permit dispatch to this cell from the fictitious source.

5. In Kargil, the heavy fighting is going on with terrorists at four hill stations D_1, D_2, D_3 and D_4. The food is to be supplied to

	D_1	D_2	D_3	D_4	a_i
S_1	3	1	2	4	28
S_2	1	4	4	6	23
S_3	2	4	5	2	19
b_j	5	15	20	25	

these stations by three aircrafts S_1, S_2 and S_3. The following cost matrix is given

In this cost matrix, let a_i denote the total number of trips that aircraft S_i can make in one day, b_j denote the trips required at the hill station D_j in one day, c_{ij} denote the amount of food grain that aircraft S_i can to carry to hill station D_j. How many trips by the aircrafts S_i should make to hill stations D_j so that maximum amount of food can be supplied.

6. Consider the following transportation problem

5	9	13	2	50
11	18	20	3	50
14	13	16	3	50
30	40	55	25	

Find the initial BFS using N-W rule, LCM, VAM. Modify the VAM by calculating the penalties as differences of the lowest cost and highest cost for each row and column, and then compute the initial BFS using allocation rules of VAM. Which heuristic gives the better BFS.

7. A transportation model has m sources and n destinations. Given that supply is more than demand. How many basic feasible solutions does this problem has?

8. The following table shows all the necessary information on the available supply from each warehouse, the requirement of each market and the unit transportation cost in each N-W entry. The shipping manager has worked out the following schedule from experience: 12 units from A to II, 1 unit from A to III, 9 units from A to IV, 15 units from B to III, 7 units from C to I and 1 unit from C to III.

	I	II	III	IV	
A	5	2	4	3	22
B	4	8	1	6	15
C	4	6	7	5	8
	7	12	17	9	

(a) Verify whether the manager has made the optimal schedule.

(b) If the manager is approached by a carrier of route C to II who offers to reduce his rate in the hope of getting some business, by how much the rate be reduced by the carrier before the manager should consider giving him business.

(c) If the supply from warehouse B reduces to 12 units and simultaneously the requirement of market III reduces further to 10 units, will the original optimal optimal schedule be affected? Write the optimal solution in this situation.

9. The following is the optimal table of a transportation problem.

		25	
20			10
	20	0	
		5	25

The dual variables corresponding to the destinations are $-1, 2, 1, 3$, respectively and those corresponding to sources are $1, 2, 0, 1$, respectively. Find the optimal transportation cost in two different ways.

10. In the following transportation model, if a unit is not shipped, a storage must incur. Let the storage cost per unit at sources S_1, S_2 and S_3 be 3, 2 and 1, respectively, then find the optimal solution.

	D_1	D_2	D_3	
S_1	2	1	4	20
S_2	3	1	2	40
S_3	5	6	7	50
	30	25	45	

11. A company has plants at A, B and C which have capacities to assemble 300, 200 and 500 cars per day. The assembling cost (per car) in these plants are $70, 60, and 66, respectively. Four retail customers have placed orders on the following pattern

Consumer	Cars required per day	Price offered ($ per car)
I	400	100
II	250	100
III	350	100
IV	150	103

Shipping costs in rupees per car from plants to customers are given in the following table.

	I	II	III	IV
A	3	5	4	6
B	8	11	9	12
C	4	6	2	8

Find the optimal schedule of the above problem.

12. A company dealing in international trading buys no-band textile outlets from China, India and the Philippines, ships to either Singapore or Taiwan for packing and labeling, and then ships to the USA and France for sale. The transportation costs per ton in $ between sources and destinations are given in the following table:

	China	India	Philippines	USA	France
Singapore	50	90	70	150	180
Taiwan	60	95	50	130	200

Suppose the company purchased 60 tons of no-bland from China, 45 tons from India and 30 tons from Philippines. The USA market demands 80 tons of labeled products and the France market 55 tons. Assume that the packaging and labeling do not change the weight of textile products.

(a) If both Singapore and Taiwan have unlimited packaging and labeling capacity, formulate a linear programme to help the company minimize the shipping cost.

(b) If Singapore can process at most 60 tons of no-bland, what will be change in your formulation?

(c) If Singapore can process at most 60 tons of no-bland and Taiwan can process at most 75 tons, what will happen to your formulation?

(d) Under condition (c), try to reduce the linear program to two independent transportation problems.

13. An industry has three assembly plants for scooters at cities 1, 2 and 3. The scooters are shipped to retail customers 6, 7 and 8 from storage centers located at cities 4 and 5. The shipping cost per scooter (in $100) are shown on the arcs of the following network.

(a) Solve the problem as transshipment model, and find the optimal solution.

(b) Suppose that the distribution centre 4 can sell 200 scooters in the local market directly to the customers. Find the new optimal solution.

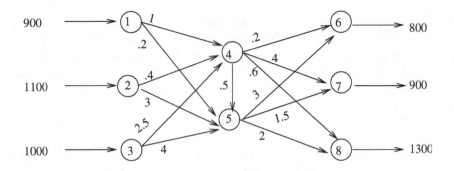

14. There are six skilled workers and six machines. In the following matrix the efficiency of each worker to work on each machine is given. Formulate the problem so that maximum efficiency is utilized.

	M_1	M_2	M_3	M_4	M_5	M_6
W_1	5	5	7	4	8	1
W_2	6	5	8	3	7	2
W_3	6	8	9	5	10	3
W_4	7	6	6	3	6	4
W_5	6	7	10	6	11	5
W_6	3	2	9	6	5	7

15. Find the minimum number of man hours required to complete the jobs J_1, J_2, J_3 and J_4 by persons M_1, M_2, M_3 and M_4 under the restriction that job J_3 can not be assigned to person M_2. The cost matrix is given as

	J_1	J_2	J_3	J_4
M_1	4	4	10	5
M_2	6	6	13	8
M_3	5	8	2	2
M_4	9	13	7	7

16. The secretary of a school is taking bids on the city's four school bus routes. Four companies have made the bids (in rupees) as

Routes

	1	2	3	4
1	4000	5000	—	—
2	—	4000	—	4000
3	3000	—	2000	—
4	—	—	4000	5000

Companies

detailed in the following table: Suppose each bider can be assigned only one route. Use the assignment model to minimize the school's cost of running the four bus routes.

17. In the modification of a plant layout of a factory four new machines M_1, M_2, M_3, M_4 are to be installed in a workshop. There is provision at five places A, B, C, D, E for installation. Because of the space limitations the machine M_2 can not be installed at the position A. The cost of installation machine i at jth place are shown below.

Locations

	A	B	C	D	E
M_1	9	11	15	10	11
M_2	—	11	7	10	9
M_3	13	11	16	14	15
M_4	14	8	19	7	13

Machines

Suggestion. This is unbalanced assignment problem. Introduce one fictitious machine in fifth row and assign zero cost to each cell in this row. Now, the problem is amenable to the Hungarian algorithm.

18. A cycle company has four distribution centres and four salesmen to operate at these centres. The distribution centres are

not having equal potential for sales in the open market. It is estimated that a typical salesman operating in each distribution centre would bring the following annual revenue

Distribution centre	I	II	III	IV
Annual sales ($)	126000	105000	84000	63000

The four salesman also differ in their ability. It has been observed that, working under the same conditions, their annual sales would be proportional as follows:

Salesman	A	B	C	D
Proportion	7	5	5	4

If the criteria is to maximize total sales, then the intuitive answer is to assign the best salesman to richest distribution centre, the next best salesman to the second richest, and so on, verify this answer by assignment by assignment technique.

Suggestion To avoid fraction values of annual sales of each salesman at each distribution centre, for convenience consider their annual sales as 21 (the sum of all proportions), taking $1,000 as one unit. The following cost matrix is obtained

		Distribution centre				
		I	II	III	IV	Sales proportion
	A	42	35	28	21	7
	B	30	25	20	15	5
Salesman	C	30	25	20	15	5
	D	24	20	16	12	4
Sales (in $1,000)		6	5	4	3	

Solve the problem as a maximization assignment problem.

19. A small airline company, owing five planes operates on all seven days of a week. Flights between three cities A, B and C according to the following schedule is given below.

Flight no.	From	Departure time (hours)	To	Arrival time (hours)
1	A	09,00	B	12.00
2	A	10.00	B	13.00
3	A	15.00	B	18.00
4	A	20.00	C	Midnight
5	A	22.00	C	02.00
6	B	04.00	A	07.00
7	B	11.00	A	14.00
8	B	15.00	A	18.00
9	C	07.00	A	11.00
10	c	15.00	A	19.00

Find how the planes should be assigned to the flights so as to minimize the total layover cost. It is assumed that that the layover cost on a route in both directions is equal to the sum of the squares of the layover times between the flights.

20. An airline that operates seven days a week has time-table as shown below. Crew must have minimum layover of 5 hours between the flights. Obtain the pairing of flights that minimizes the layover time away from home. For any given pairing, the crew will be based at the city that results in smaller layover.

Delhi-Jaipur

Flight no.	Departure	Arrival
1	7.00AM	8AM
2	8.00AM	9.00AM
3	1.30PM	2.30PM
4	6.30PM	7.30PM

Jaipur-Delhi

Flight no.	Departure	Arrival
101	8.00AM	9.15AM
102	8.30AM	9.45AM
103	12.00 Noon	1.15PM
104	5.30PM	6.45PM

For each pair also mention the town where the crew should be based.

21. **Multi-dimensional Transportation Problem.** A shoe manufacturer has ℓ factories in various parts of the country. Each of the ℓ plants can manufacture m different types of shoes. The shoes are to be transported from the factories to n different areas. The following information is available.

a_{ik} = required number of units to be transported from the factory i to the area k;

b_{jk} = required number of units of type j to be transported to the area k;

d_{ij} = number of units of type j available at the factory i;

x_{ijk} = amount of the jth type made in the ith plant transported to the kth area;

c_{ijk} = cost of transportation of one unit of the jth type from the ith plant to the kth area $x_{ijk} \geq 0$.

Formulate the problem and state the procedure to solve it.

Chapter 8

Network Analysis

This chapter is devoted to different type of networks which occur in real life problems. The analysis of these networks have been done from different angles. The main topics are the Minimal spanning tree algorithm, the Shortest route problem, and the Maximum flow problem.

8.1 Introduction to Networks

A network consists of a set of nodes linked by arcs. In Chapter 6, the transportation and assignment are also network problems. Here, we study the network in more generalized context. First, we define certain terms in reference to networks. The simple notation for a network is (N, A), where N is the number of nodes and A is the set of arcs. For example, in Fig 8.1:

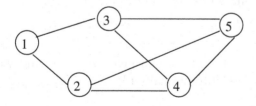

Figure 8.1

$N = \{1, 2, 3, 4, 5\}$, $A = \{(1, 3), (1, 2), (2, 4), (2, 5), (3, 4), (3, 5), (4, 5)\}$.

Each arc is assigned a capacity which may be flow, cost, distance etc. The capacity of an arc is finite or infinite. An arc is directed (oriented) if it allows positive flow in one direction and zero flow in the opposite direction. Hence, by a directed network we mean that all arcs are directed.

Again, we define a path as a sequence of distinct arcs that join two nodes of the flow irrespective of the direction of flow in each arc. A path is said to be cyclic if it connects a node to itself. In Fig. 8.1, arcs $(3,4)$, $(4,5)$ and $(5,3)$ form a closed loop (cycle). In particular, a directed cycle is a loop in which all arcs are directed in the same direction. Nodes which are connected without loop are said to be acyclic.

A connected network means that any pair of distinct nodes can be connected by at least one path. A tree is a connected network without loop. In the next section, we shall analyze such type of networks.

8.2 Minimal Spanning Tree Algorithm

This algorithm calculates the minimum distance between nodes connected directly or indirectly. Such type of problems arise when cable network is to be set up for certain areas in a town or city. Here, we are interested that each area is connected, and the length of the cable is minimum. Another famous example is about the construction of paved roads between villages so that each village is connected by the paved road, and distance is minimum, i.e., such linkage is economical.

Let $\{1, 2, \ldots, N\}$ be the set of nodes of the network. Define

C_k = Set of all nodes that have been permanently connected at iteration k.

$\overline{C}_k$ = Set of all nodes yet to be connected.

The steps of the algorithm proceed as follows.

Step 1. Start with any node i from given n nodes and write $C_1 = \{i\}$ which renders automatically $\overline{C}_1 = n - \{i\}$.

Set $k = 2$.

Step k. Select a node j, in the unconnected set $\overline{C}_{k-1}$ that gives the shortest distance to a node in C_{k-1}. Include node j permanently to C_{k-1}, and remove it from $\overline{C}_{k-1}$ to have

$$C_k = C_{k-1} + \{j\}, \qquad \overline{C}_k = \overline{C}_{k-1} - \{j\}$$

If the set of unconnected nodes $\overline{C}_k$ is empty, stop. Otherwise, set $k = k + 1$, and repeat the above step.

Example 1. A TV company is requested for providing cable services to six new housing development areas. Figure 8.2 depicts the potential TV linkage. The cable length in miles is mentioned on each branch.

Our objective is to determine the most economical network.

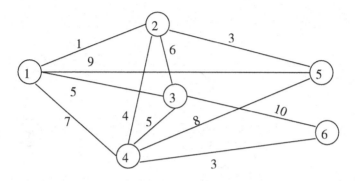

Figure 8.2

The algorithm starts at node 1 (any other node will do as well), which gives

$$C_1 = \{1\}, \qquad \overline{C}_1 = \{2, 3, 4, 5, 6\}.$$

For second iteration, choose the node from $\overline{C}_1$ which is nearest to 1. This is node 2. Hence, connect node 1 to 2 by dashed line, see Figure 8.3, and construct the sets

$$C_2 = \{1, 2\}, \qquad \overline{C}_2 = \{3, 4, 5, 6\}.$$

For third iteration, choose the node from $\overline{C}_2$ which gives the shortest distance to 1 or 2. This is node 5 which is nearest to 2. Hence, connect node 2 to 5 by dashed line, and write the sets

$$C_3 = \{1, 2, 5\}, \qquad \overline{C}_3 = \{3, 4, 6\}.$$

Similarly, connect node 2 to 4 by dashed line, and construct the sets

$$C_4 = \{1, 2, 5, 4\}, \qquad \overline{C}_4 = \{3, 6\}.$$

Now, connect node 6 to 4 by dashed line, and write

$$C_5 = \{1, 2, 5, 4, 6\}, \qquad \overline{C}_5 = \{3\}.$$

In the last iteration two branches are there which give the same short-est distance, i.e., connect node 3 to 4 or 3 to 1 by dashed line. Thus,

$$C_6 = \{1, 2, 5, 4, 6, 3\}, \qquad \overline{C}_6 = \emptyset.$$

Now dashed line structure gives the shortest length of the cable.

$$\text{Optimal length} = 1 + 3 + 4 + 3 + 5 = 16 \text{ miles.}$$

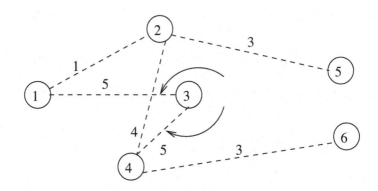

Figure 8.3

Matrix method. The above algorithm can easily be carried out on the distance matrix. Execute the following steps.

Step 1. Represent the network by distance matrix. If two nodes are not connected directly, then take ∞ in the cell concerned.

Step 2. Let Q be initially a null set whose elements at later stage will be taken as the row numbers.

Step 3. Select row one and include this in Q. Then delete Column 1 of the distance matrix.

Step 4. Find the smallest entry in all the rows contained in Q and box it. In case of tie, break it arbitrarily.

Step 5. Identify the column of boxed entry of Step 4, and let let it be K. Include row K in Q.

Step 6. Delete column K identified in Step 5.

Step 7. Check whether all the columns of the matrix are deleted. If so, go to Step 8, otherwise go to Step 4, i.e., find the smallest entry of the rows contained in Q and box it.

tep 8. Show the arcs in the spanning tree corresponding to the boxed entries and total of the boxed entries gives the minimum distance required for spanning tree.

Remark. We may start with any row in Step 3.

Let us rework out the above example by matrix method.

Write the distance matrix from the network as shown in Table 1, initially $Q = \emptyset$.

Table 1

	1	2	3	4	5	6
1	—	1	5	7	9	∞
2	1	—	6	4	3	∞
3	5	6	—	5	∞	10
4	7	4	5	—	8	3
5	9	3	∞	8	—	∞
6	∞	∞	10	3	∞	—

eration 1. Set $Q = \{1\}$ (any other node will work well) and write the distance matrix after deleting the first column, see Table 2. Star first row. Each entry of Q will be shown starred. The smallest entry in 1*-row is 1. Box it.

Table 2

	2	3	4	5	6
1*	[1]	5	7	9	∞
2	—	6	4	3	∞
3	6	—	5	∞	10
4	4	5	—	8	3
5	3	∞	8	—	∞
6	∞	10	3	∞	—

Iteration 2. The boxed entry of Iteration 1 is in 1*-row. This is in second column. For Iteration 2, delete the second column of Table 1 and star second row, see Table 3(a). The smallest entry in rows 1* and 2* is 3 and box it. $Q = \{1, 2\}$.

Table 3

(a)

	3	4	5	6
1*	5	7	9	∞
2*	6	4	[3]	∞
3	—	5	∞	10
4	5	—	8	3
5	∞	8	—	∞
6	10	3	∞	—

(b)

	3	4	6
1*	5	7	∞
2*	6	[4]	∞
3	—	5	10
4	5	—	3
5*	∞	8	∞
6	10	3	—

(c)

	3	6
1*	5	∞
2*	6	∞
3	—	10
4*	5	[3]
5*	∞	∞
6	10	—

(d)

	3
1*	5
2*	6
3*	—
4*	[5]
5*	∞
6*	10

Iteration 3. The boxed entry of Iteration 2 is in 2*-row and falls in Column 5. For executing Iteration 3 delete Column 5 of Table 3(a) and star fifth row, see Table 3(b). The smallest entry in all starred rows is 4. Box 4. $Q = \{1, 2, 5\}$

Iteration 4. The boxed entry of Iteration 3 is in 2*-row and falls in fourth column. To perform Iteration 4, delete Column 4 of Table 3(b) and star row 4, see Table 3(c). The smallest entry in starred rows (done so far) is 3. Box it. $Q = \{1, 2, 5, 4\}$.

Iteration 5. The boxed entry of previous iteration is in sixth column and hence delete the sixth column star sixth row to perform the next iteration, see Table 3(d). The smallest entry in starred rows is available at two positions, i.e., row 1* and row 4*. Thus there is a tie. Break the tie arbitrarily by taking row 4 and box the smallest entry which is 5, see Table 3(d). $Q = \{1, 2, 5, 4, 6\}$

Iteration 6. After Iteration 5, all columns have been deleted and algorithm stops, and $Q = \{1, 2, 5, 4, 3\}$. To write the minimum distance of the spanning tree add all boxed entries, and connect nodes corresponding to them. This will give minimum spanning tree. The sum of boxed entries is $1 + 3 + 4 + 3 + 5 = 16$. For spanning tree, see Fig. 8.3.

Remark. The tie in Iteration 5 indicates that there exists an alternate minimal spanning tree.

8.3 Shortest Path Problem

In this section, we analyze the minimum distance between two specified nodes connected through various feasible paths, directly or indirectly. The network may be cyclic or acyclic. To solve such type of problems two algorithms are in common use, viz.,, (i) Dijkstra's algorithm, (ii) Floyd's algorithm.

Dijkstra's algorithm. A network is given with different nodes connected directly or indirectly. This algorithm finds shortest distance between a source (given) and any other node in the network. The algorithm advances from a node i to an immediately successive node j using a labeling procedure. Let u_i be the shortest distance from node 1 to node i, and $d_{ij}(\geq 0)$ be the length of (i, j)th arc. Then, the label for node j is defined as

$$[u_j, i] = [u_i + d_{ij}, i], \quad d_{ij} \geq 0.$$

Here label $[u_j, i]$ means we are coming from node i after covering a distance u_j from the starting node. The node label in this algorithm are of two types: temporary and permanent. A temporary label can be replaced with another temporary label, if shorter path to the same node is detected. At the stage when it is certain that no better route can be found, the status of temporary node is changed to permanent.

The steps of the algorithm are summarized as follows.

Step 1. Label the *source* node (say node 1) with the permanently level $[0, -]$. Set $i = 1$.

Step i. (a) Compute the temporary labels $[u_i + d_{ij}, i]$ for each node j that can be reached form i, provided j is not permanently labeled. If node j is already labeled as $[u_j, k]$ through another node k, and if $u_i + d_{ij} < u_j$, replace $[u_j, k]$ with $[u_i + d_{ij}, i]$.

(b) If all the nodes have permanent labels, stop. Otherwise, select the label $[u_r, s]$ with shortest distance $(= u_r)$ from among all the temporary labels (break the tie arbitrarily). Set $i = r$ and repeat step i.

Remark. At each iteration among all temporary nodes, make those nodes permanent which have smallest distance. Note that at any iteration we can not move to permanent node, however, reverse is possible. After all the nodes have been label and only one temporary node remains, make it permanent.

Example 2. In the following network, five towns are connected through permissible routes. This distance in miles between any towns is given on the arc connected these towns. Find the shortest distance between town 1 and any other town.

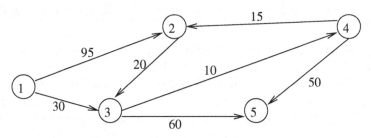

Iteration 0. Assign the permanent label $[0, -]$ to node 1.

Iteration 1. Nodes 2 and 3 can be reached from (the last permanently labeled) node 1. Thus, label $[95, 1]_{(1)}$ and $[30, 1]_{(1)}$ to nodes 2 and 3, respectively. Since $[30, 1] < [95, 1]$, make label $[30, 1]_{(1)}$ as permanent node, see Figure 8.4. The superscript (1) denotes the number of iterations.

Iteration 2. Nodes 4 and 5 can be reached from node 3, and hence label $[40, 3]_{(2)}$ and $[90, 3]_{(2)}$ to nodes 4 and 5, respectively. Since $[40, 3] < [90, 3]$, make label $[40, 3]_{(2)}$ permanent.

Iteration 3. So far, nodes 2 and 5 are temporary. Node 2 can also be reached from permanent node 4 and its label is changed to $[55, 4]_{(3)}$. Since no other permanent node exists from where we can reach at node 2 and $[55, 4] < [95, 2]$, make label $[55, 4]_{(3)}$ as permanent. Note that node 5 can also be reached from 4 with label $[90, 4]_{(3)}$.

Iteration 4. Only node 3 can be reached from node 2, but node 3 is already labeled permanently, it can not be re-labeled. The only temporary node is 5 and this does not lead to any other node, its status is changed to permanent.

The shortest route between the source node 1 and other node in the network is determined by starting at the desired destination and backtracking through the nodes using the information given by permanently labels, see Fig. 8.4. For example at node 2 we reach as

$$2 \longleftarrow 4 \longleftarrow 3 \longleftarrow 1$$

The desired route is $1 \longrightarrow 3 \longrightarrow 4 \longrightarrow 2$ with a total distance of 55 units.

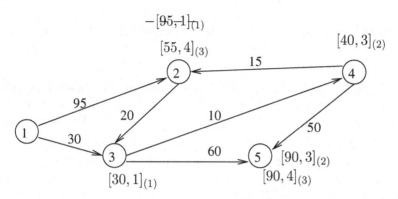

Figure 8.4

Remark. It is advised to work on all iterations on the network.

Floyd's algorithm. This is an efficient, simple programmed, and widely used algorithm that finds the shortest path between all nodes, all at once, while in Dijkstra algorithm we have to specify in advance at least least one node. The algorithm works on n-node network as a square matrix with n rows and n columns. Entry (i, j) of the matrix gives the distance d_{ij} from node i to node j, which is finite if i is connected directly to j, and infinite otherwise.

The logic of Floyd's algorithm is simple. Given three nodes i, j and k, see Fig. 8.5 with the weights (distances) shown on the three arcs. It is shorter to reach k from i passing through j if

$$d_{ij} + d_{jk} < d_{ik}$$

In this situation, it is optimal to replace the direct route from $i \longrightarrow k$ with the indirect route $i \longrightarrow j \longrightarrow k$. This **triangle operation** is applied to the network in a systematic way.

The Floyd algorithm is based on the following proposition.

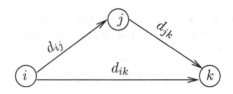

Figure 8.5

Proposition. If we perform a triangle operation for successively values $k = 1, 2, \ldots, n$, each entry d_{ij} becomes equal to the length of the shortest path from i to j, assuming the weights $c_{ij} \geq 0$.

Proof. We shall show by induction that after the triangle operation for $j = j_0$ is executed, d_{ik} is the length of the shortest path from i to k with intermediate nodes $v \leq j_0$, for all i and k. The reasoning for $j_0 = 1$ is clear.

Assume then that the inductive hypothesis is true for $j = j_0 - 1$. The proof by induction is complete, if it is true for $j = j_0$. Consider the triangle operation for $j = j_0$

$$d_{ik} = \min\{d_{ik}, d_{ij_0} + d_{j_0 k}\}.$$

If the shortest path from i to k with intermediate nodes $v \leq j_0$ does not pass through j_0, d_{ik} will remain unchanged by this operation, the first argument in min-operation will be selected, and d_{ik} will satisfy the inductive hypothesis.

On the other hand, if the shortest path from i to k with intermediate nodes $v \leq j_0$ does pass through j_0, d_{ik} will be replaced by $d_{ij_0} + d_{j_0 k}$. By assumption, d_{ij_0} and $d_{j_0 k}$ are both optimal distances with intermediate nodes $v \leq j_0 - 1$, so $d_{ij_0} + d_{j_0 k}$ is optimal with intermediate nodes $v \leq j_0$. This completes the induction.

The steps of the algorithm are performed as follows.

Step 0. Define the staring distance matrix D^0 and node sequence matrix S^0 as given subsequently. The diagonal elements are given are marked with $(-)$ to indicate that they are blocked. Set $k = 1$.

Step k. Define row k and column k as pivot row and pivot column. Apply the triangle operation to each entry d_{ij} in D^{k-1}, for all i and j, if the condition

$$d_{ik} + d_{kj} < d_{ij} \ (i \neq k, j \neq k, \ i \neq j)$$

is satisfied, make the following changes:

(a) Construct D^k by replacing d_{ij} in D_{k-1} with $d_{ik} + d_{kj}$;

(b) Construct S^k by replacing s_{ij} in S_{k-1} with k.

Set $k = k + 1$, and repeat step k till no changes are given by triangle operation.

Step k of the algorithm is well illustrated by representing D^{k-1} in Figure 8.6.

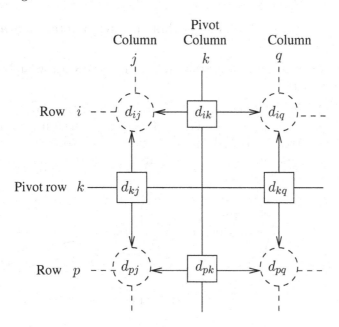

Figure 8.6

Here, row k and column k define the current pivot row and column. Row i represents any of the rows $1, 2, \ldots$, and $k - 1$ and row p represents any of the rows $k + 1, k + 2, \ldots$, and n. Similarly, column j represents any of the columns $1, 2, \ldots, k - 1$, and column q represents any of the columns $k + 1, k + 2, \ldots$, and n. The triangle operation is applied as follows.

If the sum of the entries on the pivot row and pivot column (shown by squares) is smaller than the associated intersection element (shown by circle), then it is optimal to replace the intersection distance by the sum of the pivot distances.

Thus, after n steps, we can determine the shortest route between nodes i and j from the matrices D^n and S^n using the following rules:

1. From D^n, d_{ij} gives the shortest distance between nodes i and j.

2. From S^n, determine the intermediate node $k = s_{ij}$ which yields the root $i \longrightarrow k \longrightarrow j$. If $s_{ik} = k$ and $s_{kj} = j$, stop; all the intermediate nodes of the roots are detected. Otherwise repeat the procedure between nodes i and k, and nodes j and k.

Remarks. 1. The Floyd's algorithm requires a total of $n(n-1)^2$ comparisons.

2. At each iteration the entries on the current pivot row and column are left unaffected.

Example 3. For the network in Figure 8.7, find the shortest routes between every pair of nodes. The distances (in miles) are mentioned on each arc. Arcs $(2,5)$ and $(4,5)$ are directional so that no passage is permitted from node 5 to nodes 2 and 4. All the other arcs allow passage in both directions.

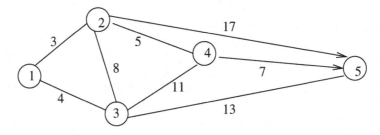

Figure 8.7

Iteration 0. The matrix D^0 and S^0 give the initial net work. D^0 is symmetric except that $d_{52} = \infty$ and $d_{54} = \infty$, because no passage is allowed from node 5 to nodes 2 and 3.

Iteration 1. In first iteration (k=1), the first row and first column are pivoted (shaded) as shown in D^0-matrix. Apply triangle operation on every entry (except entries on the shaded region). For doing this, select any entry and find the entries of the corresponding row and column of shaded region and take their sum if it is smaller than the selected entry, then box it so that it may be replaced

D⁰

	1	2	3	4	5
1	—	3	4	∞	∞
2	3	—	8	5	17
3	4	8	—	11	13
4	∞	5	11	—	7
5	∞	∞	13	∞	—

S⁰

	1	2	3	4	5
1	—	2	3	4	5
2	1	—	3	4	5
3	1	2	—	4	5
4	1	2	3	—	5
5	1	2	3	4	—

by this lower sum, leave other unchanged. For example, entry $c_{23} = 8$ has 3 and 4 in row and column, respectively of shaded region and their sum is 7 which is smaller than 8. Box it so that before applying next iteration it is reduced to 7. The only entries affected by triangle operation are c_{23} and c_{32}. These have been boxed, and the corresponding changes are incorporated in D^1. Because this is Iteration 1, the corresponding entries of node matrix S^0 are replaced by 1.

D¹

	1	2	3	4	5
1	—	3	4	∞	∞
2	3	—	7	5	17
3	4	7	—	11	13
4	∞	5	11	—	7
5	∞	∞	13	∞	—

S¹

	1	2	3	4	5
1	—	2	3	4	5
2	1	—	1	4	5
3	1	1	—	4	5
4	1	2	3	—	5
5	1	2	3	4	—

Iteration 2. Set k=2 after incorporating the changes of the first iteration, see D^1 and S^1. The second row and column, being pivoted are shaded. Apply the triangle operation for all entries in unshaded region, and make corresponding changes in node matrix to have matrices for next iteration.

Iteration 3. Set $k = 3$ after incorporating the changes due to Iteration 2, the third row and column are pivoted. Make shaded these as shown by matrix D^2. Apply the triangle operations in D^2 and make corresponding changes in S^1 to get matrices D^2 and S^2.

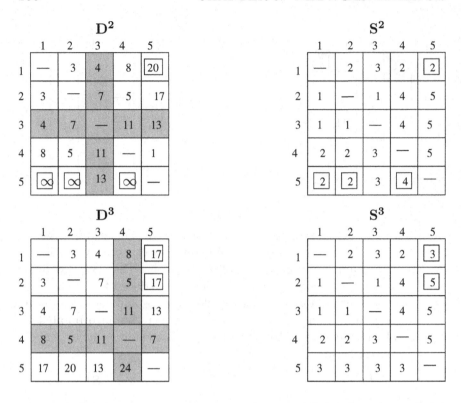

D^2

	1	2	3	4	5
1	—	3	4	8	20
2	3	—	7	5	17
3	4	7	—	11	13
4	8	5	11	—	1
5	∞	∞	13	∞	—

S^2

	1	2	3	4	5
1	—	2	3	2	2
2	1	—	1	4	5
3	1	1	—	4	5
4	2	2	3	—	5
5	2	2	3	4	—

D^3

	1	2	3	4	5
1	—	3	4	8	17
2	3	—	7	5	17
3	4	7	—	11	13
4	8	5	11	—	7
5	17	20	13	24	—

S^3

	1	2	3	4	5
1	—	2	3	2	3
2	1	—	1	4	5
3	1	1	—	4	5
4	2	2	3	—	5
5	3	3	3	3	—

Iteration 4. Set $k = 4$, as shown by the shaded region in D^3. The new matrices are D^3 and S^3. Apply triangle operation to get matrices D^4 and S^4 for next iteration.

D^4

	1	2	3	4	5
1	—	3	4	8	15
2	3	—	7	5	12
3	4	7	—	11	13
4	∞	5	11	—	7
5	17	20	13	24	—

S^4

	1	2	3	4	5
1	—	2	3	2	4
2	1	—	1	4	4
3	1	1	—	4	5
4	2	2	3	—	5
5	3	3	3	3	—

Iteration 5. Set $k = 5$, as shown in shaded region of D^4. By applying triangle operation, we observe that no further improvements are possible. Thus, matrices D^4 and S^4 are final and have all the information which we need.

To determine the shortest route between any two nodes, recall that a segment (i, j) represents a direct link only if $s_{ij} = j$, otherwise i and j are linked through at least one intermediate node.

For instance, suppose we are interested to find the shortest route between nodes 1 and 5. Note that $s_{15} = 4$. This means 1 and 5 are not directly connected. These are to be connected through node 4. Again, see

$$s_{14} = 2, \quad s_{45} = 5$$

The partial route is $1 \longrightarrow 4 \longrightarrow 5$, i.e., node 4 is connected directly, but $s_{14} \neq 4$ implies that 1 and 4 are not connected directly, and there is at least one intermediate node 2 to connect 1 and 4. Again, note

$$s_{12} = 2, \quad s_{24} = 4.$$

Stop, because 1 and 2 are connected directly and 2 and 4 are connected in the same manner. Hence the desired route between 1 and 5 is

$$1 \longrightarrow 2 \longrightarrow 4 \longrightarrow 5, \text{ minimum distance } d_{15} = 15 \text{ miles}$$

Following the above reasoning we can find route between any other two nodes.

8.4 Maximal Flow Problem

This is a very famous problem which is closely associated with the determination of optimal solution of a transportation problem. Originally, the problem was invented by Ford and Fulkerson who gave its solution by the labeling algorithm. We have two nodes called the sink and source. From sink to source some fluid flows through intermediate nodes. The capacity of an arc here is assigned as the rate of flow, and the rate of flow in each arc may be unidirectional or bidirectional.

Suppose there are two wells which supply crude oil to the refineries. The intermediate nodes are boosters and pipelines. Our objective is find the maximum flow in the network. For this we need one combined net work. This requirement is satisfied if we assume infinite capacity to supply at wells and infinite demand at refineries.

Assume that the capacity of each branch is known in both directions. Let (i, j), $i < j$ be an arc. Define c_{ij} as the rate of flow in the direction from i to j and d_{ji} is the rate of flow in opposite direction, see Fig. 8.8.

Figure 8.8

Algorithm. The algorithm presented here is the labeling technique. The idea of maximum flow algorithm is to find a breakthrough path with positive net flow that connects the source and sink nodes. Take an arc (i, j) with initial capacities (c_{ij}, d_{ji}). As the computations of the algorithm proceed, portions of these capacities will be committed to the flow in the arc. The excess capacities of the arcs are then changed accordingly. For excess capacities on the arc (i, j), we use the notation (c_{ij}, d_{ji}). The network with the updated excess capacities will be referred to as the residual network. Define

$$[a_j, i] = \text{Flow } a_j \text{ from node } j \text{ to node } i.$$

The source node is numbered 1 and the algorithm proceeds as follows:

Step 1. Let the index j refer to all nodes that can be reached directly from source node 1 by arcs with positive excess capacities, i.e., $c_{1j} > 0$ for all j. On the diagram of the network, we label nodes j with two numbers $[a_j, 1]$, where a_j is the positive excess capacity, and 1 means we have come from node 1. If in doing this we label the sink N, so that there is a branch of positive excess capacity from source to the sink, then the maximum flow along the path is given by $f_1 = c_{1N}$, and the excess capacity due to this breakthrough path is determined by f_1 in the direction of the flow and is increased by f_1 in the reverse direction. This means that for source nodes 1 and sink node N the excess flow is changed from the current (c_{1N}, d_{N1}) to

$$(c_{1N} - f_1, d_{N1} + f_1)$$

Step 2. In case in Step 1, the sink is not labeled, Choose the smallest index j of the labeled nodes and search the unlabeled nodes which can be reached from j by arcs of positive excess capacities. If there are no such nodes we move to the next lowest index j and repeat the process. If sink is labeled, we immediately compute

$$f_1 = \text{minimum of the excess capacities on the path to the sink.}$$

Subtract f_1 from excess capacity on the arc in the direction of path and add f_1 from the excess capacity in reverse direction. In this way we get fresh excess capacities. Even, if the sink is not labeled, some unlabeled nodes (other than sink) can be reached, then using the general index k, we label each as follows $[a_k, j]$, and compute f_1.

tep 3. Steps 1 or 2 give first breakthrough. Compute freshly excess capacities of all arcs which are changed due to first breakthrough.

The process is repeated from Step 1 to 3 until, in a finite number of steps, we reach the state so that no additional nodes can be labeled to reach sink. This is no breakthrough. The maximum flow is computed by

$$f = f_1 + f_2 + \cdots + f_p,$$

where p is the number of iterations to get no breakthrough.

The optimal flow in the arc (i, j) is computed as

$$(\alpha, \beta) = (c_{ij} - d'_{ij}, c_{ji} - d'_{ji}),$$

where c_{ij} and c_{ji} are the initial capacities, and d'_{ij} and d'_{ji} are the final excess capacities. If $\alpha > 0$, the optimal flow from i to j is α. Otherwise, if $\beta > 0$, the optimal flow from j to i is β. Note that α and β can not be positive together.

Remark. During labeling process, we are not worried about whether the orientation of the arc is in the direction of our move form j to k. We need that the arc must have positive residual.

The algorithm is illustrated by working out an example of maximum problem flow problem.

Example 4. Consider the network in Figure 8.9. The bidirectional capacities are shown on the respective arcs. For example, for arc $(3, 4)$, the flow limit is 10 units from 3 to 4 and 5 units from 4 to 3. Determine the maximum flow in this network between source 1 and sink 5.

The algorithm is applied in the following manner.

teration 1. At the first step, find the nodes that can be reached directly from the source by arcs of positive excess capacity $c_{ij} > 0$. These nodes are $2, 3, 4$. Label these nodes with the ordered pair of numbers $[a_j, 1]$, where $a_j = c_{1j}$ and 1 means we have reached

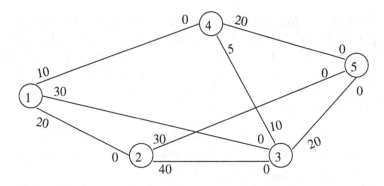

Figure 8.9

from node 1. For instance, node 2 is labeled $[20, 1]$ and likewise 3 and 4 as $[30, 1]$ and $[10, 1]$, respectively. Still sink is not labeled. Thus, select a node with lowest index from the labeled nodes, i.e., node 2. Again, find nodes that can be reached from node 2 by arcs of positive excess capacity. These nodes are $3, 5$. Ignore node 3 that has already been labeled and label sink 5 by the ordered pair $[30, 2]$. Now, sink is reached and labeling process stops as we have got first breakthrough. The flow in the network can be increased by

$$f_1 = \min\{20, 30\} = 20.$$

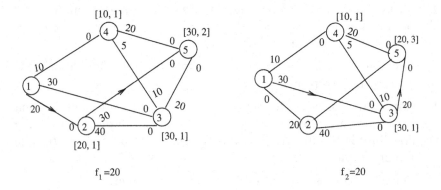

The value of f_1 indicates that increase of 20 units can be made along the path traced out in a move from source to sink. We can easily work backward to find the path. The label on the sink shows that we came from node 2. From node 2, it is seen that we came from node 1. The path is $1 \longrightarrow 2 \longrightarrow 5$. Before starting the second iteration, we first set the network showing the effect of first breakthrough. To have fresh excess capacities, subtract $f_1 = 20$ from the direction of flow in the path, and add this amount in the opposite direction. Compute the residual capacities of each arc on the path. After first iteration, arc $(1, 2)$ has residual in direction of flow $20 - 20 = 0$, and has in opposite direction $0 + 20 = 20$. Similarly, arc $(2, 5)$ has 10 and 20 in direction of flow and in opposite direction.

ration 2. Again, start from source node 1 and find the nodes that can be reached directly from the source by arcs of excess capacities. This time, we can not go to node 2 from source node 1. The nodes $3, 4$ can be reached from source 1. Label these nodes and note that sink is still not reached. Choose lowest labeled index 3 and find nodes that can be reached from 3. These are nodes $2, 4, 5$. But node $2, 4$ can not reached as arc $(3, 2)$ has no excess capacity in the direction $3 \longrightarrow 2$ and node 4 is already labeled. Label sink 5. Now the augmented path is

$$1 \longrightarrow 3 \longrightarrow 5, \quad f_2 = \min\{30, 20\} = 20.$$

Again, before starting third iteration write fresh excess capacities affected due to second breakthrough path.

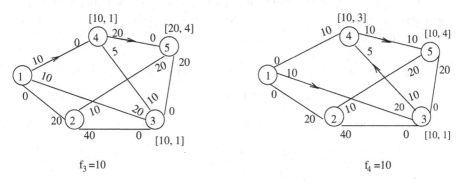

$$f_3 = 10 \qquad\qquad f_4 = 10$$

ration 3. In Iteration 3, we label $3, 4$. Choose lowest index from 3 and 4. This is 3. Now, nodes $2, 4, 5$ can not be reached from 3, since 4 is already labeled and $(3, 2)$ and $(3, 5)$ have no excess capacities. Move to higher index 4. Nodes $3, 5$ can be reached from 4. But 3 is already labeled. So label sink 5 and the augmented path is

$$1 \longrightarrow 4 \longrightarrow 5, \quad f_3 = \min\{10, 20\} = 10.$$

Iteration 4. Similarly, for the renewed network after iteration 4, the augmented path is

$$1 \longrightarrow 3 \longrightarrow 4 \longrightarrow 5, \quad f_4 = 10.$$

More iterations are not possible after iteration 4 as there is no way out to reach at sink from source. The maximum flow is

$$f = f_1 + f_2 + f_3 + f_4 = 20 + 20 + 10 + 10 = 60.$$

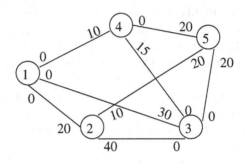

No breakthrough path

The optimal flow in different arcs is computed by subtracting the last excess capacities (when no breakthrough path is available) from the initial excess capacities. This is done in the table:

Arc	$(c_{ij}, d_{ji}) - (c_{ip}, d_{pi})$	Flow amount	Direction
$(1,2)$	$(20,0) - (0,20) = (20,-20)$	20	$1 \to 2$
$(1,3)$	$(30,0) - (0,30) = (30,-30)$	30	$1 \to 3$
$(1,4)$	$(10,0) - (0,10) = (10,-10)$	10	$1 \to 4$
$(2,3)$	$(40,0) - (40,0) = (0,0)$	0	$2 \to 3$
$(2,5)$	$(30,0) - (10,20) = (20,-20)$	20	$2 \to 5$
$(3,4)$	$(10,5) - (0,15) = (10,-10)$	10	$3 \to 4$
$(3,5)$	$(20,0) - (0,20) = (20,-20)$	20	$3 \to 5$
$(4,5)$	$(20,0) - (0,20) = (20,-20)$	20	$4 \to 5$

The solution of the problem is shown by the following network.

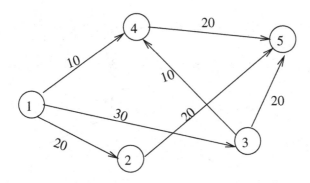

Optimal flow in different arcs

The arrows show the direction of flow and the number on each arc gives the optimal flow in that arc.

Problem Set 8

1. Determine (i) Minimal spanning tree; (ii) Shortest route from node 1 to node 7 using Dijkstra algorithm for the following network.

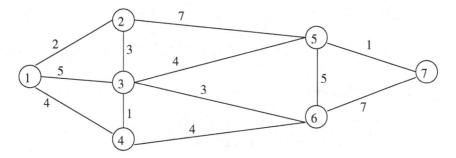

2. Apply Floyd's algorithm to the network depicted below. Determine the shortest distance between the following pairs of nodes.

   (a) From node 1 to node 7.

   (b) From node 7 to node 1.

   (c) From node 2 to node 7.

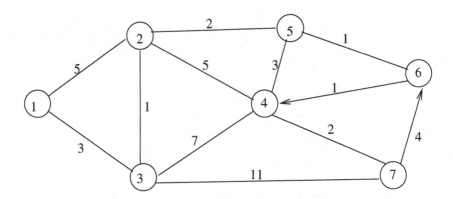

3. Determine the maximum flow and the optimal flow in each arc
for the following network

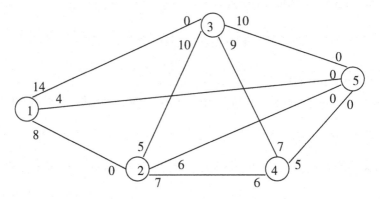

Also, determine

(a) the surplus capacities for all arcs;

(b) amount of flow through nodes 2, 3, and 4.

(c) Can the network flow be increased by increasing the capacities in the direction $3 \rightarrow 5$ and $4 \rightarrow 5$.

(d) Does the problem has alternate optimal solution? if yes, find it.

4. Three refineries send a LPG (liquid petroleum gas) to two distribution terminals through a pipeline network. Any demand that can not be satisfied through network is acquired from other source. The pipeline network is served by three pumping stations as shown in the figure below. The LPG flows in the network in the direction of arrows. The capacity of each pipe is mentioned on each arc in million bbl per day. To match the maximum capacity of the network, determine the following:

(a) The daily production at each refinery.

(b) The daily demand at each terminal.

(c) The daily capacity of each pump.

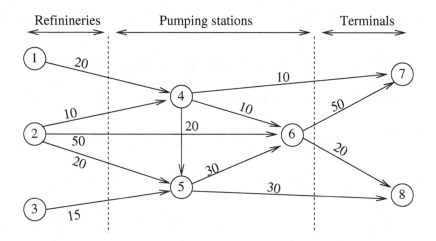

5. Suppose that the maximum daily capacity of pump 6 in the network of Figure 8.10 is limited to 60 million bbl. Remodel the network to include this restriction. Then determine the maximum capacity of the network.

6. The academic council of the Thapar Institute is seeking representations from six students who are affiliated with four academic societies. The academic council representation includes three areas: mathematics, management and engineering. At most two students in each area can be on the council. The following table shows the membership of the six students in the four academic societies:

Society	Affiliated students
1	1,2,3
2	1,3,5
3	3,4,5
4	1,2,4,6

The students who are skilled in the areas of mathematics, management and engineering are shown in the following table:

Area	Skilled students
Mathematics	1,2,4
Management	3,4
Engineering	4,5,6

A student who is skilled in more than one area must be assigned exclusively to one are only. Can all academic societies be represented on the academic council.

Chapter 9

Project Management

This chapter introduces the critical path method (CPM), a technique used for governing the successful completion of a project. The next section is devoted to learning the technique how the duration of a project can be reduced economically. In the end the project evaluation review techniques (PERT) is introduced.

9.1 Introduction

A project consists of interrelated activities which are to be executed in a certain order before the entire task is completed. The activities are interacted in a logical sequence which is known as precedence relationship. The work on a task can not be started until all its preceding activities are completed. Project management is generally applied for constructing items of public conveniences, large industrial projects etc. Every project is represented by a network for the purpose of analytical treatment. There are mainly two main techniques critical path method (CPM), a deterministic approach and the other one is project evaluation review techniques (PERT), a nondeterministic model.

In the next two sections we introduce how to sketch the network and find critical path. This is sensitive as the further analysis may give absurd results provided the network is not correct. These two studies are also essential preliminary steps for project evaluation review technique.

9.2   Critical Path Method

The CPM is a technique designed to assist in planning, scheduling and controlling the projects. A project is defined to be a task which has a definite beginning and definite end time and consists of several activities or jobs. A time duration is assigned to each activity. These activities (jobs) must be performed in a specified sequence to complete the project in successful manner. In planning, we design the problem jobs and their durations and order sequence, while in scheduling the time schedule for the project is determined. The controlling comes into play when certain changes in the planning and scheduling are proposed.

The CPM technique requires a network. Before discussing construction of a network, we are supposed to be acquainted with certain terms.

Network. The network of a project is the graphical representation of project operations (activities).

Activity. In each network an activity is shown by an arrow. The arrow head indicates the direction of progress of activity. The number on the arrow indicates the duration time of activity.

Node. The points at which an activity starts or ends are called nodes. Usually, these will be shown by circles in a network. Node number denotes an event.

The problem is given in the form of data information. The first step is how to make the network. Each network should satisfy the following conditions:

(a) Each activity must be represented by one and only one arrow.

(b) Two activities can not have same initial and same terminal nodes, i.e., each activity must be identified by two distinct nodes. The following graph is not permissible.

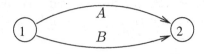

Figure 9.1

To avoid the above situation, a dummy activity is introduced between such nodes. It is assigned '0' time duration. In figure

9.1, activities A and B have the same initial and terminal nodes and hence we introduce a dummy activities in either of the styles shown in Figs. 9.2 and 9.3.

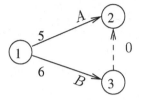 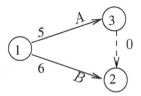

Figure 9.2

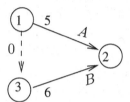 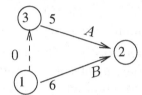

Figure 9.3

Such type of situations come in the way when some of the activities are emanating from the same node and preceding the same activity. These must be separated by using the dummy activities.

(c) Arrows should not form closed loop.

(d) Starting and ending nodes should be unique.

(e) To maintain the exact precedence relationship, we must note that (i) What activities must immediately precede the current activity? (ii) What activities must follow the current activity? (iii) What activities must occur concurrently with the current activities?

Suppose we are given

1. Activity C can start immediately after A and B are completed.

2. Activity E can start immediately after only B is completed.

Note that Figure 9.4(a) is incorrect, because it requires that A and B must be completed before E can start. Fig. 9.4(b) is correct representation of precedence by introducing a dummy activity D.

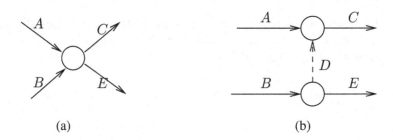

(a)   (b)

Figure 9.4

Remark. When some activity precedes two or more activities with different nodes, the precedence relation is maintained by introducing dummy one or more dummy activities to connect these different nodes.

Example 1. Construct a network where the activities satisfy the requirements: (i) A and B are the first activities of the project to start simultaneously; (ii) A and B precede C; (iii) B precedes D and E; (iv) A and B precede F; (v) F and D precede G and H; (vi) C and G precede I; (vii) E, H and I are the terminal activities.

The duration of activities A, B, C, D, E, F, G, H and I are 2, 3, 5, 2, 7, 4, 6, 11, 3, respectively.

With the data information available from the example, the network is depicted in Fig. 9.5. A dummy activity $(2, 3)$ is used to produce unique end nodes for the concurrent activities A and B. The numbering of the nodes is done in a manner that reflects the direction of the progress.

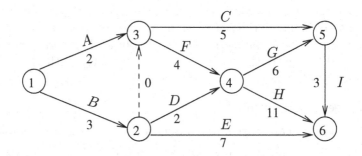

Figure 9.5

9.3   Critical Path Determination

Once the network is available, we determine the time of completion of the project. This time is not the sum of duration of individual activities, because some of the activities can be processed simultaneously. The total time of completion of a project depends on certain activities known as critical activities. The path connecting these activities is known as the critical path. The sum of durations of activities along a critical path determines the completion time of the project.

Thus, our next phase of working is to find the critical path. Once the critical path is at hand we are supposed to answer some queries: (i) Has the due time of completion of the project met? (ii) How, most economically, the duration of the project is reduced? (iii) If an activity is delayed, whether will it delay the completion of the project.

To construct the critical path define the notations:

t_{ij} = Time duration of activity emanating from node i and ending at node j;

ES_i = Earliest starting time of activities emanating from node i;

$= \max_k \{ES_k + t_{ki}\};$ $ES_1 = 0$ (by convention);

LC_i = latest completion time of activities coming to node i;

$= \min_j \{LC_j - t_{ij}\}.$

Also, $ES_1 = LC_1 = 0$ and $LC_\ell = ES_\ell$, ℓ stands for the last node.

Critical Activity. An activity (i, j) (emanating from i and terminating at node j) is called critical activity if and only if

(i) $ES_i = LC_i$;

(ii) $ES_j = LC_j$;

(iii) $ES_j - ES_i = LC_j - LC_i = t_{ij}$.

Example 2. Find the critical path of the network of the example in previous section by CPM.

Compute ES_i for each node and insert into the rectangle kept below the triangle. These computations are obtained using the above notations as

$ES_1 = 0$, by convention
$ES_2 = \max\{ES_1 + 3\} = 0 + 3 = 3$
$ES_3 = \max\{(ES_1 + 2), (ES_2 + 0)\} = \max\{2, 3\} = 3$
$ES_4 = \max\{(ES_2 + 2), (ES_3 + 4)\} = \max\{5, 7\} = 7$
$ES_5 = \max\{(ES_3 + 5), (ES_4 + 6)\} = \max\{8, 13\} = 13$
$ES_6 = \max\{(ES_2 + 7), (ES_4 + 11), (ES_5 + 3\} = \max\{10, 18, 16\} = 18$

Now, compute LC_j for each node and place them in the triangle kept above rectangle. In this case we move backward from the last node and compute as

$LC_6 = 18 = ES_6$, by definition
$LC_5 = \min\{LC_6 - 3\} = 15$
$LC_4 = \min\{(LC_5 - 6), (LC_6 - 11)\} = \min\{9, 7\} = 7$
$LC_3 = \min\{(LC_5 - 5), (LC_4 - 4)\} = \min\{10, 3\} = 3$
$LC_2 = \min\{(LC_3 - 0), (LC_4 - 2), (LC_6 - 7)\} = \min\{3, 5, 11\} = 3$
$LC_1 = 0 = ES_1$, by definition

The critical path is sketched by joining the nodes with double lines (critical activities), i.e., satisfying the conditions (i), (ii) and (iii). The critical path is depicted in the adjoining figure.

$$1 \longrightarrow 2 \longrightarrow 3 \longrightarrow 4 \longrightarrow 6$$

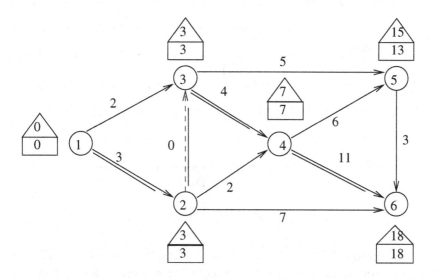

Note that $(3,5)$ is not the critical activity, since (ii) is not satisfied. Also, $(2,4)$ is not the critical activity as (iii) is not satisfied.

Remark. The critical path has longest duration between initial node and the terminal node.

9.4   Optimal Scheduling by CPM

Every project involves two type of costs, viz., , the direct and indirect costs. The direct costs are associated with individual activities. The direct cost of the project increases, if the duration of an activity is reduced by employing more resources, man power, instruments etc. The indirect costs are associated with the overhead items such as supervision etc. The indirect cost decreases, if the duration of an activity is reduced, since less time of supervision is needed for the execution of the project. The optimal scheduling by CPM means the studies of the cost-duration relationships. First, we study the direct cost-duration relationships. The graph of the direct cost relationship is depicted in Fig. 9.6.

Here T_n and C_n represent the duration and direct costs of an activity when it is executed under normal conditions. As already mentioned, the duration of an activity can be reduced by employing additional resources. However, this will increase the direct cost of the project. An activity can be reduced to a certain minimum duration.

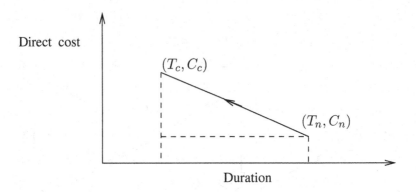

Direct cost

(T_c, C_c)

(T_n, C_n)

Duration

Figure 9.6

This minimum duration is called crash duration, denoted by T_c, and the corresponding cost involved is called crash cost, denoted by C_c. Note that an activity with crash duration can not be further reduced by employing additional resources. Obviously, crash cost of an activity should be larger than normal cost.

The slope of crashing an activity is given by

$$\text{slope} = \frac{C_c - C_n}{T_n - T_c}.$$

The slope of crashing an activity means the increment cost of expediting that activity per unit period.

The direct cost of project increases and indirect cost decreases if the duration of the project is reduced. A project manager is always interested to find duration T_m of the project with minimum cost schedule (direct cost + indirect cost), see Fig 9.7.

For direct cost relationship we use the FF (free floats) limit method. For this purpose, we define free floats as

$$FF \text{ of } (i, j)\text{th activity} = (ES_j - ES_i) - t_{ij}.$$

Clearly, the FF of an critical activity is zero. However, if $FF = 0$ for any activity, it may not necessarily be an critical activity, i.e., a noncritical activity may have $FF = 0$. Also,

Crash limit of an activity = Current duration − Crash duration;

FF limit = minimum of the positive FF's of the noncritical activities

Compression limit = min{Crash limit, FF limit}.

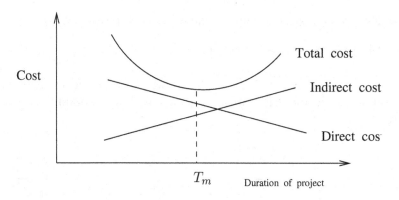

Figure 9.7

Remark. The other parameter called total float (TF) of an activity is defined as

$$\text{TF of } (i,j)\text{th activity} = (LC_j - ES_i) - t_{ij}.$$

1. FF limit method. We work out a problem to explain how the duration of a project is reduced by using free floats of the noncritical activities.

Example 3. Consider the project for which details are mentioned below

Activity	Normal duration	Normal cost	Crash duration	Crash cost
(i,j)	T_n	C_n	T_c	C_c
$(1,2)$	15	600	12	1200
$(1,3)$	8	700	5	1600
$(2,5)$	12	750	6	1500
$(3,4)$	15	650	12	1400
$(3,5)$	18	700	13	1450
$(4,5)$	8	500	5	950

(a) Find by FF limit method the minimum cost schedule, if the project is to be completed in 28 days;

(b) Find the most economical schedule for 23 days;

(c) Find the crash cost schedule.

Figure 9.8 depicts the network. It also shows the critical path by double lines, and FF's are also mentioned corresponding to each activity.

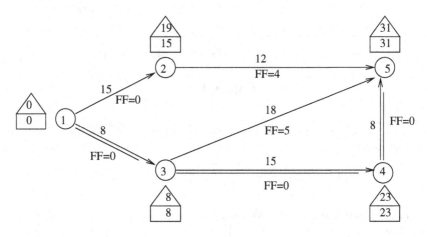

Figure 9.8

The critical path: $1 \longrightarrow 3 \longrightarrow 4 \longrightarrow 5$. Normal cost: 3900, which is the sum of all entries in the column of normal costs. The normal duration of the project: 31 days.

Further, we are interested to reduce the time duration of the project. For doing this, the slope of each activity is required for comparison purposes. Using the formula for slope we compute

Activity	Slope
(i, j)	m
$(1, 2)$	200
$(1, 3)$	300
$(2, 5)$	125
$(3, 4)$	250
$(3, 5)$	150
$(4, 5)$	150

For part (a), we have to compute the optimal cost schedule for 28 days. As the normal duration of the project is 31 days, we have to reduce it by 3 days. The reduction in the duration of critical activity will cause only change in the cost scheduling.

The question arises which critical activity should be reduced?

We proceed as follows:

Reduce the critical activity on the critical path for which the slope is minimum. Among all critical activities, the activity $(4,5)$ has minimum slope, and hence, it is best candidate for reduction. To compute FF limit for $(4,5)$ we shall take minimum of the FF's of $(2,5)$ and $(3,5)$. Thus,

FF limit $= \min\{4,5\} = 4$

Crash limit of activity $(4,5) = 8 - 5 = 3$

Compression limit for activity $(4,5) = \min\{3,4\} = 3$

Hence, reduce the duration of $(4,5)$ by 3 days. The new schedule is given in Figure 9.9.

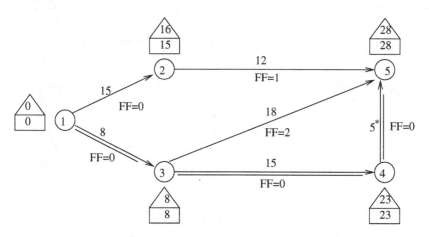

Figure 9.9

The minimum cost schedule for 28 days is given by

normal cost of the project + reduced time × slope.

Hence,

$$\text{minimum cost schedule for 28 days } = 3900 + 3 \times 150 = 4350.$$

Remark. If the most economical cost schedule for 29 days is required, then reduce the critical activity (4,5) by two days.

For part (b), a further reduction of 5 days is desired. Activity, $(4,5)$ can not be reduced as it has reached its crash duration. In Fig. 9.9, activity with asterisk means it has reached at its crash duration.

The next candidates for reduction are $(1,3)$ and $(3,4)$. Since the slope of $(3,4)$ is minimum. Hence, we reduce activity $(3,4)$. Thus, for activity (3,4),

$$\text{Crash limit} = 15 - 12 = 3$$

$$\text{FF limit} = \min\{1, 2\} = 1$$

$$\text{Compression limit} = \min\{1, 3\} = 1$$

Hence, activity $(3,4)$ is reduced by 1 day. The cost schedule for 27 days is depicted in Fig. 9.10.

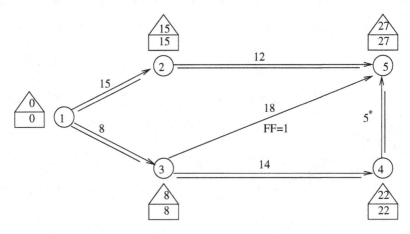

Figure 9.10

Minimum cost schedule for 27 days $= 4350 + 250 = 4600$.

At this stage, two critical paths have developed, and hence, reduce the time of both critical paths simultaneously. For critical path $(1, 2, 5)$ the best candidate for compression is $(2, 5)$, and for critical path $(1, 3, 4, 5)$ the best candidate is $(3, 4)$.

$$\text{Crash limit for activity } (2, 5) = 12 - 6 = 6$$

$$\text{Crash limit for activity } (3, 4) = 14 - 12 = 2$$

Therefore, the crash limit for two paths is $\min\{2, 6\} = 2$. Now, find FF limit for both paths separately. While considering one critical path, all other activities not on this path (including activities on other critical path) are considered noncritical activities. Hence,

$$FF \text{ limit for activity } (2, 5) = \min\{1\} = 1$$

$$FF \text{ limit for activity } (3, 4) = \min\{1\} = 1$$

Thus,

$$FF \text{ limit for two paths } = \min\{1, 1\} = 1.$$

Note that $FF = 0$ for $(2, 5)$ and $(3, 4)$, even then we have taken FF limits as 1, since, by definition positive FF is to be taken.

The compression limit is $\min\{1, 2\} = 1$. This means that both candidates must be reduced by 1 day, i.e., activities $(2, 5)$ and $(3, 4)$ are reduced. By doing so, we get the minimum cost schedule for 26 days in Figure 9.11.

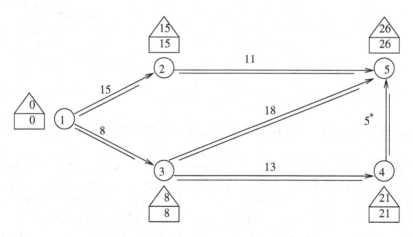

Figure 9.11

Minimum cost schedule for 26 days $= 4600 + 250 + 125 = 4975$.

Now, all the activities have turned to be critical activities, and hence, all FF's are zero. This indicates that FF limit is infinite. Again, the best candidates are activities $(2,5)$, $(3,5)$ and $(3,4)$, respectively for the critical paths $(1,2,5)$, $(1,3,5)$ and $(1,3,4,5)$, respectively.

When FF's are infinite, we compress by taking minimum of the crash limits.

$$\text{Compression limit} = \min\{11-6, 18-13, 13-12\} = 1.$$

Inspite of reducing the duration by one day for each of the activities $(3,5)$ and $(3,4)$, we reduce the common activity $(1,3)$, since, slope of $(1,3)$ < slope of $(3,5)$ + slope of $(3,4)$. The minimum cost schedule for 25 days is shown in Figure 9.12.

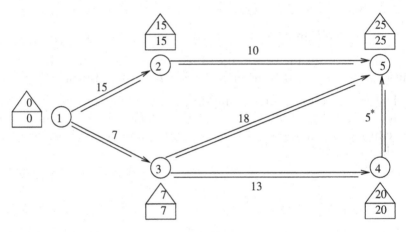

Figure 9.12

Minimum cost schedule for 25 days $= 4975 + 125 + 300 = 5400$.

Again, best pairs for compression are activities $(2,5)$, $(3,5)$ and $(3,4)$ on three different critical paths. Observe that $(1,3)$ is common to critical paths $(1,3,5)$ and $(1,3,4,5)$. Hence, compression in activity $(1,3)$ will further reduce both critical paths $(1,3,5)$ and $(1,3,4,5)$. This gives better schedule than reducing activities $(3,5)$ and $(3,4)$, simultaneously. In view of this

$$\text{Compression limit} = \min\{10-6, 7-5\} = 2.$$

The minimum cost schedule for 23 days is shown in Figure 9.13.

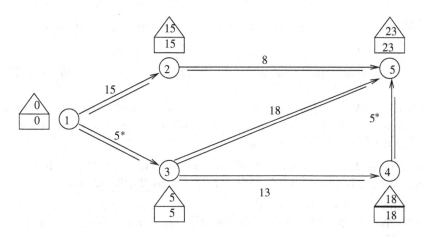

Figure 9.13

Minimum cost schedule for 23 days $= 5400 + 2 \times 125 + 2 \times 300 = 6250$.

As the common critical activity $(1,3)$ has reached its crash duration. Further, reduction in this activity is not possible. The best candidate pairs for reduction are $(2,5)$, $(3,5)$ and $(3,4)$. It can be verified that compression limit is 1. Hence, reduce all these activities by 1 day. The most economic schedule for 22 days is shown in Figure 9.14.

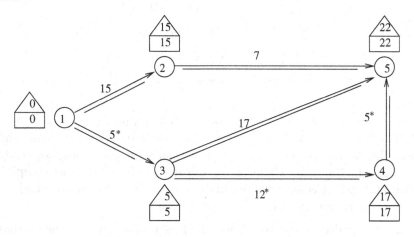

Figure 9.14

(c) Further, reduction is not possible as all the activities along one of the critical paths: $1 \longrightarrow 3 \longrightarrow 4 \longrightarrow 5$ have reached their crash duration.

The most economic crash schedule $= 6250 + 125 + 150 + 250 = 6775$.

Remark. When there are more than one critical activities as best candidates pairs for reduction is advisable to select the activity to be reduced which has larger crash limit.

Example 4. In Example 3, if the indirect cost is $50 per day then find the minimum cost schedule for $28, 27, 26, 25, 23, 22$ days.

Since indirect costs are the supervisory costs, we get the minimum cost schedule for 28 days by adding the total supervision for this period. Hence,

minimum cost schedule for 28 days $= 4350 + 50 \times 28 = 5750$.

Similarly, most economical schedule for 27 days$= 4600 + 27 \times 50 = 5950$, and so on. The following table gives all details for different durations is

Duration	Total cost
28	$4350+28\times50=5750$
27	$4600+27\times50=5950$
26	$4975+26\times50=6275$
25	$5400+25\times50=6650$
23	$6250+23\times50=7400$
22	$6775+22\times50=7875$

We have just described the optimal scheduling by CPM based on FF limit method and this method is preferred when the time and cost relationship is nearly linear. For nonlinear situations, we study optimal scheduling by CPM using another method called as stepping stone method. Because, in nonlinear case a best candidate at given iteration may not be best for next iteration.

2. Stepping step method. In this method we compress that activity along critical path which has lowest slope by only one unit duration at a time. Note that we do not use free floats for this purpose. The sensitive point is that when we compress two critical activities on different critical paths, the compression is done in such a way that both path remain critical. We may continue till all activities along some critical reach at their crash duration.

9.5 Project Evaluation and Review Technique

Most often, PERT is used as an abbreviation for Project (Programming) Evaluation and Review Technique. The first step is to determine the critical path. However, in PERT duration of an activity is not fixed but is a random variable. For an activity (i, j), the time duration T_{ij} is a random variable. It is found by experiments that probability density function of T_{ij} has a graph of either the shapes, see Fig. 9.15.

Define

$$t_0 = \text{optimistic time, if execution goes well;}$$

$$t_p = \text{pessimistic time, if execution goes badly;}$$

$$t_m = \text{most likely time: if execution goes normal.}$$

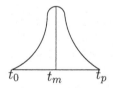 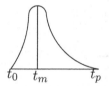 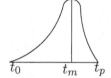

Figure 9.15

In PERT, Beta distribution is taken as the probability density function of T_{ij}

$$f(t) = \begin{cases} kt^{p-1}(1-t)^{q-1}, & 0 \le t < 1, \\ 0, & \text{otherwise,} \end{cases}$$

where

$$k = \frac{\Gamma(p+q)}{\Gamma(p)\Gamma(q)}, \quad \Gamma \text{ being gamma function.}$$

The density function $f(t)$ can be made to take any of the three shapes of Fig. 9.15 by assigning p and q appropriate values. The expected value and variance of beta distribution are computed by usual methods. However, for the purpose of PERT computations we approximate expected value and variance as follows. Assume that $(t_0 + t_p)/2$ weighs half as much as most likely t_m. Thus the expected value (mean) of the random variable T_{ij} is computed using

$$E\left(T_{ij}\right) = \frac{\frac{t_0+t_p}{2} + 2t_m}{3} = \frac{t_0 + 4t_m + t_p}{6}. \tag{9.1}$$

To ensure that ninety percent of data falls within the interval $[t_0, t_p]$, we utilize the Chebyshev inequality to find V_{ij}, variance of the random variable T_{ij}

$$V_{ij} = \left(\frac{t_p - t_0}{6}\right)^2. \tag{9.2}$$

The critical path is found as in CPM by using $E(T_{ij})$, expected value of T_{ij}.

$$T(\text{shortest completion time}) = \sum_{\substack{(i,j) \\ \text{critical} \\ \text{activity}}} T_{ij}$$

$$E(T) = \sum_{\substack{(i,j) \\ \text{critical} \\ \text{activity}}} E(T_{ij})$$

$$V(T) = \sum_{\substack{(i,j) \\ \text{critical} \\ \text{activity}}} V_{ij}.$$

The earliest starting time of event i is denoted by Z_i, which is a random variable. Suppose $E(Z_i)$ and $V(Z_i)$ denote respectively the expected value and variance of the random variable Z_i. If two or more paths lead to the same i, then we take $E(Z_i)$ to be the one corresponding to the path having largest value of $E(Z_i)$.

If there is a tie among the paths having largest $E(Z_i)$, we select the one which corresponds to the larger value of the variance, because it reflects greater uncertainty. It is very much obvious that $E(Z_i) = ES_i$ (earliest starting time of the activities emanating from node i. The path is determined by $E(Z_i)$, and then the variance $V(Z_i)$ is calculated for this path.

Also, it is significant to mention that Z_i is the sum of many independent random variables T_{ij}, hence, by central limit theorem the random variable

$$Z = \frac{Z_i - E(Z_i)}{\sqrt{V(Z_i)}}$$

has the standard normal distribution $N(0, 1)$.

Example 5. Consider the data

Activity	t_0	t_p	t_m
$(1, 2)$	2	8	5
$(1, 3)$	1	7	4
$(2, 3)$	0	0	0
$(2, 4)$	2	6	4
$(2, 6)$	5	12	7
$(3, 4)$	3	10	7
$(3, 5)$	3	3	6
$(4, 5)$	2	8	5
$(4, 6)$	4	10	6
$(5, 6)$	2	6	4

(a) Find the probability that the earliest (starting) occurrence time of event 4 is less than or equal to 13 days;

(b) Find the probability that the project is completed in 23 days.

Using equations (9.1) and (9.2), we compute $E(T_{ij})$ and $V(T_{ij})$ and write data in the form

Activity	$E(T_{ij})$	$V(T_{ij})$
$(1, 2)$	5	1
$(1, 3)$	4	1
$(2, 3)$	0	0
$(2, 4)$	4	0.44
$(2, 6)$	7.5	1.36

Activity	$E(T_{ij})$	$V(T_{ij})$
$(3,4)$	6.83	1.36
$(3,5)$	5	0
$(4,5)$	6.33	1
$(4,6)$	5	1
$(5,6)$	4	0.44

The network is shown in Figure 9.16.

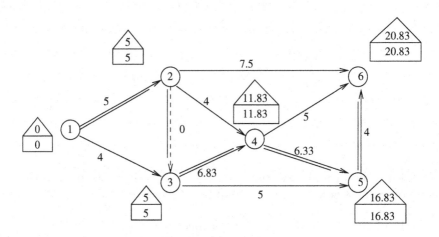

Figure 9.16

The critical path and its expected normal duration $E(T)$ and variance $V(T)$, respectively are

$$1 \longrightarrow 2 \longrightarrow 3 \longrightarrow 4 \longrightarrow 5 \longrightarrow 6,$$

$$E(T) = 20.83, \quad V(T) = 1 + 0 + 1.36 + 1 + 0.44 = 3.8.$$

Event i	$E(Z_i)$	V_{Z_i}
1	0	0
2	5	1
3	5	1
4	11.83	2.36
5	16.83	3.36
6	20.83	3.8

(a) $P[Z_4 \leq 13] = P\left[\dfrac{Z_4 - E(Z_4)}{\sqrt{V(z_4)}} \leq \dfrac{13 - E(Z_4)}{\sqrt{V(z_4)}}\right]$

$\qquad = P\left[Z \leq \dfrac{13 - 11.83}{\sqrt{2.36}}\right]$

$\qquad = P(Z \leq 0.7616) = 0.7794$

(b) $P[T \leq 23] = P\left[\dfrac{T - E(T)}{\sqrt{V(T)}} \leq \dfrac{23 - E(T)}{\sqrt{V(T)}}\right]$

$\qquad = P\left[Z \leq \dfrac{23 - 20.83}{\sqrt{3.8}}\right]$

$\qquad = P(Z \leq 1.113) = 0.8665,$

The probabilities in (a) and (b) have been written from the table of normal distribution.

Problem Set 9

1. A statue is to be erected in a village square on a stone platform which is to be built on cement concrete foundation. The statue is to be prepared at another place, moved and erected. The various operations of the project are given below. Construct the network of the project. Write clearly A, B, C, etc. on the activities.

   A : Make statue

   B : Shift statue

C : Erect statue

D : Lay foundation of the platform

E : Construct platform

2. Draw the network of the project for which (i) activities A and B start simultaneously; (ii) activities A, C, B, F, G precede C, D, E, H, I, respectively, (iii) E precedes G and H (iv) D, H, I precede J, a terminal activity.

3. Construct the project network comprised of activities A to P that satisfies the following precedence relationships:

   (i) A, B, and C, the first activities of the project, can be executed concurrently;

   (ii) D, E and F follow A;

   (iii) I and G follow both B and D;

   (iv) H follows both C and D;

   (v) K and L follow I;

   (vi) J succeeds both E and H;

   (vii) M and N succeed F, but can not start until both E and H are completed;

   (ix) O succeeds M and I;

   (x) P succeeds J, L and O

   (xi) K, N, and P are the terminal activities of the project.

4. Write the following activities by network :

Activity $(i\text{-}j)$	Activity time (days)
$A - B$	3
$A - C$	4
$B - C$	5
$B - D$	4
$C - F$	4
$D - E$	13
$E - G$	4
$E - H$	6
$F - H$	9
$G - H$	0

There is a constraint that activity $F - H$ cannot start till the activity $D - E$ is considered. Determine the Critical Path and normal duration of the project.

Suggestion. Here, $A, B, \ldots, H$ are events (nodes).

5. A project consists of a series of tasks labeled A, B, $\ldots$, I with the relationship : $X < Y, Y$ means X and Y can not start until W is complete, X, $Y < W$ means W can not start until both X and Y are complete. Construct the diagram using this notation when

$$A < D, \ E; \ B < D < F, \ C < G; \ B < H; \ F, \ G < I.$$

Also, find the minimum time of completion of the project when the time in (days) for each task is as follows:

A	B	C	D	E	F	G	H	I
23	8	20	16	24	18	19	4	10

6. A reactor and storage tank are interconnected by a 7.5 cm insulated process line that needs periodic replacement. There are valves along the lines as well as at the terminals and they need replacement as well. No pipe and valves are in stock. Their drawings are accurate and are available. The line is overhead

and requires scaffolding. Pipe sections can be fabricated at the plant. An adequate craft labour is available. The plant methods standard section has furnished the following data:

Activity		Time (Hrs.)	Predecessor(s)
A	Develop required material list	10	-
B	Procure pipe	200	A
C	Erect scaffold	15	-
D	Remove scaffold	5	H, L
E	Deactivate line	10	-
F	Prefabricate sections	40	B
G	Place new pipes	35	F, K
H	Fit up pipe and valves	10	G, J
I	Procure valves	220	A
J	Place valves	10	I, K
K	Remove old pipe and valves	40	C, E
L	Insulate	25	G, J
M	Pressure test	8	H
N	Clean-up and start-up	5	D, M

(a) Draw the arrow diagram of this project plan.

(b) Find the critical path and its duration.

7. A small project consists of the following activities where duration is in days and cost is in rupees:

		Normal		Crash	
Activity	Precedence	duration	cost	duration	cost
A	-	6	300	5	400
B	-	8	400	6	600
C	A	7	400	5	600
D	B	11	1000	4	1350
E	C	8	500	5	800
F	B	7	400	6	500
G	D, E	5	1000	3	1400
H	F	8	500	5	950

(a) Draw the network and find the normal duration;

(b) Find the most economical schedule if the project is to be completed in 21 days;

(c) Find the minimum project duration and total cost involved.

8. Consider the data of a project with duration in weeks and costs in rupees as shown in the following table

	Normal		Crash	
Activity	duration	cost	duration	cost
(1,2)	13	700	9	900
(1,3)	5	400	4	460
(1,4)	7	600	4	810
(2,5)	12	800	11	865
(3,2)	6	900	4	1130
(3,4)	5	1000	3	1180
(4,5)	9	1500	6	1800

If the indirect cost per week is $160, find the most economical schedule for the completion of the project.

9. Show by an example that, having two critical path and common activity, that at times it may be economical to compress a non-best common critical activity rather than to compress respective best activities along two critical paths.

Suggestion. Observe schedule for 25 days in Example 3.

10. A project consists of six activities. The time estimates in days of the activities are as follows.

Activity	Optimistic	Most likely	Pessimistic
(1,2)	3	7	9
(1,3)	3	4	7
(2,4)	2	4	6
(2,5)	1	3	4
(3,5)	6	8	9
(4,5)	5	7	9

The schedule date for the event 5 is given as 19 days. What is the probability of this schedule being achieved? If event 4 is delayed by 2 days, then find the probability of the project being completed in time.

11. Assuming that the expected times are normally distributed, find the probability of meeting the schedule date for the following network. The estimation of duration is given in weeks.

Activity	Optimistic	Most likely	Pessimistic
(1,2)	3	6	15
(1,3)	10	13	16
(2,4)	6	15	18
(3,4)	3	6	9
(4,5)	7	7	13
(3,5)	9	18	21

The schedule date for the project completion is 31 days. Find the date on which the project manager can complete the project with a probability of 0.90.

12. The following data is available for a project. The estimation of duration is given in days.

Activity	Least time	Greatest time	Most likely time
(1,2)	6	11	9
(1,3)	19	23	21
(1,4)	27	41	34
(2,5)	17	21	19
(2,6)	14	24	21
(3,6)	7	13	10
(4,7)	6	11	12
(5,7)	8	10	9
(6,7)	4	6	5

Find the following:

(a) Expected task time and their variance;
(b) The earliest and latest expected times to reach each node;
(c) The critical path;
(d) The probability of node occurring at the proposed completion date if the original contract time for completion of project is 42.5 weeks.

13. A civil engineering firm has to bid for the construction of a dam. The activities and time estimates are as follows:

	Duration (in weeks)		
Activity	Optimistic	Most likely	Pessimistic
(1,2)	14	17	25
(2,3)	14	18	21
(2,4)	13	15	18
(2,8)	16	19	28
(3,4)	0	0	0
(3,5)	15	18	0

(4,6)	13	17	21
(5,7)	0	0	0
(5,9)	14	18	20
(6,7)	0	0	0
(6,8)	0	0	0
(7,9)	16	20	41
(8,9)	14	16	22

The policy of the firm with respect to submitting bids is to bid the minimum amount that will give 0.95 probability of at best at breaking even. The fixed costs for the project are eight lacks and the variable cost are $9,000 every week spent working on the project.

What amount should the firm bid under this policy?

14. A promoter is organizing a sports meeting. The relationship among the activities and time estimates are shown in the following table:

Activity	Task	Immediate predecessor	Duration (days)		
			t_0	t_m	t_p
A	Prepare draft program	–	3	7	11
B	Send to sports authority and wait for comments	A	14	21	28
C	Obtain promoters	A	11	14	17
D	Prepare and sign document for stadium hire	A, C	2	2	2
E	Redraft program and request entries	B	2	3.5	8
F	Enlist Officials	B	10	14	21

G	Arrange accommodation for tourist teams	E	3	4	5
H	Prepare detailed programme	E, F	4	4.5	8
I	Make final arraignments	G, H	1	2	4

Chapter 10

Sequencing Problems

This chapter deals in which order (sequence) a finite number of jobs should be processed on finite number of machines so that the elapsed time (total time involved in completion of all jobs) is minimized.

10.1 Introduction

The selection of an appropriate order for finite number of different jobs to be done on a finite number of machines is called sequencing problem. In a sequencing problem we have to determine the optimal order (sequence) of performing the jobs in such a way so that the total time (cost) is minimized. Suppose n jobs are to be processed on m machines for successful completion of a project. Such type of problems frequently occur in big industries. The sequencing problem is to determine the order (sequence) of jobs to be executed on different machines so that the total cost (time) involved is minimum. Since these arrangements are large in number, in particular, 5 jobs on 4 machines can be processed in $(5!)^4 = 207360000$ ways. Thus, by enumeration it is impossible to solve these problems even for small number of jobs and machines. It motivates to devise some technique which can help to locate the specific arrangement in limited number of iterations so that the elapsed time is minimized.

Before developing the algorithm we define certain terms as

M_{ij} = processing time required by ith job on the jth machine ($i = 1$ to n, $j = 1$ to m).

T_{ij} = idle time on machine j from the completion of $(i-1)$th job to the start of ith job.

T = elapsed time (including idle time) for the completion of all the jobs.

The problem is to determine a sequence $i_1, i_2, \ldots, i_n$, where $i_1, i_2, \ldots, n$ is a some permutation of the integers $1, 2, \ldots, n$ that minimizes the total elapsed time T. Each job is processed on machine M_1 and then on machine M_2, and we say jobs functioning order is $M_1 M_2$. Before developing the algorithm in the next section we make certain assumptions.

(i) No Machine can process more than one job at a time.

(ii) Each job once started on a machine must be completed before the start of new job.

(iii) Processing times M_{ij}'s are independent of the order of processing the jobs.

(iv) Processing times M_{ij}'s are known in advance and do not change during operation.

(v) The time required in transferring a job from one machine to other machine is negligible.

(vi) There is only one of each type of machine.

10.2   Problem of n Jobs and 2 Machines

We start with the simple problem of processing n jobs on two machines. The procedure illustrated below is due to S. M. Johnson. Let A_i and B_i be the processing time of ith job on machines M_1 and M_2, respectively. To find the optimal sequence of jobs on the machine M_1 and M_2 in the order $M_1 M_2$, we write the problem in the format:

Jobs	Processing Time on Machine M_1	Processing Time on Machine M_2
1	A_1	B_1
2	A_2	B_2
⋮	⋮	⋮
i	A_i	B_i
⋮	⋮	⋮
n	A_n	B_n

We adopt the following procedures:

Algorithm. The steps of an iterative procedure to get the optimal sequence for the problem of n jobs on two machines are as follows:

Compute

$$\min_i \{A_i, \ B_i\}$$

Step 1. If this processing time is in column A say A_r, then process the job r first, if this processing time is in column B, say B_s, then process the sth job in the last.

Step 2. In the case of a tie among the smallest processing timings in columns A and B, then proceed as

(a) If the smallest value falls in both columns (there is a tie), then schedule the job in column A first and the job of column B in the last.

(b) If all smallest values fall in column A, select one which corresponds to the lowest entry in column B and place in the sequence from the left. Here all tied jobs are are processed looking for the lowest, next to the lowest and so on in column B.

(c) If all smallest values fall in column B, select one which corresponds to the lowest entry in column A and place in the sequence from the right. Here all tied jobs are are processed looking for the lowest, next to the lowest and so on in column A.

Step 3. Cross all the jobs assigned by Steps 1 and 2 and again find the minimum of the remaining A_i and B_i. Repeat the above procedure till all the jobs are assigned.

Remark. If the smallest processing time for any job is same for both machines, then this can be placed from the left or right. This situation results in alternate optimal solution.

The above procedure is illustrated by solving some examples.

Example 1. There are six jobs each of which is to be processed through the machines M_1 and M_2 in order $M_1 M_2$. The processing time is given in hours. Determine the sequence of these jobs which minimizes the total elapsed time.

Job number	1	2	3	4	5	6
Processing time on M_1	5	9	4	7	8	5
Processing time on M_2	8	4	8	3	6	6

Write the data in the format

Jobs	Processing Time on M_1 (in hours)	Processing Time on M_2 (in hours)
1	5	8
2	9	4
3	4	8
4	7	3
5	8	6
6	5	6

The minimum of all entries (processing time) in column A and B is 3, and this is in column B and correspond to job 4. Hence, process job 4 in last, i.e.,

					4

As job 4 has been placed cross the row containing job 4. Calculate the minimum of the remaining entries. The minimum is 4 and falls in column A and B both. This minimums corresponds to job 3 for and job 2 for column B. Hence, place job 3 first and job 2 in last (i.e., before already filled space). This gives

3				2	4

Calculate minimum of all entries ignoring the entries in row 2, 3 and 4. The minimum is 5 and there is a tie as this corresponds to jobs 1 and 6. But the lowest entry in column B is for job 6. Hence, process job 6 first and then job 1. The next arrangement is

3	6	1		2	4

All jobs are assigned except job 5. Fill up this in the space left. The optimal sequence is

3	6	1	5	2	4

To compute the elapsed time and the idle time for machines M_1 and M_2, construct the table

Job	Machine M_1		Machine M_2		Idle time
	Time in	Time out	Time in	Time out	for M_2
3	0	4	4	12	4
6	4	9	12	18	0
1	9	14	18	26	0
5	14	22	26	32	0
2	22	31	32	36	0
4	31	38	38	41	2

From the table, we write

$$\text{Total elapsed time} = 41 \text{ hours}$$
$$\text{Idle time for } M_2 = 4 + 2 = 6 \text{ hour}$$
$$\text{Idle time for } M_1 = 41 - 38 = 3 \text{ hours}$$

Example 2. Find the sequence that minimizes the total elapsed time required to complete the following tasks. The table below gives the processing time in hours. Also, calculate the idle time for both machines.

Job	1	2	3	4	5	6	7
M_1	1	5	2	4	2	4	4
M_2	4	4	2	2	4	3	5

As usual write the problem in column format, see Example 1, and note that 1 is minimum of all entries and this corresponds to job 1. Hence, process job 1 first. Delete the row containing job 1.

Compute fresh minimum of all the entries in both columns, and this is 2. But there is a tie for 2 to appear in both columns and also at two places in the same column. First, break the tie according to Step 2 (i) of algorithm. This finds placement for job 3 after job 1 and job 4 in the last. At most two jobs may be placed at a time. Delete rows for jobs 1, 3 and 4.

The next minimum is 2 in the first column and corresponds to job 5. Place job 5 just after job 3. Delete all rows containing the jobs assigned so far.

The next smallest value is 5 and there is a tie for the corresponding lowest entry in both columns. This implies that jobs 2 and 7 can be placed in any order. This indicates alternate optimal solution. Suppose, we first place 7 and then 2.

The only left space is filled by the only left job 6. The optimal sequence is

1	3	5	7	2	6	4

Construct the table to find minimum elapsed time as

Job	Machine M_1		Machine M_2		Idle time
	Time in	Time out	Time in	Time out	for M_2
1	0	1	1	5	1
3	1	3	5	7	0
5	3	5	7	11	0
7	5	9	11	16	0
2	9	14	16	20	0
6	14	18	20	23	0
4	18	22	23	25	0

From the table, we conclude

$$\text{Total elapsed time} = 25 \text{ hours}$$
$$\text{Idle time for } M_2 = 1 \text{ hour}$$
$$\text{Idle time for } M_1 = 25 - 22 = 3 \text{ hour}$$

Remark. Example 2 has alternate optimal solution as jobs 2 and 7 can also be placed in the order job 2 and then job 7.

10.3   Problem of n Jobs and m Machines

Let there be n jobs to be processed through m machines M_1, M_2, ..., M_m in that order. If either or both of the conditions hold, then the problem can be related to n jobs and two machines.

$$\min_i M_{i1} \geq \max_i M_{ij}, \quad j = 2, 3, \ldots, m - 1 \qquad (10.1)$$
$$\min_i M_{im} \geq \max_i M_{ij}, \quad j = 2, 3, \ldots, m - 1. \qquad (10.2)$$

Note. If neither of the above conditions are satisfied, then no solution procedure is available.

Suppose at least one of the conditions is available as valid. Then the algorithm proceeds as

Algorithm.

Step 1. Let there be two fictitious machines F_1 and F_2, Calculate

$$F_1 = M_{i1} + M_{i2} + \cdots + M_{i,m-1}, \quad i = 1, 2, \ldots, n, (10.3)$$
$$F_2 = M_{i2} + M_{i3} + \cdots + M_{im}, \quad i = 1, 2, \ldots, n, \quad (10.4)$$

where M_{ij} is the processing time for ith ($i = 1$ *textto* n) job on machines M_j ($j = 1$ to m).

Step 2. Find the optimal sequence for n jobs with two machines F_1 and F_2 in the order $F_1 F_2$. This optimal sequence is also the optimal sequence for the given problem with n jobs and m machines.

Example 3. Suppose there are five jobs each of which must go through machines M_1, M_2 and M_3 in the order M_1 M_2 M_3. Processing times are given below. Determine the optimal sequence and total elapsed time.

Jobs	1	2	3	4	5
M_1	8	5	4	6	5
M_2	6	2	9	7	4
M_3	10	13	11	10	12

Since, $\min M_{i1} = 4$, $\max M_{i2} = 9$, $\min M_{i3} = 10$, and condition (10.2) is satisfied, the algorithm works. Introduce fictitious machines F_1 and F_2, and compute processing for each job on machines F_1 and F_2 using relations (10.3) and (10.4). This procedure yields the following equivalent sequencing problem.

Jobs	1	2	3	4	5
F_1	14	7	13	13	9
F_2	16	15	20	17	16

Now, solve the problem as five jobs and two machines. By usual method of Examples 1 and 2, the optimal sequence is

2	5	4	3	1

Once the optimal sequence is available, we use the original data of the problem to find the total elapsed time and idle time for all machines. Suppose I_d denotes the idle time in the table constructed below.

Job	Machine M_1		Machine M_2			Machine M_3		
	Time in	Time out	Time in	Time out	I_d	Time in	Time out	I_d
2	0	5	5	7	5	7	20	7
5	5	10	10	14	3	20	32	0
4	10	16	16	23	2	32	42	0
3	16	20	23	32	0	42	53	0
1	20	28	32	38	0	53	63	0

From the data in the above table, we get

$$\text{Total elapsed time} = 63 \text{ hours}$$
$$\text{Idle time for } M_1 = 63\text{-}28 = 35 \text{ hours}$$
$$\text{Idle time for } M_2 = 10+(63\text{-}38) = 35 \text{ hours}$$
$$\text{Idle time for } M_3 = 7 \text{ hours}$$

Similarly, the problem can be done for 4 or 5 machines.

Remark. In addition to (10.1) and (10.2) conditions, if the following condition also holds

$$M_{i2} + M_{i3} + \cdots + M_{i,m-1} = \text{ constant}; \quad i = 1, 2, \ldots, n$$

then the problem can be solved as n jobs 2 machines but in a more simplified way, i.e., we have to get the optimal sequence for the first and the last machines only. For more illustration see the next example.

Example 4. Solve the following sequencing problem

$$\begin{array}{cccccc}
\text{Jobs} & 1 & 2 & 3 & 4 & 5 \\
M_1 & 7 & 10 & 8 & 9 & 7 \\
M_2 & 2 & 1 & 4 & 0 & 5 \\
M_3 & 5 & 6 & 3 & 7 & 2 \\
M_4 & 8 & 7 & 12 & 10 & 9
\end{array}$$

Note that both conditions (10.1) and (10.2) are satisfied. In addition, we also have

$$M_{i2} + M_{i3} = 7, \quad i = 1 \text{ to } 5$$

Hence, by the above remark this problem is equivalent to the following two machines $P_1 \equiv M_1$ and $P_2 \equiv M_4$ problem

Jobs	1	2	3	4	5
P_1	7	10	8	9	7
P_2	8	7	12	10	9

By the usual method the optimal sequence is

1	5	3	4	2

With this optimal sequence at hand construct the table to find total elapsed time and the idle time for different machines. Suppose I, O and I_d stands for time in, time out and idle time. Then

Job	Machine M_1		Machine M_2			Machine M_3			Machine M_4		
	I	O	I	O	I_d	I	O	I_d	I	O	I_d
1	0	7	7	9	7	9	14	9	14	22	14
5	7	14	14	19	5	19	21	5	22	31	0
3	14	22	22	26	3	26	29	5	31	43	0
4	22	31	31	31	5	31	38	2	43	53	0
2	31	41	41	42	10	42	48	4	53	60	0

From the table we compute

$$\text{Total elapsed time} = 60 \text{ hours}$$

$$\text{Idle time for } M_1 = 19 \text{ hours}$$

$$\text{Idle time for } M_2 = 48 \text{ hours}$$

$$\text{Idle time for } M_3 = 37 \text{ hours}$$

$$\text{Idle time for } M_4 = 14$$

10.4 Two Jobs on Ordered m Machines

Here we consider job 1 and job 2 to be processed on machines M_1, M_2, ..., M_m. The technological order of these machines to execute each job is specified in advance. Certainly, the ordering is different for both jobs, otherwise this problem will reduce to 2 jobs on m machines already discussed.

The exact or expected processing time on machines M_1, M_2, ..., M_m are given. The problem is to minimize the total elapsed from the start of the first job on the first machine to the completion on the last machine. The elapsed time for job 2 will turn to be same. The optimal sequence can be obtained graphically for which the algorithm is given as follows.

Algorithm. The algorithm proceeds as

Step 1. Draw two perpendicular lines, one representing the processing time for job 1, while job 2 remains idle, and vertical line representing the processing time for job 2, while job remains idle.

Step 2. Mark the processing tome for job1 and job 2 on the horizontal and vertical line respectively according to the specified order of machines.

Step 3. Start from the origin (start point), construct various blocks by pairing the same machines until the end point.

Step 4. Sketch the line from origin to the end point by moving horizontally,, vertically and diagonally at 45° with the horizontal base. The horizontal segment of this line indicates that the first job is under process while the second job is idle. Similarly, the vertical segment of the line indicates that the second job is under process while the first job is idle. The diagonal segment denotes that both jobs are under process.

5. The path that minimizes the idle time is the optimal path for both of the jobs. Thus, the path on which the diagonal movement is maximum is to be selected.

6. The total elapsed time is calculated by adding the idle time for either of the job to the completion of that job on all machines.

The following example will make this graphical approach more illustrative.

Example 5. Use the graphical method to minimize the time required to process the following jobs on two machines. Find the job which should be processed first. Also, calculate the total elapsed time to complete both jobs.

$$\text{Job 1} \begin{cases} \text{order} & A \quad B \quad C \quad D \quad E \\ \text{Time} & 3 \quad 4 \quad 2 \quad 6 \quad 2 \end{cases}$$

$$\text{Job 2} \begin{cases} \text{order} & B \quad C \quad A \quad D \quad E \\ \text{Time} & 5 \quad 4 \quad 3 \quad 2 \quad 6 \end{cases}$$

Draw the lines (horizontal and vertical) OX and OY representing the processing time of jobs 1 and 2. Mark the processing time of these jobs as shown in Fig. 10.1. Here, $v_1 + v_2 = 2 + 3 = 5$ (idle time for job 1), and $h_1 = 2$ (idle time for job 2).

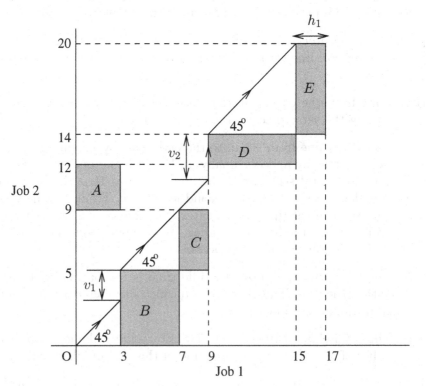

Figure 10.1

Now, draw the shaded rectangular blocks corresponding to each machine. Take A horizontal vs A vertical to complete the block A. Similarly, do for other machines. Start from the origin O, we move on doing jobs avoiding the shaded rectangular blocks until the finish point is reached. Here it is important to note that we shall try to move as such as we can along a line having an angle 45° to the horizontal. Whenever it is not possible to move along the line having 45° degree with horizontal, we shall move along horizontally and vertically as required.

Both jobs can not be processed simultaneously, this means that that the diagonal movement through blocked out ares is not permitted. The best path is shown by arrows in Fig 10.1. From the graph we conclude

Job 1 before job 2 on machine A

Job 2 before job 1 on machine B

Job 2 before job 1 on machine C

Job 1 before job 2 on machine D

Job 2 before job 1 on machine E

Total elapsed time = Processing time for job 1 + Idle time for job 1
$$= 17 + 2 + 3 = 22 \text{ hours}$$

or

Total elapsed time = Processing time for job 2 + Idle time for job 2
$$= 20 + 2 = 22 \text{ hours}$$

Example 6. Using the graphical method calculate the minimize time needed to process the job 1 and job 2 on five machines A, B, C, D, E, i.e., on each machine. Find the job which must be done first. Also, determine the total elapsed time to complete both jobs.

Job 1 $\begin{cases} \text{order} & A \quad B \quad C \quad D \quad E \\ \text{Time} & 1 \quad 2 \quad 3 \quad 5 \quad 1 \end{cases}$

Job 2 $\begin{cases} \text{order} & C \quad A \quad D \quad E \quad B \\ \text{Time} & 3 \quad 4 \quad 2 \quad 1 \quad 5 \end{cases}$

Draw the lines (horizontal and vertical) OX and OY representing the processing time of jobs 1 and 2. Mark the processing time of these jobs as shown in Fig. 10.2.

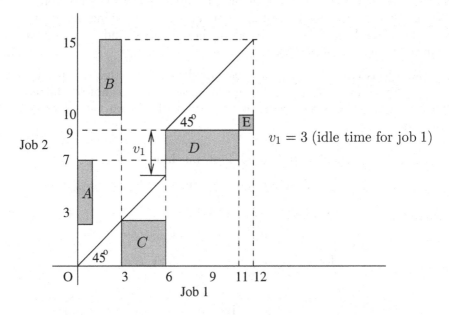

Figure 10.2

Now, draw the shaded rectangular blocks corresponding to each machine. Take A horizontal vs A vertical to complete the block A. Similarly, do for other machines. Start from the origin O, we move on doing jobs avoiding the shaded rectangular blocks until the finish point is reached. Here it is important to note that we shall try to move as such as we can along a line having an angle 45° to the horizontal. Whenever it is not possible to move along the line having 45° degree with horizontal, we shall move along horizontally and vertically as required

Both jobs can not be processed simultaneously, this means that that the diagonal movement through blocked out areas is not permitted. The best path is shown by arrows in Fig 10.2. From the graph we conclude

Job 1 before job 2 on machine A

Job 1 before job 2 on machine B

Job 2 before job 1 on machine C

Job 1 before job 2 on machine D

Job 2 before job 1 on machine E

Total elapsed time = Processing time for job 1 + Idle time for job 1
$$= 12 + 3 = 15 \text{ hours}$$

or

Total elapsed time = Processing time for job 2 + Idle time for job 2
$$= 15 + 0 = 15 \text{ hours}$$

Remark. To decide the precedence relation of two jobs on any machine we observe the coordinate of the south-west corner of each square. Suppose coordinate of any south-west corner are (x, y), $x < y$. Then job 1 precedes job 2 on that machine.

Problem Set 10

1. The following table shows the machine time (in days) for 5 jobs to be processed on two different machines M_1 and M_2 in order $M_1 M_2$:

Job	1	2	3	4	5
M_1	3	7	4	5	7
M_2	6	2	7	3	4

   Assuming passing is not permitted, find the optimal sequence of jobs to be processed and corresponding total idle time for the machines in this period.

2. A book binder has one printing press, one binding machine, and the manuscripts of a number of different books. The time (in hours) required to perform the printing and binding operations for each book are shown below. It is required to determine the

order in which books should be processed so as to minimize the total time needed to turn out all the books.

Books	1	2	3	4	5	6
Printing time	35	130	60	30	90	120
Binding time	80	100	90	60	30	20

3. Find the optimal sequence that minimizes the total elapsed time (in hours) required to complete the following tasks:

Tasks	A	B	C	D	E	F	G
Processing time on M_1	4	9	8	5	8	9	7
Processing time on M_2	5	3	2	4	1	3	5
Processing time on M_3	7	6	8	12	6	5	11

4. Find the optimal sequence that minimizes the total elapsed time required to complete the following tasks. Each job is processed in the order $M_1 M_3 M_2$.

Job	1	2	3	4	5	6	7
Processing time on M_1	13	5	6	12	6	8	7
Processing time on M_2	8	9	8	4	8	9	4
Processing time on M_3	4	5	2	5	3	4	6

5. A ready-made garment manufacturer has to process seven items through two stages of production, i.e., cutting and sewing. The time taken for each of these items at the different stages are given below in hours.

Items	1	2	3	4	5	6	7
Cutting processing time	6	8	4	5	7	6	14
Sewing Processing Time	3	7	6	4	10	6	9

(a) Find an optimal order (sequence) of items to be processed through these stages so as to minimize the total processing time.

(b) Suppose the third stage, i.e., pressing and packing is added. The processing time for this stage are as follows:

Items	1	2	3	4	5	6	7
Pressing & Packing Time	10	12	11	13	12	10	11

Find the optimal order of the items to be processed as well as minimum total processing time.

6. Find the optimal sequence for processing 4 jobs, A, B, C, D on four machines M_1, M_2, M_3, M_4 in the order $M_1 M_2 M_3 M_4$. The processing time for jobs (in hours) are given below.

Job/machine	M_1	M_2	M_3	M_4
A	14	6	5	15
B	13	3	10	13
C	12	4	6	16
D	16	3	0	20

7. Suppose that we have four jobs J_1, J_2, J_3, J_4 which must be processed through machines M_1, M_2, M_3, M_4 in the following order.

$$J_1 : M_1 M_2 M_3 M_4$$
$$J_2 : M_1 M_3 M_2 M_4$$
$$J_3 : M_2 M_3 M_4 M_1$$
$$J_4 : M_2 M_4 M_3 M_1$$

Each processing of job on a machine takes an hour. Find a feasible job sequence, and compute the corresponding elapsed time T from the start of the first job to the completion of the last job finished.

8. Two jobs are to be processed on four machines M_1, M_2, M_3, M_4. The technological order for these jobs on machines are as follows:

$$J_1 : M_1 M_2 M_3 M_4$$
$$J_2 : M_4 M_2 M_1 M_3$$

The processing time for jobs (in hours) are given below:

Job/machine	M_1	M_2	M_3	M_4
J_1	5	6	8	3
J_2	3	7	4	8

Find the order of jobs to be processed on each machine. Also, calculate the total elapsed time.

9. A manufacturing company processes 6 different jobs on 2 machines A and B in that order The number of units of each job and its processing time (in minutes) on A and B are given in the following table:

Job	Units of each job	Processing time Machine A	Machine B
1	3	5	8
2	4	16	7
3	2	6	11
4	5	3	5
5	2	9	7.5
6	3	6	14

Find the optimal sequence, the total elapsed time and idle time for each machine.

Chapter 11

Integer Programming

In practical problems decision variables may not be continuous. For studying problems where the decision variables take positive integral values, the integer linear programming has been developed. The procedures involved in the integer linear programming are explained in this chapter and in the end we emphasize on its two applications, viz., the traveling salesman problem and Cargo loading problem.

11.1 Introduction

In many real life problems the decision variables make sense only if they are integers. For example, it is often necessary to assign activities to men, machines, and vehicles in integer quantities. Let us define the integer linear programming in a general case.

Definition 1. An integer linear programming problem (ILPP) is a LPP where some or all the decision variables are restricted to be integer valued.

Remarks. 1. An ILPP is said to be pure if all decisions variables are restricted to be integers.

2. An ILPP is said to be mixed if some of the decision variables are restricted to be integers while others can assume fractional or continuous values.

The general ILPP is written as

opt $x_0 = C^T X$

s.t. $a_{i1}x_1 + a_{i2}x_2 + \cdots + a_{in}x_n \geq, =, \leq b_i, \quad i = 1, 2, \ldots, m$

$X \geq 0.$

Some or all of the $\{x_1, x_2, \cdots, x_n\}$ may be integers. Separate these out and write

opt $x_0 = C^T X$

s.t. $a_{i1}x_1 + a_{i2}x_2 + \cdots + a_{in}x_n \geq, =, \leq b_i, \ i = 1, 2, \ldots, m$

$x_1, x_2, \ldots, x_k$ are integers and $x_{k+1}, x_{k+2}, \ldots, x_n \geq 0.$

Now, we develop a method widely known as branch and bound algorithm or Dakin's method for solving the integer linear programming problem.

11.2   Branch and Bound Algorithm

We would explain the Branch and Bound algorithm by an example as follows:

Consider the problem

$$\text{opt}\quad x_0 = 3x_1 + 2x_2$$
$$\text{s.t.}\quad x_1 \leq 2$$
$$x_2 \leq 2$$
$$x_1 + x_2 \geq 7/2$$
$$x_1, x_2 \geq 0 \text{ and integers.}$$

(a) The initial step is to solve the ILPP by ignoring the restriction on the decision variables of being integers.

(b) In the optimal solution, x_2 turns to be fractional. But this is not an optimal solution in terms of integers. By restricting x_2 to be an integer optimal solution may worse but certainly not better as this is done by adding the constraints. If, in optimal table $x_2 = \alpha$ (not positive integer), then we add the constraint $x_2 \leq [\alpha]$, where $[\alpha]$ denotes the greatest integer contained in α.Again solve the problem for optimality. If this table contains both x_1 and x_2 as positive integers, then stop at this node and compute z, the objective function value.

(c) Further, we see the effect of adding $x_2 \geq [\alpha] + 1$, and solve for optimality, if an integer solution is available or if infeasibility occurs then stop. In case of the integer solution compare the value z with the value computed in (b) and consider the solution which gives better value.

(d) In steps (b) and (c), if neither infeasibility exists nor a integer solution is obtained, then we continue with further additions of constraints until all nodes are fathomed.

Definition 2. A node is said to be fathomed if it satisfies one of the conditions.

1. The LPP optimal solution at that node is integer valued.

2. The LPP problem is infeasible.

3. The optimal value of z for the LPP is not better than current lower bound (for max problem) and current upper bound (for min problem).

In view of the above discussion, continue branching from those nodes only which has larger value (for max) smaller value (for min) of z then obtained in the recent integer solution, i.e., from unfathomed nodes.

The solution of the above problem by the graphical method gives five vertices, viz., $(0,0)$, $(2,0)$, $(0,2)$, $(2,1.5)$, $(1.5,2)$. Obviously, the maximum occurs at the vertex $(2,1.5)$, and the maximum value: $z = 9$. Mention this at Node 1. Since the optimal solution $x_1 = 2, x_2 = 1.5$ is not integer valued at x_2. Hence, we add the constraints $x_2 \leq [1.5]$ and $x_2 \geq [1.5] + 1$, i.e., $x_2 \leq 1$ and $x_2 \geq 2$ to the given problem and proceeds as follows:

(i) Now, with the addition of $x_2 \leq 1$, the vertices are $(0,0)$, $(1,0)$, $(2,1)$, $(0,1)$. The maximum occurs at $x_1 = 2, x_2 = 1$, and the maximum value: $z = 8$. Write this solution at Node 2.

(ii) However, the addition of $x_2 \geq 2$ gives the vertices as $(1.5,2)$, $(1,2)$. The maximum occurs at $x_1 = 1.5, x_2 = 2$ with the maximum value $z = 8.5$. Write this solution at Node 3.

(iii) The optimal solution at Node 1 is integer valued, while at Node 3 improves in spite of the fact it is noninteger valued. Hence, Node 2 is fathomed, and we continue at Node 3.

(iv) Like Steps (i) and (ii), we add the constraints $x_1 \le 1$ and $x_1 \ge 2$. For the addition of $x_1 \le 1$, we get the optimal solution at $x_1 = 1, x_2 = 2$ with the optimal value $z = 7$, see Node 4. This is fathomed as we have got integer valued solution. But the addition of $x_1 \ge 2$ turns the problem infeasible, see Node 5, and hence it is fathomed.

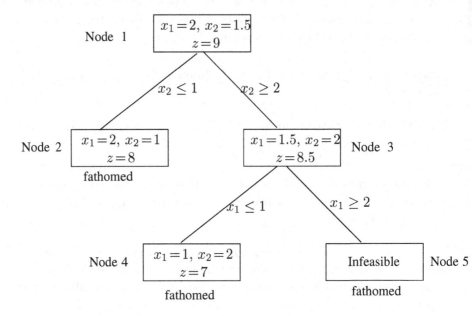

Figure 11.1

From all fathomed nodes, we pick up the best solution, and this is $x_1 = 2, x_2 = 1$ with maximum value $z = 8$, see Fig. 11.1.

Example 1. Find the optimal solution of the ILPP by Branch and Bound technique. Draw the diagram showing the solution inside nodes.

$$\max \quad z = x_1 - 2x_2 + 3x_3$$

$$\text{s.t.} \quad 2x_1 + 3x_2 - x_3 \le 8$$

$$x_1 - x_2 + x_3 \ge 5$$

$$x_1, x_3 \ge 0, \text{ and } x_2 \text{ is integer.}$$

First, we solve the related linear programming problem by the simplex method (sometimes the optimal table of the related LPP may also be given). The related LPP is

$$\max \quad z = x_1 - 2x_2 + 3x_3$$
$$\text{s.t.} \quad 2x_1 + 3x_2 - x_3 + s_1 = 8$$
$$x_1 - x_2 + x_3 + s_2 = 5$$
$$x_1, x_2, x_3, s_1, s_2 \geq 0.$$

The optimal solution of the related LPP by the simplex method is

B V	x_1	x_2	$x_3 \downarrow$	s_1	s_2	Soln
x_0	-1	2	-3	0	0	0
s_1	2	3	-1	1	0	8
$\leftarrow s_2$	1	-1	$\boxed{1}$	0	1	5
x_0	2	$-1 \downarrow$	0	0	3	15
$\leftarrow s_1$	3	$\boxed{2}$	0	1	1	13
x_3	1	-1	1	0	1	5
x_0	$7/2$	0	0	$1/2$	$5/2$	$43/2$
x_2	$3/2$	1	0	$1/2$	$1/2$	$13/2$
x_3	$5/2$	0	1	$1/2$	$3/2$	$23/2$

Since the above optimal solution is fractional at $x_2 = 13/2$, while we need a positive integer to the above optimal table. We add the constraint $x_2 \leq 6$. The effect of the addition of the constraint is shown in Table 1.

Table 1

B V	x_1	x_2	x_3	$s_1 \downarrow$	s_2	s	Soln
x_0	$7/2$	0	0	$1/2$	$5/2$	0	$43/2$
x_2	$3/2$	1	0	$1/2$	$1/2$	0	$13/2$
x_3	$5/2$	0	1	$1/2$	$3/2$	0	$23/2$
$\leftarrow s$	$-3/2$	0	0	$\boxed{-1/2}$	$-1/2$	1	$-1/2$
	0	1	0	0	0	1	6

Table 1 (Contd.)

B V	x_1	x_2	x_3	$s_1 \downarrow$	s_2	s	Soln
x_0	2	0	0	0	3	4	21
x_2	0	0	0	0	1	1/2	6
x_3	1	0	1	0	1	1/2	11
s_1	3	0	0	1	1	-2	1

As per Table 1, the optimal solution is integer valued. However, in case it is not integer valued, say variable x_2 in the solution, we must have proceeded further.

Next, we investigate the effect of adding the constraint $x_2 \geq [13/2]+ 1 = 7$ as described in Table 2.

Table 2

B V	x_1	x_2	x_3	s_1	s_2	s	Soln
x_0	7/2	0	0	1/2	5/2	0	43/2
x_2	3/2	1	0	1/2	1/2	0	13/2
x_3	5/2	0	1	1/2	3/2	0	23/2
$\leftarrow s'$	3/2	0	0	1/2	1/2	1	-7
	0	1	0	0	0	-1	7

Thus, the addition of $x_2 \geq 7$ makes the solution infeasible.

From Tables 1 and 2, we sketch the flow diagram. Node 1 is the solution of the related LPP which is fractional in x_2. On adding the constraint $x_2 \leq 6$, we get an integer valued solution as $x_1=0$, $x_2=6$, $x_3=11$ with the optimal value $x_o=43/2$. This is shown on Node 2. Thus, Node 2 is fathomed.

Next, we add the constraint $x_2 \geq 7$ to the last iteration table and note that the resulting solution is infeasible, see Node 3. In this way, Node 3 is also fathomed. The flow diagram is shown in Fig. 11.2. Obviously, from the diagram Fig. 11.2, the optimal solution is

$$x_1 = 0, x_2 = 6, x_3 = 11, \quad \text{max value } x_0 = 21.$$

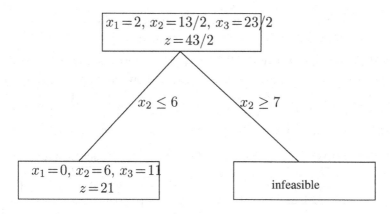

Figure 11.2

11.3 Traveling Salesman Problem

In a traveling salesman problem, a traveling salesman wishes to visit n cities and starts the journey from some city and visits each of $(n-1)$ cities once and only once and returns to the start. It is desired to find the order in which he should visit all the cities to minimize the total distance traveled. For the distance travelled, we may consider time, cost and other measure of effectiveness as desired. The distance, cost, and time etc., between any two cities are pre-assumed known.

The problem is well known because of its easy statement but difficult solution procedure. The difficulty is entirely computational since the solution obviously holds. There are $(n-1)!$ possible tours one or more of which may give minimum cost. The minimum cost could be conceivably infinite. It is conventional to assign infinite cost between the city pairs that have no direct connection. To solve the traveling salesman problem (TSP), we need a reduced cost matrix (see Section 7.6) at various iterations.

Example 2. The cost between city pairs are given as

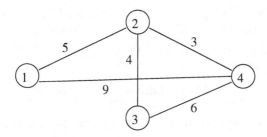

Find the tour of the traveling salesman so that cost of travel is minimum. Given that he starts journey from city 2.

Solution procedure. Write the problem in the matrix form, and get its reduced matrix by performing row operations followed by column operations (if necessary). Write the amount of total reduction on the bottom of right corner of the reduced matrix S. Both matrices are shown in the following table

Cost matrix

∞	5	∞	9
5	∞	4	3
∞	4	∞	6
9	3	6	∞

Reduced matrix

∞	0	∞	4
0	∞	0	0
∞	0	∞	2
4	0	2	∞

18

In the reduced matrix, city pairs with cost '0' are candidates to be considered in the tour. From the reduced matrix $(1,2)$, $(2,1)$, $(2,3)$, $(2,4)$, $(3,2)$, $(4,2)$ are the candidate pairs.

The least cost of exclusion (LCE) of (i,j) pair is the sum of the smallest cost in ith row and jth column (excluding (i,j) entry).

Step 1. The candidate pair for which the LCE is the largest should be retained in the tour (in case of a tie choose arbitrarily) or this is convenient to move from the top most row or the left most column.

Step 2. $\overline{(i,j)}$ means (i,j) pair is not included in the tour. The lower bound (LB) on $\overline{i,j}$ is computed using the formula:

$$\text{LB on } \overline{(i,j)} = \text{ LB on parent node } + \text{ LCE of } (i,j).$$

Step 3. To determine LB on (i,j), write the matrix of parent node by deleting ith row and jth column. Inclusion of (i,j) pair makes the cell (j,i) infeasible. Also check other infeasibility if exists. After this get the reduced matrix and calculate

$$\text{LB on } (i,j) = \text{ LB on parent node } + \text{ amount of reduction.}$$

Steps 2 and 3 are shown in the following cost matrices:

S

∞	0 $\boxed{4}$	∞	4
0 $\boxed{4}$	∞	0	0 $\boxed{2}$
∞	0 $\boxed{2}$	∞	2
4	2	2 $\boxed{2}$	∞

18

$(1,2)$

—	—	—	—
∞	—	0	0
∞	—	∞	$\not{1}$ 0
$\not{4}$ $\not{1}$ 0	—	$\not{1}$ 0	∞

24

Now, we execute the following iterations as follows:

Iteration 1. The LCE of each candidate pair is mentioned in East-South corner (ES) of the concerned cells. Note that, according to Step 1, the pairs $(1,2)$ and $(2,1)$ have the largest LCE, and hence must be be retained, see matrix S. Break the tie arbitrarily, and for convenience we retain (1,2), i.e., row-wise. Also,

$$\text{LB on } \overline{(1,2)} = 22.$$

Now, the set of all tours S is partitioned into two classes:

(a) Node $\boxed{(1,2)}$ means all tours in which salesman travels from city 1 to 2.

(b) Node $\boxed{\overline{(1,2)}}$ means all tours in which salesman does not travel from city 1 to 2.

LB on $(1,2)$: Write the matrix of $(1,2)$ by deleting the first row and second column of the parent reduced matrix S. Put ∞ in $(2,1)$ cell. After these changes, reduce the matrix, if required, see matrix $(1,2)$.

$$\text{LB on } (1,2) = 18 + 6 = 24.$$

Now, branching is done from the node which has the smallest LB, and hence in this case, branching is done from node $\overline{(1,2)}$. One iteration is over with outcome shown in the following graph.

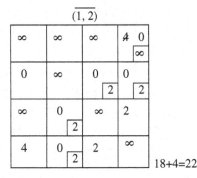

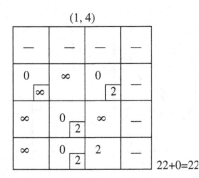

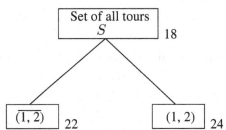

Iteration 2. To go for the second iteration, write the matrix $\overline{(1,2)}$ from the parent reduced matrix S, i.e., put ∞ in cell $(2,1)$, and find the candidate pairs to be included in the tour. We have to be careful while branching is done from any node the matrix must be in the reduced form. By Step 1, $(1,4)$ has the largest LCE, see matrix $\overline{(1,2)}$.

$$\text{LB on } \overline{(1,4)} = \infty.$$

Include $(1,4)$ in the tour.

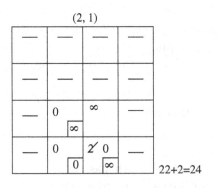

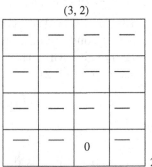

<u>LB on $(1,4)$</u>: Write the matrix of $(1,4)$ from the parent matrix $\overline{(1,2)}$ by deleting the first row and fourth column, and put ∞ at

$(4, 1)$ position. Reduce it, if necessary, see the table of matrix $(1, 4)$.

$$\text{LB on } (1, 4) = 22 + 0 = 22.$$

What we have done in the second iteration is depicted in the graph:

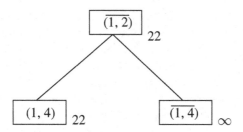

Iteration 3. Branching is done from $(1, 4)$ and $(2, 1)$ is the candidate pair with

$$\text{LB on } \overline{(2, 1)} = \infty.$$

LB on $(2, 1)$: Write the matrix $(2, 1)$ by deleting the second row and first column from its parent matrix $(1, 4)$ and making the cell $(2, 1)$ infeasible.

$$\text{LB on } (2, 1) = 22 + 2 = 24.$$

The outcome of the third iteration is

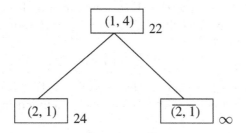

Iteration 4. Now, branching can be done from either $(2, 1)$ or $(1, 2)$. Let us continue from $(2, 1)$. Find the candidate pair to be included in the tour. This is $(3, 2)$. Also,

$$\text{LB on } \overline{(3, 2)} = 24 + \infty = \infty.$$

LB on $(3,2)$: Delete the third row and second column from the matrix of $(2,1)$ and put ∞ on $(2,3)$ position. After reduction, this ensures

$$\text{LB on } (3,2) = 24.$$

The outcome of the fourth iteration is

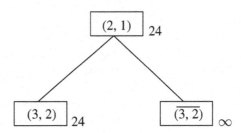

Branching stops when there is a single entry, and include this entry in the tour, i.e., $(4,3)$. Connecting all the graphs, we get the left side branching of Fig. 11.3.

We have observed that in the fourth iteration branching was also possible from node $(1,2)$. This will generate another tour. So, start branching from $(1,2)$ and follow the same steps as explained above. Observe the matrix of $(1,2)$ which is already available in previous branching. From this matrix, note that cell $(3,4)$ has the largest LCE, and hence it is to be retained in the tour. Also,

$$\text{LB on } \overline{(3,4)} = \infty.$$

LB on $(3,4)$: Delete the third row and fourth column of matrix $(1,2)$. Put ∞ in cell $(4,3)$. Then get it reduced to have the matrix of $(3,4)$.

$$\text{LB on } (3,4) = 24.$$

Again, branching is done from Node $(3,4)$. From the matrix of $(3,4)$, it is obvious that $(2,3)$ is to be included in the tour. Also,

$$\text{LB on } \overline{(2,3)} = \infty.$$

LB on $(2,3)$: Write the matrix of $(2,3)$ by deleting the second row and third column from matrix $(3,4)$ and make position $(3,2)$ infeasible (but it is already infeasible).

$$\text{LB on } (2,3) = 24.$$

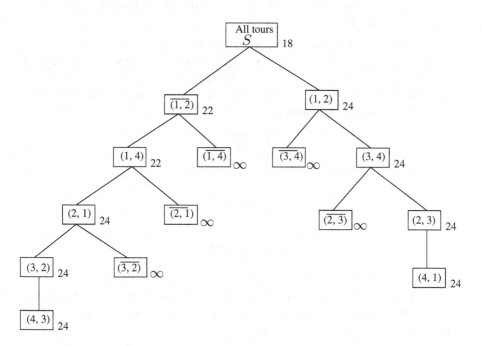

Since the matrix of $(2,3)$ contains only single entry in cell $(4,1)$, we include this cell in the tour. Combining all branching from node $(1,2)$, we get the left side of the graph in Fig. 11.3. From the complete graph, we conclude:

Figure 11.3

The first tour retains the candidates pairs $(1,4), (2,1), (3,2), (4,3)$, and the tour is constituted as

$$2 \longrightarrow 1 \longrightarrow 4 \longrightarrow 3 \longrightarrow 2, \text{ optimal cost of travel} = 24.$$

The second tour retains the candidates pairs $\{(1,2), (3,4), (2,3), (4,1)\}$, and the tour is constituted as

$$2 \longrightarrow 3 \longrightarrow 4 \longrightarrow 1 \longrightarrow 2, \text{ optimal cost of travel} = 24.$$

Remarks. 1. While solving TSP by the branch and bound technique, the reader is advised to start branching on a separate paper and the remaining work looking at the branching.

2. Note that the second tour is also obtained by reversing the arrows. This happens as the cost matrix is symmetric. For an asymmetric cost matrix this may not happen.

3. Sometimes it is convenient and the solution converges rapidly if we reduce the matrix by applying the column reduction first and then row reduction. Any way, select the reduced matrix which has higher starting LB. This reduces the length of a tree diagram, only such choice of reduction is effective in the asymmetric TSP, see problem 7 of Problem set 11.

4. For the cost matrix of order greater than 4, one must be very cautious to find LB on non bar type nodes. Because at the second and third stage of branching, the infeasibility may enter, and hence determination is sensitive. Consider the situation for some problem given in the following graph:

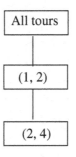

LB on $(2,4)$: Write the matrix $(1,2)$, delete 2nd row and 4th column. Put $(4,2)$ position infeasible. Also $(4,1)$ position turns out to be infeasible, otherwise the tour

$$1 \longrightarrow 2 \longrightarrow 4 \longrightarrow 1$$

is not complete.

Formulation of TSP as Assignment Problem: A traveling salesman has to visit n cities. He has to start from any one city and visit each city once and only once, and return to the starting point. Suppose he starts from kth city and the last city he visited is m. Then journey from m to k is mandatory for the tour. Let c_{ij} be the cost of travel from city i to city j. Then the traveling salesman problem can be formulated as an assignment problem given by

$$\min \quad z = \sum_{i=1}^{m} \sum_{j=1}^{n} c_{ij} x_{ij} \tag{11.1}$$

$$\text{s.t.} \quad \sum_{i=1}^{n} x_{ij} = 1, \; j = 1, 2, \ldots, n, \; i \neq j, \; i \neq m \tag{11.2}$$

$$\sum_{j=1}^{n} x_{ij} = 1, \; i = 1, 2, \ldots, n, \; i \neq j, \; j \neq m \tag{11.3}$$

$$x_{mk} = 1 \tag{11.4}$$

$$x_{ij} = 0 \text{ or } 1.$$

Solution procedure. Solve the problem as an assignment problem and find all assignment solutions. Any cyclic assignment solution is a solution of the traveling salesman problem. In case cyclic assignments are not available, then include the lowest entry (other than 0) for the assignment. Delete the row and column of this lowest entry and find '0' assignment in the remaining matrix. Again, observe whether cyclic assignment are available, otherwise include the next higher entry (other than the above lowest entry) in the tour and delete the row and column of this choice, and find '0' assignment in the remaining matrix. Continue the above procedure till at least one cyclic assignments is available.

Remark. If more than one cyclic assignment solutions are available at any stage, the TSP has alternate optimal solutions.

Example 3. Solve Example 2 for traveling salesman by the assignment technique.

Using the Hungarian method, the optimal solution is given by the assignment matrix shown below. The optimal assignment are

$$1 \longrightarrow 2, \; 2 \longrightarrow 1, \; 3 \longrightarrow 4, \; 4 \longrightarrow 3.$$

However, these assignments are not cyclic. Hence, consider the lowest entry 2 of cell $(4, 1)$ or $(1, 4)$, other than '0'. Thus, there is a tie for

∞	$\boxed{0}$	∞	2
$\boxed{0}$	∞	$\cancel{0}$	$\cancel{0}$
∞	$\cancel{0}$	∞	$\boxed{0}$
2	$\cancel{0}$	$\boxed{0}$	∞

the lowest cost entry. Break the tie arbitrarily. Let us consider 2 in cell $(4,1)$ and make an assignment in this cell. Delete the row and column containing this entry and make '0' assignment in the remaining submatrix, see following matrix

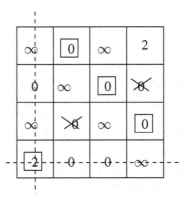

The next optimal assignment are $(1,2)$, $(2,3)$, $(3,4)$, $(4,1)$, and this is cyclic. Thus, the required tour is

$$1 \longrightarrow 2 \longrightarrow 3 \longrightarrow 4 \longrightarrow 1, \quad \text{Travel cost} = 24.$$

The alternative tour is obtained if the tie is broken by taking 2 in cell $(1,4)$. The required tour in this case is

$$1 \longrightarrow 4 \longrightarrow 3 \longrightarrow 2 \longrightarrow 1, \quad \text{Travel cost} = 24.$$

11.4   Cargo Loading Problem

The cargo loading problem is also known as Knapsack problem. We describe this problem by an example as follows:

Trieste food corporation plans to dispatch a truck with food items to another city. There are five items the corporation plans to dispatch with the restriction that the weight of the items on the truck must not exceed 115 units. The details of the food items are as

Item no.	Weight	Value	Value per unit weight
1	80	40	1/2
2	20	50	5/2
3	30	60	2
4	55	55	1
5	40	60	3/2

What should be the loading pattern so that the value of the truck is maximum? The divisibility of the food items is not permissible.

Write the problem in order of increasing value per unit weight, and assign the index number correspondingly.

Item no.	Index no.	Weight	Value	Value/weight
2	1	20	50	5/2
3	2	30	60	2
5	3	40	60	3/2
4	4	55	55	1
1	5	80	40	1/2

Node $\boxed{(\bar{1})}$ means item with index no. 1 is completely excluded;

Node $\boxed{(1)}$ means item with index no. 1 is completely included.

Assuming divisibility for the last index to be included, find UB (Upper Bound) as

UB on ($\overline{1}$)		
index no.	weight	value
2	30	60
3	40	60
4	35	35
	105	155

UB on (1)		
index no.	weight	value
1	20	50
2	30	60
3	40	60
4	15	15
	105	185

Start branching from the node having the largest UB value, and hence in this case from node (1),

UB on (1, 2)		
index no.	weight	value
1	20	50
2	30	60
3	40	60
4	15	15
	105	185

UB on (1, $\overline{2}$)		
index no.	weight	value
1	20	50
3	40	60
4	45	45
	105	155

UB on (1, 2, 3)		
index no.	weight	value
1	20	50
2	30	60
3	40	60
4	15	15
	105	185

UB on (1, 2, $\overline{3}$)		
index no.	weight	value
1	20	50
2	30	60
4	55	55
	105	165

UB on (1, 2, 3, 4)			UB on (1, 2, 3, $\overline{4}$)		
index no.	weight	value	index no.	weight	value
1	20	50	1	2	50
2	30	60	2	30	60
3	40	60	3	40	60
4	55	55	5	15	7.5
	145	225		105	177.5
	infeasible				

The node $(1, 2, 3, 4)$ is infeasible, we continue branching from $(1, 2, 3, \overline{4})$ as

UB on (1, 2, 3, $\overline{4}$, 5)			UB on (1, 2, 3, $\overline{4}, \overline{5}$)		
index no.	weight	value	index no.	weight	value
			1	2	50
	infeasible		2	30	60
			3	40	60
				90	170

Hence, the optimal loading pattern is

Item	Weight	Value
2	20	50
3	30	60
5	40	60
	90	170

All the analysis is shown by a tree diagram depicted in Fig. 11.4.

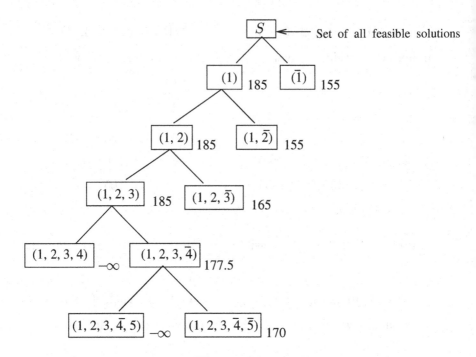

Figure 11.4

Problem Set 11

1. Solve the following ILPP

$$\min \quad x_0 = 2x_1 - x_2$$
$$\text{s.t.} \quad x_1 + 3x_2 = 13$$
$$x_1, x_2 \geq 0 \text{ and integers}$$

Suggestion. In solving the related LPP, x_1 will function to give an identity matrix.

2. Solve the following ILPP whose related LPP's optimal table is also given.

$$\min \quad x_0 = 6x_1 + 5x_2$$
$$\text{s.t.} \quad x_1 + 2x_2 \geq 20$$
$$3x_1 + 2x_2 \geq 50$$
$$x_1, x_2 \geq 0 \text{ and integers}$$

B V	x_1	x_2	s_1	s_2	Soln
x_0	0	0	$-3/4$	$-7/4$	$205/2$
x_2	0	1	$-3/4$	$1/4$	$5/2$
x_1	1	0	$1/2$	$-1/2$	15

3. The optimal table of the related LPP of an ILPP is given below. Find the optimal solution of the ILPP

B V	x_1	x_2	s_1	s_2	Soln
x_0	0	0	$5/3$	$4/3$	$38/3$
x_1	1	0	$1/3$	$-1/3$	$4/3$
x_2	0	1	$1/3$	$2/3$	$10/3$

Suggestion. Branch off from the value of x_1.

4. Solve TSP for which the effectiveness matrix is given below

∞	3	2	7
3	∞	4	2
6	13	∞	5
15	7	3	∞

5. Solve the following TSP

∞	2	5	6	19
5	∞	3	4	17
1	3	∞	2	11
6	3	7	∞	18
11	17	18	7	∞

under the condition that the traveling salesman is asked to travel from city 1 to city 4 and the remaining journey is left at his choice.

Suggestion. Put ∞ in the cells $(1, 2), (1, 3)$ and $(4, 1)$, and then solve the problem.

6. Using branch and bound technique solve the traveling sales problem for which the cost of direct travel between city pairs is given in the table

∞	80	70	50
20	∞	60	40
30	100	∞	30
70	50	40	∞

Suggestion. To get the reduced matrix, if we reduce row-wise followed by column-wise then $LB = 150$. However, it is interesting to note that if the process is reversed, then $LB = 170$. Hence, for computational efficiency use the sequence of reduction for which LB is higher.

7. A machine operator processes 4 type of items on his machine each week and is interested to find a sequence for them. The setup cost per change depends on the item presently on the machine and item to be made according to the following table

		To item			
		A	B	C	D
From item	A	∞	4	7	3
	B	4	∞	6	3
	C	7	6	∞	7
	D	3	3	7	∞

If he produces each type of item once and only once, each week, how should he sequence the items on the machine in order to minimize the total setup cost.

Suggestion. This is to be solved as traveling salesman problem. If we solve by assignment method, then it is obvious how the assignment method to solve TSP is freak. To get cyclic assignment when we go from 0 assignment to next higher assignment at 1 we get cyclic assignments but with different costs. Take that cyclic assignment which corresponds to minimum cost.

8. The wandering salesman problem (WSP) is a TSP, except that the salesman can start wherever he wishes and does not have to

return to the starting city after visiting all the cities. Show how to transform in polynomial time any instance of the TSP to an equivalent instance of the WSP.

Suggestion. By equivalent we mean that the optimal tour of one can be easily derived from the original tour of the other.

9. Write the formulation of the cargo loading problem solved in Section 11.4

10. A hiker plans to go on an excursion and wishes to take four items with him. However, he is not in a position to carry more than 60 lbs weight. To assist in selection process he has assigned a value to each item in ascending order of significance as

Item	1	2	3	4
Weight	52	23	35	7
Value	100	60	70	15

What items should he take to maximize the total value without exceeding weight restrictions?

Chapter 12

Dynamic Programming

Dynamic programming is introduced. Next its applications to solve various well known real life problems have been demonstrated by offering various numerical problems. The last section is devoted to give an exposition of linear vis a vis dynamic programming.

12.1 Introduction

Dynamic Programming was invented by Richard E. Bellman. This technique became so popular that it took the place of well known simplex method for linear systems. Both type of the systems, i.e., linear and non-linear problems have been simultaneously dealt with the help of dynamic programming. Dynamic programming is related to branch and bound in the sense that it performs an intelligent enumeration of all feasible points of a problem, but it does in an entirely different way. The idea is to work backward from the last decision to the earlier ones. However, it works in forward direction also.

Suppose, we have to make a sequence of n decisions to solve a problem, say $D_1, D_2, \ldots, D_n$. If the sequence is optimal, the last k decisions

$$D_{n-k+1}, D_{n-k+2}, \ldots, D_n$$

must be optimal. That is completion of an optimal sequence of decisions must be optimal. This approach is often referred to as the *principal of optimality*.

The usual application of dynamic programming entails breaking down the problem into stages (subproblems) at which the decisions take place and finding a recurrence relation that takes us backward from one stage to the previous stage. The optimal solution of one stage (subproblem) is used as an input to the next stage (subproblem). By the time we reach at the last stage, we will have at hand the optimal solution for the entire problem. The subproblems are linked together by some common constraints. As we move from one subproblem to the next, we must be careful for the feasibility of these constraints.

We shall explain the method by an example, stating the shortest path problem for layered networks, in which the sequence of decisions from the last to first is clear.

Remark. The dynamic programming programming is really simpler than other techniques to solve linear and nonlinear models. However, computationally it is not so efficient as compared to other techniques.

12.2   Formulation

Consider the following network (maps of various routes in going from A to I). The distances between the cities are marked on the connecting arcs as depicted in Fig. 12.1.

Objective. A and I are connected through different routes with specified distances. Our objective is to select the route along which it requires the minimum distance to reach I from A.

Backward Dynamic Programming. The problem is solved in stages:

Stage 1. Consider the cities G and H which are connected to destination I. We say G and H are on the last stage $j = 1$. We are coming from A. Suppose we have reached G (How we reached is insignificant). We have only one route from G to I with distance 5 units.

Similarly, if we are at H, then we have only one route from H to I with distance 7 units.

Thus, we have exhausted all states of the stage $j = 1$ ($x_1 = G, H$).

Stage 2. Now come backward to stage $j = 2$. The state variable can take three values $x_2 = D, E, F$.

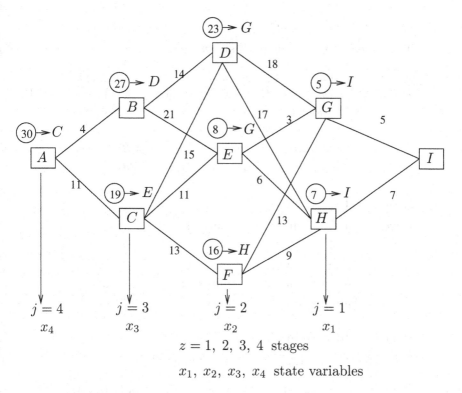

$$z = 1, 2, 3, 4 \text{ stages}$$

$$x_1, x_2, x_3, x_4 \text{ state variables}$$

Figure 12.1

If we are at D (how we reached at D is the least concerned), then there are two choices:

Going through G requires $18 + 5 = 23$ units;
Going through H requires $17 + 7 = 24$ units.

Hence, go to G from D as it takes the minimum distance. Similarly, from E, we have 2 choices:

Going through G requires $3 + 5 = 8$ units;
Going through H requires $6 + 7 = 13$ units.

Hence, go to G from E to travel the minimum distance. Similarly, from F, we have 2 choices:

Going through G requires $13 + 5 = 18$ units;
Going through H requires $14 + 7 = 21$.

Hence, go to G from F.

Thus, we have exhausted all states of the stage $j = 2$.

Stage 3. Again, come backward to the stage $j = 3$, x_3 can take two values B and C. If we are at B (how we reached, again least worried), we have two choices:

Going through D requires $14 + 23 = 27$ units;
Going through E requires $21 + 8 = 29$

Hence, go to E from D.

If we are at C, we have three options:

Going through D requires $15 + 23 = 38$ units;
Going through E requires $11 + 8 = 19$ units;
Going through F requires $13 + 16 = 29$ units

Hence, go to E from C.

All states of the stage $j = 3$ are exhausted

Stage 4. Come backward to stage $j = 4$. x_4 can take only one value A.

Going through B requires $27 + 4 = 31$ units;
Going through C requires $11 + 19 = 30$ units

Hence, go to C from A.

Finally, the optimal tour can be read from the diagram looking at circular indications from node A to node I. The optimal route is

$$A \longrightarrow C \longrightarrow E \longrightarrow G \longrightarrow I; \text{ minimum distance} = 30\text{units}.$$

Remark. Decision variables are: $y_1 = 5$, 7; $y_2 = 18$, 17, 3, 6, 13, 9; $y_3 = 14$, 21, 15, 11, 4, 11; and the decision variables giving optimal policy are:
$$y_1 = 5, \ y_2 = 3, \ y_3 = 11, \ y_4 = 11.$$

Forward Dynamic Programming. The above problem can also be solved by an another way called forward dynamic programming. The procedure is as follows:

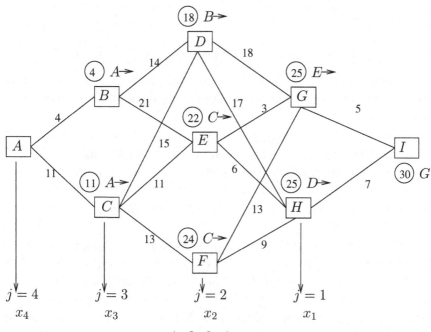

$z = 1, 2, 3, 4$ stages

x_1, x_2, x_3, x_4 state variables

Figure 12.2

Let us take the same map of routes as shown in Fig. 12.2.

Consider states B and C of stage $j = 3$. To reach at B, there is only one choice, i.e., $A \longrightarrow B$ (4 units). Also, to reach at B again one choice is available with distance $A \longrightarrow C$ as 11 units.

The stage $j = 3$ is exhausted, and we move forward to $j = 2$ (with respect to A). Here, state variable x_2 can take three values, D, E, F.

If we are at D (no matter how we reached), we have two choices:

Coming through B requires $4 + 14 = 18$ units;
Coming through C requires $11 + 15 = 26$ units

Prefer to come through B. At E, there are two choices:

Coming through B requires $21 + 4 = 25$ units;
Coming through C requires $11 + 11 = 22$ units

Prefer to come from C. At F, there is only one choice:

Coming through C requires $16 + 11 = 27$ units;

Thus, stage $j = 2$ is exhausted. Move forward to $j = 1$. Here, state variable x_1 can take values G, H. At G, there are three choices:

Coming through D requires $18 + 18 = 36$ units;
Coming through E requires $22 + 3 = 25$ units;
Coming through F requires $27 + 13 = 40$ units

Prefer to come from E. At H, there are three choices:

Coming through D requires $18 + 17 = 35$ units;
Coming through E requires $22 + 6 = 28$ units;
Coming through F requires $24 = 9 = 33$ units.

Prefer to come from E. Now, $j = 1$ is exhausted and we are at destination I. To come to I, we have two choices:

Coming through G requires $25 + 5 = 30$ units;
Coming through H requires $33 + 7 = 40$ units.

Prefer to come from G. Finally, tracing back from I, the shortest route is

$$A \longrightarrow C \longrightarrow E \longrightarrow G \longrightarrow I; \text{ minimum distance} = 30 \text{ units}$$

Remark. The nodes corresponding to the same stage can be directly connected. For such type of networks duplicate nodes may be introduced, see Problem 1.

Bellman's Optimal Principle. An optimal policy (or a set of decisions) has the property that whatever the initial state and initial decision are the remaining decisions must constitute an optimal policy with regard to the state resulting from the preceding decision.

Observe that route $D \longrightarrow H$ was rejected. No matter what the previous choice have been. If specified that we are at D, we should do our best to do with it. The decision variable $y_2 = 17$ and 18 will decide the optimal policy irrespective of G and H, and so when we are at D we take $y_2 = 18$.

Thus, Bellman's optimality principle states that the problem of decision making must be split up in stages and the optimal decisions should be taken sequentially. The iterative procedure is depicted in Fig. 12.3.

12.3 Recursive Relations

Many problems can be solved by dynamic programming using Bellman's principle of optimality. If a problem can be decomposed into stages and the decisions can be taken stage-wise using the principle of optimality, then it can be solved by dynamic programming.

Suppose the objective function can be written as

$$F = \sum_{i=1}^{N} f_i(y_i)$$

The value of decision variables are dependent on the state of the system. For example, in the previous discussion of shortest route problem by backward dynamic programming, for stage $j = 2$, we have three states, and when we consider state D only two decisions are possible $D \longrightarrow G$ or $D \longrightarrow H$, i.e., $y_2 = 18$ or 17. Divide the problem of optimizing F in a sequential manner as described in Fig. 12.4.

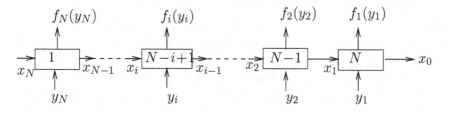

Figure 12.3

$f_i(y_i) = $ return function due to decision y_i;

$x_i = T_i(x_{i+1}, y_{i+1})$, $i = 1, 2, \ldots, N-1$ are state transformation functions.

The problem is formulated as

$$\text{opt} \quad F = f_1(y_1) + f_2(y_2) + \cdots + f_N(y_N) \qquad (12.1)$$
$$\text{s.t.} \quad a_1 y_1 + a_2 y_2 + \cdots + a_N y_N \geq, =, \leq b \qquad (12.2)$$
$$a_i > 0, i = 1, 2, \ldots, N, \ b > 0 \text{ are constraints}$$
$$y_i \geq 0 \text{ are decision variables}$$

We define the state variables as

$$x_N = a_1 y_1 + a_2 y_2 + \cdots + a_N y_N \geq, =, \leq b$$
$$x_{N-1} = a_1 y_1 + a_2 y_2 + \cdots + a_{N-1} y_{N-1} = x_N - a_N y_N = T_{N-1}(x_N, y_N)$$
$$\vdots$$

$$x_2 = a_1 y_1 + a_2 y_2 = x_3 - a_3 y_3 = T_2(x_3, y_3)$$
$$x_1 = a_1 y_1 = x_2 - a_2 y_2 = T_1(x_2, y_2)$$

Thus, by defining the state variables, we have brought the given problem in the format of Fig. 12.3. Now, Dynamic Programming can be utilized.

The first subproblem is shown in Fig. 12.4.

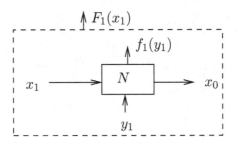

Figure 12.4

If x_1 is specified, the principle of optimality dictates that the decision variable y_1 must be selected $f_1(y_1)$ (irrespective of the preceding stages). Hence y_1 is chosen so that $f_1(y_1)$ is optimum for the state x_1. The optimum return from the first state x_1 is

$$F_1(x_1) = \text{opt}_{y_1} \sum_{i=1}^{N} f_1(y_1), \quad y_1 = \frac{x_1}{a_1} \qquad (12.3)$$

Equation (12.3) expresses the information that for a given state x_1 the optimum value f_1 is F_1 with y_1 adjusted to yield this value. This is called one stage policy.

The second subproblem is shown in Fig. 12.5.

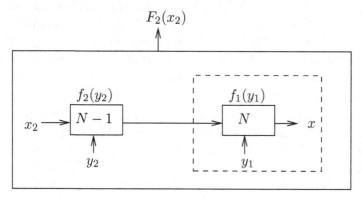

Figure 12.5

Let $F_2(x_2)$ be the optimal cumulative return from the IInd subproblem. Hence,

$$F_2(x_2) = \text{opt}_{y_1, y_2}[f_1(y_1) + f_2(y_2)] \qquad (12.4)$$

Now, principle of optimality requires that y_2 be selected to optimize f_2 for a given x_2. Therefore, two stage policy can be written as

$$F_2(x_2) = \text{opt}_{y_2}[F_1(x_1) + f_2(y_2)] \qquad (12.5)$$

Note that here, $x_1 = T_1(x_2, y_2)$ for a given point x_2.

Equation (12.5) is a function of y_1 and y_2 but the principle of optimality has allowed us to reduce this to a function of y_2 only in x_2.

In general, we can rewrite the cumulative return at ith state.

$$F_i(x_i) = \text{opt}_{y_i, y_{i-1}, \ldots, y_1}[f_1(y_1) + f_2(y_2) + \cdots + f_i(y_i)]$$

in the form

$$F_i(x_i) = \text{opt}_{y_i}[F_{i-1}(x_{i-1}) + f_i(y_i)], \ i = 2, 3, \cdots, N, \qquad (12.6)$$

with the information

$$F_1(x_1) = f_1(\frac{x_1}{a_1}), \text{ since } y_1 = \frac{x_1}{a_1}. \qquad (12.7)$$

The optimal value of the problem is $F_N(y_N)$. Here, $f_i(y_i)$ is the return from decisions y_i, and $F_{i-1}(x_{i-1})$ is the optimal cumulative return from the preceding $(i-1)$ stages. Relations (12.6) and (12.7) are the recursive formulas for the problem in format (12.1) and (12.2).

The above discussion is called backward recursion because the stage transformation is of the type $x_i = T_i(x_{i+1}, y_{i+1})$. This is convenient when x_N is specified. In the above formulation, $X_N \geq, =, \leq b$ is specified, and hence we used backward recursion. As in the shortest route problem of Section 5.1, we first fix $x_4 = A$, and then $x_3 = C$, $x_2 = E$, $x_1 = G$, $x_0 = I$.

The basic idea before dynamic programming is that we decompose any optimization problems in sub-problems. The contents of the remaining part of this chapter are optimization problems with discrete or continuous data.

12.4 Continuous Cases

In solving such type of problems calculus will be used as decision variables can take all values in a specified interest. The nature of these problems will be of the following types.

objective function	constraint
+	+
+	.
.	+
.	.

Example 1. Using the dynamic programming, find

(a) three nonnegative real numbers such that sum of the squares of these is minimum with the restriction their sum is not less than 45.

(b) the minimum of the sum of squares of these numbers whose product is 27.

For part (a), the problem is

$$\min \quad y_0 = y_1^2 + y_2^2 + y_3^2$$
$$\text{s.t.} \quad y_1 + y_2 + y_3 \geq 45$$
$$y_i > 0, \ i = 1, 2, 3$$

The problem is in appropriate format (12.1) and (12.2) and restriction on y_i's to be integers is not required, and hence calculus will be used.

The state variables are

$$x_3 = y_1 + y_2 + y_3 \geq 45$$
$$x_2 = y_1 + y_2 = x_3 - y_3$$
$$x_1 = y_1 = x_2 - y_2$$

The Ist subproblem is

$$F_1(x_1) = \min_{y_1} f_1(y_1) = y_1^2 = (x_2 - y_2)^2.$$

The IInd subproblem is

$$F_2(x_2) = \min_{y_2} \left[F_1(x_1) + f_2(y_2) \right] = \min_{y_2} \left[(x_2 - y_2)^2 + y_2^2 \right].$$

Now

$$\frac{d}{dy_2} \left[(x_2 - y_2)^2 + y_2^2 \right] = 0 \implies y_2 = \frac{x_2}{2}, \text{ and} \frac{d}{dy_2^2} > 0, \text{ at } y_2 = \frac{x_2}{2}.$$

Hence,

$$F_2(x_2) = \frac{x_2^2}{2} = \frac{(x_3 - y_3)^2}{2}.$$

The last subproblem is

$$F_3(x_3) = \min_{y_3} \left[F_2(x_2) + f_3(y_3) \right] = \min_{y_3} \left[\frac{(x_3 - y_3)^2}{2} + y_3^2 \right]$$

using $\frac{d}{dy_3} = 0$ and observing $\frac{d^2}{dy_3^2} > 0$, we find $y_3 = x_3/3$ is the minimum point. Hence,

$$F_3(x_3) = \frac{x_3^2}{3}$$

In this problem $x_N = x_3 \geq 45$ is satisfied. For $x_3 = 45$, the minimum occurs, and min $= 675$ when $x_3 = 45$. Tracing back

$$y_3 = \frac{x_3}{3} = 15, \quad \text{since } x_3 = 45$$

$$y_2 = \frac{x_2}{2} = 15, \quad \text{since } x_2 = x_3 - y_3 = 30$$

$$y_1 = x_1 = 15, \quad \text{since } x_1 = x_2 - y_2$$

$$y_0 = 675$$

For part (b) the problem is

$$\min \quad y_0 = y_1^2 + y_2^2 + y_3^2$$
$$\text{s.t.} \quad y_1 y_2 y_3 = 27$$
$$y_i > 0, \; i = 1, 2, 3$$

This is again continuous case of the type $(+, \cdot)$, and hence calculus is used.

Define state variables

$$x_3 = y_1 y_2 y_3 = 27$$
$$x_2 = y_1 y_2 = \frac{x_3}{y_3}$$
$$x_1 = y_1 = \frac{x_2}{y_2}$$

The Ist subproblem is

$$F_1(x_1) = y_1^2 = \left(\frac{x_2}{y_2}\right)^2 .$$

The IInd subproblem is

$$F_2(x_2) = \min_{y_2} \left[(x_2 y_2)^2 + y_2^2 \right] .$$

The minimum occurs at $y_2 = \sqrt{x_2}$ (can be checked by calculus as above). Hence,

$$F_2(x_2) = 2x_2 = \frac{2x_3}{y_3}$$

The last subproblem is

$$F_3(x_3) = \min_{y_3} \left[\frac{2x_3}{y_3} + y_3^2 \right]$$

Again, by derivative test, we find that the minimum occurs at $y_3 = x_3^{1/3}$. Thus,

$$F_3(x_3) = 3x_2^{2/3}.$$

Given that $x_3 = 27$ and this implies $F_3(x_3) = 27$.

Tracing back, the optimal solution is

$$y_3 = x_3^{1/3}, \qquad \text{since } x_3 = 27$$
$$y_2 = \sqrt{x_2} = 3, \quad \text{since } x_2 = \frac{x_3}{y_3} = 9$$
$$y_1 = 3, \qquad\qquad \text{since } x_1 = \frac{x_2}{y_2} = 3$$
$$y_0 = 27.$$

12.5 Discrete Cases

In such type of problems the decision variables can take only integral values. We shall study following problems which are again outcome of four types

$$(+, +), \quad (+, \cdot), \quad (\cdot, +), \quad (\cdot, \cdot).$$

The basic theme of this section is to get acquainted with some well-known real problems which can be solved by dynamic programming.

1. General Problems

2. Reliability Problems

3. Capital Budgeting Problem

4. Cargo Loading Problem

Now, we discuss these cases one by one and solve some problems concerning to each case.

General problems. Any problem will be based on any one of the cases mentioned above. We work out an example for the case $(+, \cdot)$. Other cases can be dealt with the similar approach.

Example 2. Find the maximum of sum of cubes of three positive value integers whose product does not exceed 5.

The problem is formulated as

$$\max \quad y_1^3 + y_2^3 + y_3^3$$
$$\text{s.t.} \quad y_1 y_2 y_3 \leq 5$$
$$y_i > 0, \ i = 1, 2, 3 \text{ are integers}$$

Define the state variables as

$$x_3 = y_1 y_2 y_3 \leq 5$$
$$x_2 = y_1 y_2 = x_3/y_3$$
$$x_1 = y_1 = x_2/y_2$$

Each y_i can assume any value from 1 to 5. Also, x_3 is permissible from 1 to 5. The state variables are computed in the following tables

x_3 \ y_3	1	2	3	4	5
1	1	—	—	—	—
2	2	1	—	—	—
3	3	—	1	—	—
4	4	2	—	1	—
5	5	—	—	—	1

$$x_2 = x_3/y_3$$

x_2 \ y_2	1	2	3	4	5
1	1	—	—	—	—
2	2	1	—	—	—
3	3	—	1	—	—
4	4	2	—	1	—
5	5	—	—	—	1

$$x_1 = x_2/y_2$$

The whole process is decomposed into subproblems as in Example 1. All computations have been done on the tables.

Ist subproblem

$x_1 = y_1$	1^*	2	3	4	5
$F_1(x) = f_1(y_1) = y_1^3$	1^*	8	27	64	125

The optimal solution is $y_1 = 1, y_2 = 1, y_3 = 5$; optimal value $= 127$ and this can be read by backtracking through (*) entries in all the tables of subproblems.

Reliability problem. An electric component has three components 1, 2 and 3 which are connected in a series, i.e., if one component fails the system does not work. To improve the reliability of the system each component is supplemented by 1 or 2 or 3 parallel units (in case any

IInd subproblem

y_3 \ x_3	$f_2(y_2) = y_2^3$					$F_1(x_1)$					$F_2(x_2)$
	1*	2	3	4	5	1*	2	3	4	5	add and ma:
1	1	—	—	—	—	1*	—	—	—	—	2*
2	1	8	—	—	—	8	1	—	—	—	9
3	1	—	27	—	—	27	—	1	—	—	28
4	1	8	—	64	—	64	8	—	1	—	65
5	1	—	—	—	125	125	—	—	—	1	126

IIIrd subproblem

y_3 \ x_3	$f_3(y_3) = y_3^3$					$F_2(x_2)$					$F_3(x_3)$
	1*	2	3	4	5	1*	2	3	4	5	add and ma:
1	1	—	—	—	—	2	—	—	—	—	3
2	1	8	—	—	—	9	2	—	—	—	10
3	1	—	27	—	—	28	—	2	—	—	29
4	1	8	—	64	—	65	9	—	2	—	66
5	1	—	—	—	125	126	—	—	—	2	127

component fails the supplement unit starts working and the system remains in operation). The following data gives the cost of parallel units and reliability at each component. The total money available for the improvement is \$9,000.

Number of Parallel units	Component I cost(Reliability)	Component II cost(Reliability)	Component III cost(Reliability)
1	1(0.4)	2(0.5)	3(0.7)
2	3(0.6)	4(0.6)	4(0.8)
3	4(0.8)	5(0.7)	5(0.9)

Here cost is given in thousands. Find the money spent on each component for attaching supplementary units so that reliability of the system is maximum.

Suppose y_i = money is spent on attaching supplementary units to component i, $i = 1, 2, 3$. $f_i(y_i)$ = reliability of the ith component when y_i amount is spent for attaching supplementary units to component i.

Our objective is find y_i such that reliability of the system is maximum. By the reliability principle, the reliability of the combined system is the product of reliabilities of its components. Thus, the problem is

$$\text{max} \quad f_1(y_1) \cdot f_2(y_2) \cdot f_3(y_3)$$
$$\text{s.t.} \quad y_1 + y_2 + y_3 \leq 9$$
$$y_i > 0, \ i = 1, 2, 3$$

From the given table, we have

y_1	$f_1(y_1)$
1	0.4
3	0.6
4	0.8

y_2	$f_2(y_2)$
2	0.5
4	0.6
5	0.7

y_3	$f_3(y_3)$
3	0.7
4	0.8
5	0.9

The sate variables are

$$x_3 = y_1 + y_2 + y_3 \leq 9$$
$$x_2 = y_1 + y_2 = x_3 - y_3$$
$$x_1 = y_1 = x_2 - y_2$$

From the above tables $y_1 + y_2 + y_3 \geq 6$. Hence, $6 \leq x_3 \leq 9$. The least value of $y_1 + y_2 = 3 = $ the least value of x_2. The largest value of $x_2 = $ largest value of x_3 - least value of $y_3 = 9 - 3 = 6$. Hence, $3 \leq x_2 \leq 6$ and $x_1 = y_1 = 1, 3, 4$.

Actually after getting bounds, we should decide the exact values which state variables take. The exact values are

$$x_3 = \{6, 7, 8, 9\}, \quad x_2 = \{3, 5, 6\}, \quad x_1 = \{1, 3, 4\}.$$

The computations of state variables and all subproblems are shown in the next tables. Tracing back through (*) entries, the optimal solution is given as

$$y_3 = 3, \ y_2 = 2, \ y_1 = 4, \quad \text{optimal value} = 28.$$

Hence $4, 2, 3$ thousand rupees should be spent for attaching supplementary units to 1st, IInd and IIIrd component, respectively.

y3 / x3	3	4	5
6	3	—	—
7	4	3	—
8	5	4	3
9	6	5	4

↑ x_2

y2 / x2	2	4	6
3	1	—	—
4	—	—	—
5	3	1	—
6	4	—	1

↑ x_1

Ist sub–problem

$x_1 = y_1$	1	2	4
$F_1(x_1)$	.4	.6	.8

IInd subproblem

	$f_2(y_2)$			$F_1(x_1)$			$F(x_1)f_2(y_2)$			$F_2(x_2)$
x_2 \ y_2	2	4	6	2	4	5	2	4	5	max
3	.5	—	—	.4	—	—	.2	—	—	.2
4	—	—	—	—	—	—	—	—	—	—
5	.5	.6	—	.6	.4	—	.3	.24	—	.3
6	.5	—	.7	.8	—	.4	.4	—	.28	.4

IIIrd subproblem

	$f_3(y_3)$			$F_2(x_2)$			$F(x_2)f_3(y_3)$			$F_3(x_3)$
x_3 \ y_3	2	4	6	2	4	5	2	4	5	max
6	.7	—	—	.2	—	—	.14	—	—	.14
7	.7	—	—	—	.2	—	—	.16	—	.16
8	.7	.8	.9	.3	—	.2	.21	—	.18	.21
9	.7	.8	—	.4	.3	.4	.28	.24	—	.28

Capital budgeting problem. The Punjab Supply Electricity Board intends to improve the total hydroelectric power P produced at three dams with available capital 3 units of money (in lakhs of dollars). If $f_i(u_i)$, the power generated from ith dam by investing money u_i is given as

$u_i's$	0	1	2	3
f_1	0	2	4	6
f_2	0	1	5	6
f_3	0	3	5	6

Find the optimal policy for budgeting the available capital.

The return function is defined as

$$\max \quad f_1(u_1) + f_2(u_2) + f_3(u_3)$$
$$\text{s.t.} \quad u_1 + u_2 + u_3 \leq 3$$
$$u_i \geq 0, \ i = 1, 2, 3$$

The state variables are

$$x_3 = u_1 + u_2 + u_3 \leq 3$$
$$x_2 = u_1 + u_2 = x_3 - u_3$$
$$x_1 = u_1 = x_2 - u_2$$

This implies each u_i can take values $0, 1, 2, 3$, and state variables can also take values $0, 1, 2, 3$.

The computations of state variables and solution of all subproblems are:

x_3 \ u_3	0	1	2	3
0	0	—	—	—
1	1	0	—	—
2	2	1	—	—
3	3	2	1	0

$\uparrow$ x_2

x_2 \ u_2	0	1	2	3
0	1	—	—	—
1	1	0	—	—
2	2	1	0	—
3	3	2	1	0

$\uparrow$ x_1

Ist sub–problem

$x_1 = u_1$	0	1	2
$F_1(x_1)$	0	1	2

IInd subproblem

x_2 \ u_2	$f_2(u_2)$ 0	1	2	3	$F_1(x_1)$ 0	1	2	3	$F(x_1) + f_2(u_2)$ 0	1	2	3	$F_2(x_2)$ max
0	0	—	—	—	0	—	—	—	0	—	—	—	0
1	0	1	—	—	2	0	—	—	2	1	—	—	2
2	0	1	5	—	4	2	0	—	4	3	5	—	5
3	0	1	5	6	6	4	2	0	6	5	7	6	7

Tracing back through (*) entries in all subproblems, the optimal solution is given by

$$u_3 = 1, \ u_2 = 2, \ u_1 = 0, \quad \text{max value} = 8$$

IIIrd subproblem

x_3 \ u_3	$f_3(u_3)$ 0	1	2	3	$F_2(x_2)$ 0	1	2	3	$F(x_2) + f_3(u_3)$ 0	1	2	3	$F_3(x_3)$ max
0	0	—	—	—	0	—	—	—	0	—	—	—	0
1	0	3	—	—	2	—	—	—	2	3	—	—	3
2	0	3	5	—	5	2	0	—	5	5	5	—	5
3	0	3	5	6	7	5	2	0	7	8	7	6	8

Cargo loading problem. This problem has already been discussed in Section 11.4 concerning integer programming. Here, this problem is solved using dynamic programming.

We have to load truck with three different items. The maximum allowable weight is 6. The weight and values per unit are given in the following table. It is required to find the loading pattern which maximizes value of the truck under the weight constraint.

Item	Weight per unit	Value per unit
1	1	15
2	3	70
3	2	35

Let y_1, y_2, y_3 be the number of items of Ist, IInd and IIIrd kind, respectively, which are to be loaded. Then the problem is formulated as

$$\max \quad 15y_1 + 70y_2 + 35y_3$$
$$\text{s.t.} \quad y_1 + 3y_2 + 2y_3 \le 6$$
$$y_i \ge 0, \ i = 1, 2, 3 \text{ and integers}$$

The state variables are given by the relations:

$$x_3 = y_1 + 3y_2 + 2y_3 \le 6$$
$$x_2 = y_1 + 3y_2 = x_3 - 2y_3$$
$$x_1 = y_1 = x_2 - 3y_2$$

This is a discrete problem because y_i take only integer values, and hence, calculus method is not applicable. The recursive relations are:

$$F_3(x_3) = \max_{y_3} [f_3(y_3) + F_2(x_2)]$$
$$F_2(x_2) = \max_{y_2} [f_2(y_2) + F_1(x_1)]$$
$$F_1(x_1) = f_1(y_1)$$

In view of the state variables' relations:

$y_1 = 0, 1, 2, 3, 4, 5, 6$ (put $y_2 = 0$, $y_3 = 0$ then $y_1 \leq 6$

gives these values)

$y_2 = 0, 1, 2$, put $y_1 = 0$, $y_3 = 0$ then $3y_2 \leq 6 \Rightarrow y_2 \leq 2$)

$y_3 = 0, 1, 2, 3$ (as above $2y_3 \leq 6 \Rightarrow y_3 \leq 3$)

Solving these state variable relations, we get that x_i, $i = 1, 2, 3$ can take any value from $0, 1, 2, 3, 4, 5, 6$.

The computations of state variables and all subproblems are mentioned in the following tables:

x_3 \ y_3	0	1	2	3
0	0	–	–	–
1	1	–	–	–
2	2	0	–	–
3	3	1	–	–
4	4	2	0	–
5	5	3	1	–
6	6	4	2	0

x_2

x_2 \ y_2	0	1	2
0	0	–	–
1	1	–	–
2	2	–	–
3	3	–	–
4	4	1	–
5	5	2	–
6	6	3	0

x_1

Ist sub–problem

$x_1 = y_1$	0	1	2	3	4	5	6
$F_1(x_1)$	0	15	30	45	60	75	90

IInd subproblem

x_2 $\diagdown$ y_2	$f_2(y_2) = 70y_2$			$F_1(x_1)$			$F(x_1) + f_2(y_2)$			$F_2(x_2)$
	0	1	2	0	1	2	0	1	2	max
0	0	—	—	0	—	—	0	—	—	0
1	0	—	—	15	—	—	15	—	—	15
2	0	—	—	30	—	—	30	—	—	30
3	0	—	—	45	0	—	45	—	—	45
4	0	70	—	60	15	—	60	85	—	85
5	0	70	—	75	30	—	75	100	—	100
6	0	70	140	90	45	0	90	115	140	140

IIIrd subproblem

x_3 $\diagdown$ y_3	$f_3(y_3) = 35y_3$				$F_2(x_2)$				$F(x_2) + f_3(y_3)$				$F_3(x_3)$
	0	1	2	3	0	1	2	3	0	1	2	3	max
0	0	—	—	—	0	—	—	—	0	—	—	—	0
1	0	—	—	—	15	—	—	—	15	—	—	—	15
2	0	35	—	—	30	—	—	—	30	35	—	—	30
3	0	35	—	—	45	0	—	—	45	50	—	—	45
4	0	35	70	—	60	15	0	—	85	65	70	—	85
5	0	35	70	—	75	30	15	—	100	70	85	—	100
6	0	35	70	105	90	45	30	0	140	115	100	105	140

Back tracing through (*) entries in all subproblems, the optimal is given as

$$y_1 = 0, \ y_2 = 2, \ y_3 = 0, \quad \text{max value} = 140$$

12.6 Forward Recursions

In case when x_0 is specified, then it would be convenient to reverse the direction, see shortest route path problem of Section 12.1, when solved by forward dynamic programming. This is called forward recursion, and is depicted in Fig. 12.6.

From Fig. 12.6,

$$x_{i+1} = T_{i+1}(x_i, y_i), \quad i = 1, 2, \ldots, N.$$

It is convenient, in forward recursion to start with x_1 instead of x_0. Here, the recursion formula will start with $F_N(x_N)$. In the end, given

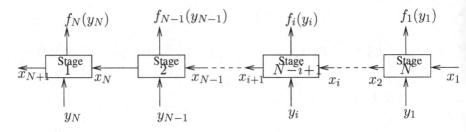

Figure 12.6

x_1, the functions are determined in order $F_1, F_2, \ldots, F_N$ (forward). Note that in backward recursion functions were determined in order $F_N, F_{N-1}, \ldots, F_1$ (backward). Here, we compute state variables for a stage from the state variable of earlier stage.

12.7　Linear Programming vs Dynamic Programming

This section concerns how a LPP can be solved with the help of dynamic programming. Consider the LPP

$$\text{opt}\quad y_0 = c_1 y_1 + c_2 y_2 + \cdots + c_n y_n \tag{12.8}$$

$$
\begin{array}{llllll}
\text{s.t.} & a_{11}y_1 & + \; a_{12}y_2 & + \; \cdots \; + & a_{1n}y_n & \leq \; b_1 \\
& a_{21}y_1 & + \; a_{21}y_2 & + \; \cdots \; + & a_{2n}y_n & \leq \; b_2 \\
& \vdots & + \; \cdots & + \; \cdots \; + & \cdots & \leq \; \vdots \\
& a_{m1}y_1 & + \; a_{m2}y_2 & + \; \cdots \; + & a_{mn}y_n & \leq \; b_m \\
& \text{Stage 1} & \text{Stage 2} & & \text{Stage n} & \text{Resources}
\end{array}
$$

$$\tag{12.9}$$

where y_j $(j = 1$ to $n)$ are nonnegative decision variables and $b_i \geq 0$ $(i = 1$ to $m)$ are the amounts available at m resources. Assume that $a_{ij} \geq 0$. We define the state variables as follows:

Let x_{1j} be the amount of resource 1 allocated to the stages $j, j +$

$1, \ldots, n$. Thus,

$$x_{11} = a_{11}y_1 + a_{12}y_2 + \cdots + a_{1n}y_n \leq b_1$$

$$x_{12} = a_{12}y_2 + \cdots + a_{1n}y_n = x_{11} - a_{11}y_1$$

$$\vdots$$

$$x_{1j} = a_{1j}y_j + \cdots + a_{1n}y_n = x_{1,j-1} - a_{1,j-1}y_{j-1} \qquad (12.10)$$

$$\vdots$$

$$x_{1n} = a_{1n}y_n = x_{1,n-1} - a_{1,n-1}y_{n-1}$$

Let x_{2j} be the amount of resource 2 allocated to the stages $j, j + 1, \ldots, n$. Thus,

$$x_{21} = a_{21}y_1 + a_{22}y_2 + \cdots + a_{2n}y_n \leq b_2$$

$$x_{22} = a_{22}y_2 + \cdots + a_{2n}y_n = x_{21} - a_{21}y_1$$

$$\vdots$$

$$x_{2j} = a_{2j}y_j + \cdots + a_{2n}y_n = x_{2,j-1} - a_{2,j-1}y_{j-1} \qquad (12.11)$$

$$\vdots$$

$$x_{2n} = a_{2n}y_n = x_{2,n-1} - a_{2,n-1}y_{n-1}$$

Similarly, we define $x_{3j}, x_{4j}, \ldots, x_{mj}$. Let us introduce the notation

$$X_j = (x_{1j}, x_{2j}, \ldots, x_{mj})^T, \quad j = 1, 2, \ldots, n. \qquad (12.12)$$

Thus, X_j denotes the state of the system at stage j. The amounts x_{1j} of different resources available at stage j denote the different states at stage j. This is depicted in Figure 12.7.

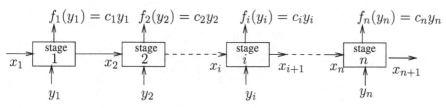

Figure 12.7

As usual $f_i(y_i)$ is the return at stage i from decision y_i. Note that $X_1 \leq (b_1, b_2, \ldots, b_m)^T$. Here, X_1 is specified, and hence we use forward dynamic programming. From (12.9) and (12.10), we have

$$X_{j+1} = T_{j+1}(X_j, y_j), \quad j = 1 \text{ to } n. \qquad (12.13)$$

where $X_j = (x_{1j}, x_{2j}, \ldots, x_{mj})^T$ are the m resources available for allocation to stages $j, j+1, \ldots$ and n, that is at stage j the maximum quantity of m resources available for allocation are $x_{1j}, x_{2j}, \ldots, x_{mj}$. At stage j, the decision variable is y_j; hence we must have

$$0 \leq a_{1j} y_j \leq x_{1j}$$
$$0 \leq a_{2j} y_j \leq x_{2j}$$
$$\vdots$$
$$0 \leq a_{mj} y_j \leq x_{mj}$$

Let

$$\min \left(\frac{x_{1j}}{a_{1j}}, \frac{x_{2j}}{a_{2j}}, \ldots, \frac{x_{mj}}{a_{mj}} \right)^T = \beta_j \qquad (12.14)$$

Then

$$0 \leq y_j \leq \beta_j. \qquad (12.15)$$

The recursive relations as developed in Section 12.6 are

$$F_n(X_n) = \operatorname{opt}_{y_n} f_n(y_n)$$
$$F_i(X_i) = \operatorname{opt}_{y_i} \{ c_i y_i + F_{i+1}(x_{i+1}) \}, \quad i = n-1, n-2, \ldots, 1,$$
$$\qquad (12.16)$$

where $0 \leq x_{ij} \leq b_i$ for all i and j. The better exposition of the theoretical ideas will be clear by solving a numerical example.

Example 3. Solve the following LPP using dynamic programming

$$\begin{aligned} \max \quad & y_0 = 28y_1 + 7y_2 \\ \text{s.t.} \quad & 4y_1 + 3y_2 \leq 12 \\ & 2y_1 + 5y_2 \leq 10 \\ & y_1, y_2 \geq 0 \end{aligned}$$

The state variables are

$$x_{11} = 4y_1 + 3y_2 \leq 12$$
$$x_{12} = 3y_2 = x_{11} - 4y_1$$

$$x_{21} = 2y_1 + 5y_2 \le 10$$

$$x_{22} = 5y_2 = x_{21} - 2y_1$$

where $X_1 = (x_{11}, x_{21})^T$, $X_2 = (x_{12}, x_{22})^T$. The recursive relations are shown with the help of (12.9),

$$F_2(X_2) = F_2(x_{12}, x_{22}) \max_{0 \le y_2 \le \beta_2} 7y_2,$$

where

$$\beta_2 = \min\left(\frac{x_{12}}{3}, \frac{x_{22}}{5}\right),$$
$$F_2(X_2) = 7\beta = 7 \min\left(\frac{x_{12}}{3}, \frac{x_{22}}{5}\right).$$

Now,

$$F_1(X_1) = F_1(x_{11}, x_{21}) = \max_{0 \le y_1 \le \beta_1} \{f_1(y_1), F_2(X_2)\}$$

$$= \max_{0 \le y_1 \le \beta_1} \left\{28y_1 + 7 \min\left(\frac{x_{11} - 4y_1}{3}, \frac{x_{21} - 2y_1}{5}\right)\right\}$$

Also,

$$\beta_1 = \min\left\{\frac{x_{11}}{a_{11}}, \frac{x_{21}}{a_{21}}\right\} = \min\left\{\frac{x_{11}}{4}, \frac{x_{21}}{2}\right\}$$

This stage, being the last stage, $x_{11} = 12$, $x_{21} = 10$, and hence $\beta_1 = 3$. Using these values, we get

$$F_1(X_1) = 28y_1 + \begin{cases} 7 \times \frac{10 - 2y_1}{5}, & 0 \le y_1 \le \frac{15}{7} \\ 7 \times \frac{12 - 4y_1}{3}, & \frac{15}{7} \le y_1 \le 3 \end{cases}$$

or

$$F_1(X_1) = \begin{cases} \frac{126}{5}y_1 + 14, & 0 \le y_1 \le \frac{15}{7} \\ \frac{56}{3}y_1 + 28, & \frac{15}{7} \le y_1 \le 3 \end{cases}$$

The largest $F_1(X_1)$ is 84, which occurs for $y_1 = 3$. Thus optimal $y_1 = 3$, i.e., $y_1^* = 3$.

To obtain optimal y_2, i.e., y_2^*, note the following

$$y_2 = \beta_2 = \min\left\{\frac{x_{12}}{3}, \frac{x_{22}}{5}\right\} = \min\left\{\frac{x_{11} - 4y_1}{3}, \frac{x_{21} - 2y_1}{5}\right\}.$$

Hence,

$$y_2^* = \min\left\{\frac{12 - 4y_1^*}{3}, \frac{10 - 2y_1^*}{5}\right\} = 0, \text{ because } y_1^* = 3.$$

Thus

optimal solution: $y_1 = 3$, $y_2 = 0$, $y_0 = F_1(X_1) = 84$.

Remark. While solving the LPP we have assumed $a_{ij} \geq 0$. Without this restriction the solution becomes cumbersome. This indicates that dynamic programming is not a good tool to solve linear programming problems.

Problem Set 12

1. Find the minimum of sum of cubes of three positive value integers whose product is 6.

2. Find the point nearest to the origin on the ellipse $x^2 + y^2 = 36$ in the first quadrant.

   Suggestion. Test the end points.

3. Find the point on the line $2y_1 + y_2 = 6$ nearest to the origin in the first quadrant.

4. Find the maximum of the sum of three positive integers whose product is 18.

5. If $y_1^2, 4y_2, 2y_3$ are the returns in millions of rupees from the plants A_1, A_2, A_3, respectively on allocating the amounts y_1, y_2, y_3 in millions of rupees to them, then use dynamic programming technique to find optimal values of y_1, y_2, y_3 which minimizes total return from the plants so that the total amount invested on them does not exceed 10 million rupees in the case when y_1 can take the values $1, 4$; y_2 can take the values $3, 5$; and y_3 can take the values $1, 2, 3$. Also, write the maximum value of the total return.

6. A company is remodeling its three factories. It has five repairmen who can be sent to three factories. Each repairman can be assigned only one factory. The table below shows the gain due to the repairs when the number of repairmen sent is 1 to 3. It is necessary that at least one repairman is sent to each factory. The problem is determine the repairmen who should be sent to each plant in order to maximize the total profit.

Factory

No. of repairmen	Delhi	Jaipur	Pilani
1	10	17	15
2	21	35	31
3	38	37	42

Use dynamic programming to solve the above problem.

7. Use dynamic programming to solve the LPP

$$\max \quad z = x_1 + 9x_2$$
$$\text{s.t.} \quad 2x_1 + x_2 \le 25$$
$$x_1 \le 11$$
$$x_1, x_2 \ge 0$$

8. Use dynamic programming to solve the LPP

$$\max \quad z = 3x_1 + 5x_2$$
$$\text{s.t.} \quad 3x_1 + 2x_2 \le 18$$
$$x_1 \le 11$$
$$x_2 \le 18$$
$$x_1, x_2 \ge 0$$

9. A student must select four elective courses from four different departments with the condition that he must choose at least one course from each department. He measures his learning ability as a function of the number of courses he takes in each department on a 100 point scale and produces the following chart. It is assumed that the course groupings satisfy prerequisites for each department.

Course

Department	1	2	3	4
I	25	50	60	80
II	20	40	90	100
III	40	60	80	100
IV	10	20	30	40

Formulate the problem as a dynamic programming problem.

Chapter 13

Nonlinear Programming

This chapter deals with nonlinear programming problems when the objective function includes several variables and any type of constraints. There are several techniques to deal with such type of problems but here we are concerned with the Lagrange multiplier method applicable to nonlinear problems with equality constraints and Kuhn Tucker theory applicable to nonlinear problems with inequality constraints. In the last two sections, the concept of separable programming and duality for nonlinear programming problems are introduced.

13.1 Introduction

First, we develop the theory of quadratic functions which appear frequently in nonlinear programming problems.

Quadratic Form. Consider the function

$$f(X) = p_{11}x_1^2 + p_{22}x_2^2 + \cdots + p_{nn}x_n^2$$
$$+ p_{12}x_1x_2 + p_{13}x_1x_3 + \cdots + p_{1n}x_1x_n$$
$$+ p_{23}x_2x_3 + \cdots + p_{2n}x_2x_n$$
$$\dots\dots\dots\dots\dots\dots\dots\dots$$
$$+ p_{n-1,n}x_{n-1}x_n.$$

The above function is called the quadric or quadratic form in n variables. Let us define

$$p_{ii} = a_{ii}, \quad \frac{p_{ij}}{2} = a_{ij} = a_{ji}, \ i \neq j.$$

Since $x_i x_j = x_j x_i$ (commutative law for numbers), the above function can be written as

$$f(X) = (x_1, x_2, \ldots, x_n) \begin{bmatrix} a_{11} & a_{12} & \cdots & a_{1n} \\ a_{21} & a_{22} & \cdots & a_{2n} \\ \vdots & \vdots & \vdots & \vdots \\ a_{n1} & a_{n2} & \cdots & a_{nn} \end{bmatrix} \begin{bmatrix} x_1 \\ x_2 \\ \vdots \\ x_n \end{bmatrix} = X^T A X,$$

where A is a real symmetric matrix of order n, because $a_{ij} = a_{ji}$, $i, j \in \{1, 2, \ldots, n\}$.

Definition 1. A quadratic form $X^T A X$ is said to be (a) positive definite, if $X^T A X > 0 \; \forall \; X \neq 0$; (b) positive semi-definite if $X^T A X \geq 0 \; \forall \; X \neq 0$, and for at least one nonzero X, $X^T A X = 0$.

For example, (i) $f(X) = x_1^2 + 2x_2^2 + 3x_3^2$ is positive definite, since $f(X) > 0$ for all $(x_1, x_2, x_3) \neq (0, 0, 0)$; (ii) $f(X) = x_1^2 + (x_2 - x_3)^2$ is positive semi-definite since for $X = (0, 1, 1)$, $X^T A X = 0$ and $X^T A X \geq 0 \; \forall \; X \neq 0$.

Similarly, we have the concept of negative definite and negative semi-definite simply by reversing the sense of inequality in (a) and (b), respectively.

A function which escapes out of this classification is termed as indefinite. For example, $f(X) = x_1^2 - 2x_2^2$ is indefinite.

Characterization of the definiteness depends on the matrix A in $X^T A X$. Hence we have the following simple test for definiteness. We define $X^T A X$ to be positive definite, positive semi-definite, negative definite, negative semi-definite or indefinite according to the matrix A.

1. Matrix minor test. For any square matrix of order n, a principal matrix minor means any submatrix of order $m(\leq n)$ which contains first m elements of the principal diagonal. These principal matrix minors decide the nature of $X^T A X$. Thus, $f(X) = X^T A X$ is

(a) positive definite if all the principal minor determinants of A are positive,

$$D_1 = a_{11} > 0, \quad D_2 = \begin{vmatrix} a_{11} & a_{12} \\ a_{21} & a_{22} \end{vmatrix} > 0,$$

$$D_3 = \begin{vmatrix} a_{11} & a_{12} & a_{13} \\ a_{21} & a_{22} & a_{23} \\ a_{31} & a_{32} & a_{33} \end{vmatrix} > 0, \cdots, \quad D_n > 0;$$

(b) positive semi-definite if $D_1 > 0$, $D_i \geq 0$, and at least one of the $D_i = 0$ for $i = 1, 2, \ldots, n$;

(c) negative definite if $D_1 < 0$, $D_2 > 0$, $D_3 < 0, \cdots, (-1)^i D_i > 0$, $i = 2, 3, \ldots, n$;

(d) negative semi-definite if $D_1 < 0$, $D_2 \geq 0$, $D_3 \leq 0, \cdots, (-1)^i D_i \geq 0$, and at least one of the $D_i = 0$ for $i = 2, 3, \ldots, n$;

(e) indefinite, if none of the above cases happen.

2. Eigenvalue test. Since matrix A is a real symmetric matrix in $X^T A X$, it follows that its eigenvalues are real. Then $X^T A X$ is

(a) positive definite if $\lambda_i > 0$, $i = 1, 2, \ldots, n$;

(b) is positive semi-definite if $\lambda_i \geq 0$, and at least one of $\lambda_i = 0$, $i = 1, 2, \ldots, n$;

(c) negative definite if $\lambda_i < 0$, $i = 1, 2, \ldots, n$;

(d) negative semi-definite if $\lambda_i \leq 0$, and at least one of $\lambda_i = 0$, $i = 1, 2, \ldots, n$;

(e) indefinite if A has both positive and negative eigenvalues.

Example 1. Decide the definiteness of the function

$$f(X) = -3x_1^2 + 2x_2^2 - 3x_3^2 - 10x_1x_2 + 4x_2x_3 + 6x_1x_3.$$

Write the function in the form $X^T A X$ as

$$f(X) = (x_1, x_2, x_3) \begin{bmatrix} -3 & -5 & 3 \\ -5 & 2 & 2 \\ 3 & 2 & -3 \end{bmatrix} \begin{bmatrix} x_1 \\ x_2 \\ x_3 \end{bmatrix}$$

Note that

$$D_1 = -3 < 0, \ D_2 = \begin{vmatrix} -3 & -5 \\ -5 & 2 \end{vmatrix} = -31 < 0,$$

$$D_3 = \begin{vmatrix} -3 & -5 & 3 \\ -5 & 2 & 2 \\ 3 & 2 & -3 \end{vmatrix} = 27 > 0,$$

and this ensures that the function is indefinite.

Hessian matrix. Let $f(X)$ be a function of n variables such that $f(X) \in C^2$, C^2 denotes the space of all real functions whose second order partial derivatives are continuous. Then the Hessian matrix of f is a $n \times n$ symmetric matrix of second order partial derivatives defined by

$$H(X) = \begin{bmatrix} \frac{\partial^2 f}{\partial x_1^2} & \frac{\partial^2 f}{\partial x_1 \partial x_2} & \cdots & \frac{\partial^2 f}{\partial x_1 \partial x_n} \\ \frac{\partial^2 f}{\partial x_2 \partial x_1} & \frac{\partial^2 f}{\partial x_2^2} & \cdots & \frac{\partial^2 f}{\partial x_2 \partial x_n} \\ \vdots & \vdots & \cdots & \vdots \\ \frac{\partial^2 f}{\partial x_n \partial x_1} & \frac{\partial^2 f}{\partial x_n \partial x_2} & \cdots & \frac{\partial^2 f}{\partial x_n^2} \end{bmatrix}$$

Since $f(X) \in C^2$, it follows that $\partial^2 f / \partial x_i \partial x_j = \partial^2 f / \partial x_j \partial x_i \ \forall \ i, j \in \{1, 2, \ldots, n\}$.

Remark. For the quadratic form $X^T A X$, $H = 2A$. This relation may not be true for other forms.

The term $\nabla f(X)$, called gradient of $f(X)$, is the column vector

$$\nabla f(X) = \left(\frac{\partial f}{\partial x_1}, \frac{\partial f}{\partial x_2}, \cdots, \frac{\partial f}{\partial x_n} \right)^T.$$

A necessary conditions for $f(X) \in C^2$ to have stationary points (critical points) is that $\nabla f(X) = 0$, i.e.,

$$\frac{\partial f}{\partial x_1} = 0, \ \frac{\partial f}{\partial x_2} = 0, \cdots, \ \frac{\partial f}{\partial x_n} = 0.$$

Solving these equations, we get the stationary points. Suppose $X^* = (x_1^*, x_2^*, \ldots, x_n^*)^T$ is a stationary point. Then minimum or maximum may occur at X^*. Note that it may be a saddle point or point of inflexion.

From the Taylor expansion of f in a neighbourhood of X^*, we have

$$f(X) = f(X^*) + (X - X^*)\nabla f(X^*) + (X - X^*)^T H(X^*)(X - X^*) + \cdots .$$

This ensures that (i) if $H(X^*)$ is positive definite, then X^* is a point of relative minimum; (ii) if $H(X^*)$ is negative definite, then X^* is a point of relative maximum; (iii) if $H(X^*)$ is indefinite, then X^* is a saddle point. In other situations the test fails, and we may have go to higher order derivatives which becomes a complicated process.

In view of the above remark, for a quadratic $X^T A X$, X^* is a point of relative minimum if A is positive definite, and is a point of relative minimum if A is negative definite. If A is indefinite then X^* is a saddle point. For other situations the test fails.

Example 2. Find stationary points and classify

$$f(X) = 2 + 2x_1 + 3x_2 - x_1^2 - x_2^2.$$

The necessary condition $\partial f / \partial x_1 = 0$ and $\partial f / \partial x_2 = 0$ gives rise to the equations

$$2 - 2x_1 = 0$$
$$2 - 3x_2 = 0$$

The solution of these equations: $x_1 = 1, x_2 = 3/2$ is the stationary point. The Hessian matrix

$$\begin{bmatrix} -2 & 0 \\ 0 & 2 \end{bmatrix}$$

is negative definite, and hence $(1, 3/2)$ is a point of the local maximum. The maximum value obtained by substituting $x_1 = 1, x_2 = 3/2$ in the function, is $21/4$.

Now, we discuss some classical methods for optimizing nonlinear functions under the influence of given set of restrictions (constraints).

13.2 Lagrange Multipliers Method

Consider the NLPP

$$\text{opt} \quad f(X)$$
$$\text{s.t.} \quad g_i(X) = 0, \quad i = 1, 2, \dots, m,$$

where $X = (x_1, x_2, \ldots, x_n)^T$, $\lambda = (\lambda_1, \lambda_2, \ldots, \lambda_m)^T$, and $f(X), g(X) \in C^1$, space of all functions whose first order partial derivatives are continuous.

Define the Lagrange function $L(X, \lambda)$ as

$$L \equiv L(X, \lambda) = f(X) + \sum_{i=1}^{m} \lambda_i g_i(X). \tag{13.1}$$

The numbers λ_i are called the Lagrange multipliers. The equations

$$\frac{\partial L}{\partial x_j} = 0 \quad \text{and} \quad \frac{\partial L}{\partial \lambda_i} = 0, \; j = 1, 2, \ldots, n; \; i = 1, 2, \ldots, m \tag{13.2}$$

give the necessary conditions for determining the stationary points of $L(X, \lambda)$. We shall prove at later stage that the unconstrained $L(X, \lambda)$ and the constrained $f(X)$ have the same set of stationary points. Moreover, a saddle point of $L(X, \lambda)$ may be an optimum point for $f(X)$.

Remark. The necessary conditions in (13.2) are also sufficient if the functions $f(X)$ and each $g_i(X)$ are convex, see Section 13.4. Also, relative minimum is global minimum, see Theorem 1, Section 13.3.

To test the nature of stationary points given by the solution of equations in (13.2), we mention without proof the sufficient condition for the Lagrange method by the use of the bordered Hessian matrix defined as

$$H^B = \begin{bmatrix} 0 & P \\ P^T & Q \end{bmatrix}_{(m+n) \times (m+n)}$$

where 0 is a $m \times m$ zero matrix and

$$P = \begin{bmatrix} \frac{\partial g_1}{\partial x_1} & \frac{\partial g_1}{\partial x_2} & \cdots & \frac{\partial g_1}{\partial x_n} \\ \frac{\partial g_2}{\partial x_1} & \frac{\partial g_2}{\partial x_2} & \cdots & \frac{\partial g_2}{\partial x_n} \\ \vdots & \vdots & \cdots & \vdots \\ \frac{\partial g_m}{\partial x_1} & \frac{\partial g_m}{\partial x_2} & \cdots & \frac{\partial g_m}{\partial x_n} \end{bmatrix} = \begin{bmatrix} \nabla g_1^T(X) \\ \nabla g_2^T(X) \\ \vdots \\ \nabla g_m^T(X) \end{bmatrix},$$

$$Q = \begin{bmatrix} \frac{\partial^2 L}{\partial x_1^2} & \frac{\partial^2 L}{\partial x_1 \partial x_2} & \cdots & \frac{\partial^2 L}{\partial x_1 \partial x_n} \\ \frac{\partial^2 L}{\partial x_2 \partial x_1} & \frac{\partial^2 L}{\partial x_2^2} & \cdots & \frac{\partial^2 L}{\partial x_2 \partial x_n} \\ \vdots & \vdots & \cdots & \vdots \\ \frac{\partial^2 L}{\partial x_n x_1} & \frac{\partial^2 L}{\partial x_n \partial x_2} & \cdots & \frac{\partial^2 L}{\partial x_n^2} \end{bmatrix}$$

Here P^T is the transpose of matrix P and Q is the Hessian matrix of the Lagrange function L. The matrix H^B is called the bordered Hessian matrix.

Given the stationary point (X^*, λ^*) for the Lagrange function L and the bordered Hessian matrix evaluated at (X^*, λ^*), then X^* is

1. a maximum point, if starting with the principal minor determinant of order $2m+1$, the last $n-m$ principal minor determinants of H^B form an alternating sign pattern starting with $(-1)^{m+1}$.

2. a minimum point, if starting with the principal minor determinant of order $2m+1$, the last $n-m$ principal minor determinants of H^B have the sign of $(-1)^m$.

Remarks. 1. These conditions are sufficient for identifying the extreme points (points of maximum or minimum). In other words a stationary point may be an extreme point without satisfying these conditions.

2. Sometimes a convenient method to find stationary points resulting from the necessary conditions is to select successive values of λ and then solve the equation for X. This is repeated until X satisfies all the constraints.

Let us workout some problems on the Lagrange multiplier method.

Example 3. Use the Lagrange multiplier method to solve the following NLPP. Does the solution maximize or minimizes the objective function?

$$\text{opt} \quad f(X) = 2x_1^2 + x_2^2 + 3x_3^2 + 10x_1 + 8x_2 + 6x_3 - 100$$
$$\text{s.t.} \quad x_1 + x_2 + x_3 = 20$$
$$x_1, x_2, x_3 \geq 0$$

The Lagrange function is

$$L = 2x_1^2 + x_2^2 + 3x_3^2 + 10x_1 + 8x_2 + 6x_3 - 100 + \lambda(x_1 + x_2 + x_3 - 20).$$

The necessary conditions for the stationary points are

$$\frac{\partial L}{\partial x_1} = 4x_1 + 10 + \lambda = 0$$

$$\frac{\partial L}{\partial x_2} = 2x_2 + 8 + \lambda = 0$$

$$\frac{\partial L}{\partial x_3} = 6x_3 + 6 + \lambda = 0$$

$$\frac{\partial L}{\partial \lambda} = x_1 + x_2 + x_3 - 20 = 0$$

From the first three equations, it follows that

$$x_1 = -\frac{10 + \lambda}{4}, \ x_2 = -\frac{8 + \lambda}{2}, \ x_3 = -\frac{6 + \lambda}{6}.$$

Substitution of these values in the last equation yields

$$-\frac{10 + \lambda}{4} - \frac{8 + \lambda}{2} - \frac{6 + \lambda}{6} = 20 \implies \lambda = -30.$$

Thus, the stationary point is $X^* = (5, 11, 4)$. For its test, write the bordered Hessian matrix

$$\begin{bmatrix} 0 & 1 & 1 & 1 \\ 1 & 4 & 0 & 0 \\ 1 & 0 & 2 & 0 \\ 1 & 0 & 0 & 6 \end{bmatrix}$$

Here $n = 3$, $m = 1$, $2m + 1 = 3$. Starting from the principal minor determinant of order 3, we check $n - m = 2$ principal minor determinants. Note that

$$|\Delta_3| = \begin{bmatrix} 0 & 1 & 1 \\ 1 & 4 & 0 \\ 1 & 0 & 2 \end{bmatrix} = -6, \quad |\Delta_4| = \begin{bmatrix} 0 & 1 & 1 & 1 \\ 1 & 4 & 0 & 0 \\ 1 & 0 & 2 & 0 \\ 1 & 0 & 0 & 6 \end{bmatrix} = -44.$$

Both $|\Delta_3|$ and $|\Delta_3|$ have sign of $(-1)^1$. Thus, X^* is a minimum point, and the minimum value of the objective function is

$$f_{\min} = 2 \times 5^2 + 11^2 + 3 \times 4^2 + 10 \times 5 + 8 \times 11 + 6 \times 4 - 100 = 281.$$

Example 4. Solve the following nonlinear programming problem by Lagrange multiplier method.

$$\max \quad f(X) = 4x_1^2 + 2x_2^2 + x_3^2 - 4x_1x_2$$
$$\text{s.t.} \quad x_1 + x_2 + x_3 = 15$$
$$2x_1 - x_2 + 2x_3 = 20$$
$$x_1, x_2, x_3 \geq 0.$$

The Lagrange function $L(X, \lambda) \equiv L$ is

$$L = 4x_1^2 + 2x_2^2 + x_3^2 - 4x_1x_2 + \lambda_1(x_1 + x_2 + x_3 - 15) + \lambda_2(2x_1 - x_2 + 2x_3 - 20).$$

The necessary conditions for the stationary points are

$$\frac{\partial L}{\partial x_1} = 8x_1 - 4x_2 + \lambda_1 + 2\lambda_2 = 0$$

$$\frac{\partial L}{\partial x_2} = 4x_2 - 4x_1 + \lambda_1 - \lambda_2 = 0$$

$$\frac{\partial L}{\partial x_3} = 2x_3 + \lambda_1 + 2\lambda_2 = 0$$

$$\frac{\partial L}{\partial \lambda_1} = x_1 + x_2 + x_3 - 15 = 0$$

$$\frac{\partial L}{\partial \lambda_2} = 2x_1 - x_2 + 2x_3 - 20 = 0$$

The solution of these simultaneous linear equations (use row reduced echelon form) yields

$$x_1 = \frac{11}{3}, \quad x_2 = \frac{10}{3}, \quad x_3 = 8, \quad \lambda_1 = -\frac{40}{9}, \quad \lambda_2 = -\frac{52}{9}.$$

Next, to check the nature of stationary points we construct the bordered Hessian matrix

$$H^B = \begin{bmatrix} 0 & 0 & 1 & 1 & 1 \\ 0 & 0 & 2 & -1 & 2 \\ 1 & 2 & 8 & -2 & 0 \\ 1 & -1 & -2 & 4 & 0 \\ 1 & 2 & 0 & 0 & 2 \end{bmatrix}$$

Here $n = 3$, $m = 2$ and $2m + 1 = 5$. Starting from the principal minor determinant of order 5, we check the sign for only one principal minor determinant $(3 - 2 = 1)$ to be $(-1)^2 = 1 > 0$. Here this is $|H^B| = 72 > 0$. Thus, $X^* = (11/3, 10/3, 8)$ is a minimum point, and the minimum value is

$$f_{\min} = 4 \left(\frac{11}{3}\right)^2 + 2 \left(\frac{10}{3}\right)^2 + 8^2 - 4\frac{11}{3} \cdot \frac{10}{3} = \frac{820}{9}.$$

Note. There is another condition which is necessary and sufficient both for classifying the extreme points. The only difficulty is of computation of the roots of polynomial arising in the analysis. Define a matrix

$$\Delta = \begin{bmatrix} 0 & P \\ P^T & Q - \mu I \end{bmatrix}_{(m+n)\times(m+n)}$$

where I is an identity matrix of order $n \times n$ and matrices P and Q have already been defined in the starting of this section.

The stationary point (X^*, λ^*) of the Lagrange function $L(X, \lambda)$ is a point of minimum (or maximum) of the constrained function $f(X)$, if the polynomial in μ given by $|\Delta| = 0$ has all its roots positive (or negative). If some roots are positive and some are negative, then X^* is not an extreme point.

Remark. It is worthwhile to mention that for the sufficient condition we are assuming that $f(X), g_i(X) \in C^2$, space of all functions whose second order partial derivatives are continuous.

13.3   Convex Nonlinear Programming Problem

First, we define the convexity of a function which facilitates the further studies on nonlinear programming problems with equality and inequality constraints.

Definition 2. Let S be a convex set in $\mathbb{R}^n$. A function $f(X)$ defined on S is said be convex if for any pair of points X_1, X_2 in S and $\forall \alpha : 0 \leq \alpha \leq 1$,

$$f((1 - \alpha)X_1 + \alpha X_2) \leq (1 - \alpha)f(X_1) + \alpha f(X_2).$$

Geometrically speaking in two dimensional plane, Definition 2 means that $f(x)$ is convex if for any two points x_1 and x_2 in S, the chord joining the points $(x_1, f(x_1))$ and $(x_2, f(x_2))$ is above $f(x)$, i.e., for any point $x \in [x_1, x_2]$, $f(x) \le PQ$, where Q is on the chord, see Fig. 13.1.

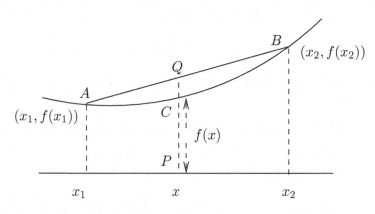

Figure 13.1

Remarks. 1. A function $f(X)$ is strictly convex if we have strict inequalities in Definition 2.

2. A function $f(X)$ is concave (or strictly concave) if $-f(X)$ is convex (or strictly convex).

3. A linear function is convex as well as concave.

Proposition 1. The sum of two convex functions is convex.

Proof. Let f_1 and f_2 be two convex functions defined on a convex set $S \subseteq \mathbb{R}^n$. Then, for any two points X_1 and X_2 in S, we have for all $\alpha, 0 \le \alpha \le 1$,

$$f_1((1 - \alpha)X_1 + \alpha X_2) \le (1 - \alpha)f_1(X_1) + \alpha f_1(X_2) \qquad (13.3)$$

$$f_2((1 - \alpha)X_1 + \alpha X_2) \le (1 - \alpha)f_2(X_1) + \alpha f_2(X_2). \qquad (13.4)$$

Using (13.3) and (13.4), we have

$$\begin{aligned}
(f_1 + f_2)((1 - \alpha)X_1 + \alpha X_2) &= f_1((1 - \alpha)X_1 + \alpha X_2) \\
&\quad + f_2((1 - \alpha)X_1 + \alpha X_2) \\
&\le (1 - \alpha)f_1(X_1) + \alpha f_1(X_2) \\
&\quad + (1 - \alpha)f_2(X_1) + \alpha f_2(X_2) \\
&= (1 - \alpha)(f_1 + f_2)(X_1) + \alpha(f_1 + f_2)(X_2).
\end{aligned}$$

Proposition 2. Let $f(X) = X^T A X$. Then $f(X)$ is convex in $\mathbb{R}^n$ if $X^T A X$ is positive semi-definite.

Proof. Given that $X^T A X \geq 0 \; \forall \; X \in \mathbb{R}^n$, and there exists at least one nonzero X such that $X^T A X = 0$. Let $X_1, X_2 \in \mathbb{R}^n$ such that $X = (1 - \alpha)X_1 + \alpha X_2$, $0 \leq \alpha \leq 1$, since $\mathbb{R}^n$ is convex. Thus, to prove the theorem we have to show that

$$(1 - \alpha)f(X_1) + \alpha f(X_2) - f((1 - \alpha)X_1 + \alpha X_2) \geq 0.$$

Since $X_1^T A X_2 = X_2^T A X_1$, being scalar, the left hand side of the above inequality turns up

$$(1 - \alpha)X_1^T A X_1 + \alpha X_2^T A X_2 - ((1 - \alpha)X_1 + \alpha X_2)^T A((1 - \alpha)X_1 + \alpha X_2)$$
$$= (1 - \alpha)X_1^T A X_1 + \alpha X_2^T A X_2 - (1 - \alpha)^2 X_1^T A X_1 - \alpha(1 - \alpha)X_2^T A X_1$$
$$\quad - \alpha(1 - \alpha)X_1^T A X_2 - \alpha^2 X_2^T A X_2$$
$$= \alpha(1 - \alpha)[X_1^T A X_1 + X_2^T A X_2 - 2X_1^T A X_2], \;\; \text{using } X_1^T A X_2 = X_2^T A X_1$$
$$= \alpha(1 - \alpha)(X_1 - X_2)^T A(X_1 - X_2)$$

Now, as $X_1 - X_2 \in \mathbb{R}^n$ where $\mathbb{R}^n$ is a linear space, it follows that

$$\alpha(1 - \alpha)(X_1 - X_2)^T A(X_1 - X_2) \geq 0.$$

This proves the proposition.

Corollary 1. Under the conditions of Proposition 2, if $f(X) = X^T A X$ is positive definite, then $f(X)$ is strictly convex.

Corollary 2. Under the conditions of Proposition 2, we have

1. $f(X) \in C^2$ is convex $\iff$ its Hessian matrix is positive semi-definite.

2. $f(X) \in C^2$ is strictly convex $\iff$ its Hessian matrix is positive definite.

The proof of Corollary 2 is a direct consequence of the fact that $H = 2A$ in a quadratic form. Note that quadratic forms and quadratic functions have different meaning.

Corollary 2 is very useful to decide convexity of any quadratic function.

Theorem 1. Let $f(X)$ be a convex function defined over a convex set S in $\mathbb{R}^n$. Then the local minimum is global minimum of $f(X)$ over S.

Proof. Let X^* be a point of local minimum. Hence,

$$f(X^*) \leq f(X^* + \delta), \text{ where } \{X : |X - X^*| < \delta, \; \delta > 0\} = N_\delta(X^*).$$

Take any point $X_1 \in N_\delta(X^*)$. There exists $\alpha : 0 \leq \alpha \leq 1$ such that $X_1 = \alpha X^* + (1 - \alpha)X$ for any X in S. Now, in view of convexity of f on S, we have

$$f(X_1) = f(\alpha X^* + (1 - \alpha)X)$$
$$\leq \alpha f(X^*) + (1 - \alpha)f(X)$$
$$\leq \alpha f(X_1) + (1 - \alpha)f(X)$$

This implies $f(X^*) \leq f(X)$ or $f(X^*) \leq f(X_1) \leq f(X)$. Since X is arbitrary point, and hence $f(X^*) \leq f(X)$ for all X in S. So, X^* is a point of the global minimum.

For developing the theory, write the problem in the format

$$\min \quad f(X) \tag{13.5}$$
$$\text{s.t.} \quad g_i(X) \leq 0, \quad i = 1, 2, \ldots, m \tag{13.6}$$

$$X \geq 0, \tag{13.7}$$

$$f(X) \text{ and } g_i(X) \text{ are convex functions.} \tag{13.8}$$

The nonlinear problem in the format (13.5)-(13.8) is called convex nonlinear programming problem (CNLPP). Note that $f(X)$ and $g_i(X)$ are convex functions over some common convex set.

Remark. Theorem 1 ensures that in a CNLPP the relative minimum or relative maximum is global minimum or global maximum.

Theorem 2. A set $S = \{X : g_i(X) \leq 0, X \geq 0\}$ of feasible solutions of CNLPP is a convex set.

Proof. In a CNLPP, $g_i(X)$ means convex functions over some convex set S. Let X_1 and X_2 be arbitrary points in S. Consider

$$Y = (1 - \alpha)X_1 + \alpha X_2, \ 0 \leq \alpha \leq 1.$$

Now $g_i(X_1) \leq 0$ and $g_i(X_2) \leq 0$ since $X_1, X_2 \in S$. Again

$$g_i(Y) = g_i((1 - \alpha)X_1 + \alpha X_2)$$
$$\leq (1 - \alpha)g_i(X_1) + \alpha g_i(X_2) \leq 0.$$

$Y \in S$, being clc of points in S, and hence S is convex.

13.4   Kuhn Tucker Theory

If an NLPP is a CNLPP and at least one of the $g_i(X)$ is an inequality constraint, then the Lagrange multiplier method has been extended for utility by Kuhn Tucker Theory as follows:

Kuhn Tucker conditions (K-T conditions). Consider the CNLPP

$$\min \quad f(X)$$
$$\text{s.t.} \quad g_i(X) \leq 0, \quad i = 1, 2, \ldots, m$$
$$X \geq 0,$$
$$f(X) \ \text{and} \ g_i(X) \ \text{are convex functions,}$$

where $f(X)$, $g_i(X) \in C^1$ over some convex set S.

Then the necessary and sufficient conditions for X^* to be the solution of the above system are

$$\frac{\partial L(X^*, \lambda^*)}{\partial x_j} \geq 0; \qquad x_j \frac{\partial L(X^*, \lambda^*)}{\partial x_j} = 0, \quad j = 1, 2, \ldots, n \tag{13.9}$$

$$\frac{\partial L(X^*, \lambda^*)}{\partial \lambda_i} = g_i(X^*) \leq 0; \qquad \lambda_i^* \frac{\partial L(X^*, \lambda^*)}{\partial \lambda_i} = 0, \quad i = 1, 2, \ldots, m \tag{13.10}$$

$$x_j^* \geq 0, \ \lambda_i^* \geq 0 \tag{13.11}$$

where $L \equiv L(X, \lambda)$ is the Lagrange function already defined in (13.1).

The conditions (13.9)-(13.11) are known as Kuhn Tucker (K-T) conditions. Note that X^* is a solution (satisfying K-T conditions) means we have minimized the function, and this is the optimal solution of CNLPP.

Remarks. 1. If the variable x_k is unrestricted in sign, then $\partial L(X^*, \lambda^*)/\partial x_k = 0$.

2. If the pth constraint is equality constraint, i.e., $g_p(X) = 0$, then the pth Lagrange Multiplier λ_p will be unrestricted.

Let us give an example of a NLPP which in not a CNLPP. Consider

$$\max \quad f(X) = 2x_1^2 + 2x_2^2 + 2x_1x_2 - 4x_1 - 6x_2 - 49$$
$$\text{s.t.} \quad x_1x_2 - 2 \le 0$$
$$x_1, x_2 \ge 0$$

This is not a CNLPP as $g_1(X) = x_1x_2 - 2 \le 0$ is not a convex function.

Example 5. Solve the following NLPP using K-T conditions:

$$\max \quad f(X) = 4x_1 + 6x_2 - 2x_1^2 - 2x_1x_2 - 2x_2^2 + 49$$
$$\text{s.t.} \quad x_1 + x_2 \le 2$$
$$x_1, x_2 \ge 0$$

Writing the problem in appropriate format

$$\min \quad f(X) = 2x_1^2 + 2x_2^2 + 2x_1x_2 - 4x_1 - 6x_2 - 49$$
$$\text{s.t.} \quad x_1 + x_2 - 2 \le 0$$
$$x_1, x_2 \ge 0$$

The above system is a CNLPP in appropriate format, and hence K-T conditions can be applied as

$$\frac{\partial L}{\partial x_1} = -4 + 4x_1 + 2x_2 + \lambda_1 \ge 0, \qquad x_1(-4 + 4x_1 + 2x_2 + \lambda_1) = 0$$
$$\frac{\partial L}{\partial x_2} = -6 + 4x_2 + 2x_1 + \lambda_1 \ge 0, \qquad x_2(-6 + 4x_2 + 2x_1 + \lambda_1) = 0$$
$$\frac{\partial L}{\partial \lambda_1} = x_1 + x_2 - 2 \le 0, \qquad \lambda_1(x_1 + x_2 - 2) = 0$$
$$x_1, x_2, \lambda_1 \ge 0$$

If there is only one constraint, then the solution of the above system can be found by inspection. We solve right side of the above system with nonnegative restrictions and check whether the solution satisfies the left side inequalities. In this procedure we try first for $\lambda_1 = 0$, and then for x_1 and x_2 if needed. For $\lambda_1 = 0$,

$$2x_1 + x_2 = 2$$
$$x_1 + 2x_2 = 3$$

Solving these equations, we have $x_1 = 1/3, x_2 = 4/3$. Now, $(1/3, 4/3, 0)$ is the solution of the system (given by K-T conditions). Hence, the optimum solution is

$$x_1 = 1/3, \ x_2 = 4/3, \ f_{\min} = -483/9, \ \text{or} \ f_{\max} = 483/9.$$

Example 6. Consider the problem

$$\min \quad z = -2x_1 - x_2$$
$$\text{s.t.} \quad x_1 - x_2 \leq 0$$
$$x_1^2 + x_2^2 \leq 4$$
$$x_1, x_2 \geq 0$$

The firs step is to check whether the problem is in appropriate format (13.5)-(13.8) or not, as K-T conditions are applicable to this format only. The Lagrange function is

$$L(X, \lambda) = -2x_1 - x_2 + \lambda_1(x_1 - x_2) + \lambda_2(x_1^2 + x_2^2 - 4)$$

The K-T conditions are

$$\frac{\partial L}{\partial x_1} = -2 + \lambda_1 + 2\lambda_2 x_1 \geq 0, \qquad x_1(-2 + \lambda_1 + 2\lambda_2 x_1) = 0$$

$$\frac{\partial L}{\partial x_2} = -1 - \lambda_1 + 2\lambda_2 x_2 \geq 0, \qquad x_2(-1 - \lambda_1 + 2\lambda_2 x_2) = 0$$

$$\frac{\partial L}{\partial \lambda_1} = x_1 - x_2 \leq 0, \qquad\qquad \lambda_1(x_1 - x_2) = 0$$

$$\frac{\partial L}{\partial \lambda_2} = x_1^2 + x_2^2 - 4 \leq 0, \qquad\quad \lambda_2(x_1^2 + x_2^2 - 4) = 0$$

$$x_1, x_2, \lambda_1, \lambda_2 \geq 0$$

We solve right side of these equations with the nonnegative restrictions. Some solutions are

$$(0, 0, 0, 0), \ (\sqrt{2}, \sqrt{2}, 1/2, 3/4\sqrt{2}), \ (0, 2, 0, 1/4).$$

Out of these only the solution $(x_1, x_2, \lambda_1, \lambda_2) = (\sqrt{2}, \sqrt{2}, 1/2, 3/4\sqrt{2})$ satisfies the left side inequalities of the above system. Hence, this is the only solution. Thus, the optimal solution is the point $(\sqrt{2}, \sqrt{2})$, since K-T conditions are necessary and sufficient. Moreover, this gives global minimum.

$$\text{Optimal solution: } x_1 = \sqrt{2}, \ x_2 = \sqrt{2}, \quad z = -3\sqrt{2}.$$

Remarks. 1. In general, the right hand side of (13.9)-(13.11) with nonnegative restrictions are solved, and then we check which solution satisfies left side of these relations. We have observed that even for two variables the computations are cumbersome. This does not lower down the utility of K-T conditions. Nevertheless, these conditions still provide valuable clues as to the identity of an optimal solution. In specialized type of problems, K-T conditions result in giving very simple expressions which can be solved by known methods. Problems of this type will be discussed in next section.

2. In format (13.9)-(13.11), we may also include constraints of the type

$$g_i(X) \geq 0, \ i = m + 1 \text{ to } p,$$

where $g_i(X)$, $i = m + 1, m + 2, \ldots, p$ are convex functions. Then the corresponding term to be added in $L(X, \lambda)$ is

$$\lambda_{m+1} g_{m+1}(X) + \cdots + \lambda_p g_p(X).$$

The additional K-T conditions are

$$\frac{\partial L(X^*, \lambda^*)}{\partial \lambda_{m+1}} = g_{m+1} \geq 0, \qquad \lambda_{m+1} \frac{\partial L(X^*, \lambda^*)}{\partial \lambda_{m+1}} = 0;$$

$$\vdots \qquad\qquad\qquad \vdots$$

$$\frac{\partial L(X^*, \lambda^*)}{\partial \lambda_p} = g_p(X) \geq 0, \qquad \lambda_p \frac{\partial L(X^*, \lambda^*)}{\partial \lambda_p} = 0;$$

$$\lambda_{m+1}^*, \ldots, \lambda_p^* \leq 0.$$

Note there is only change in the last requirement. This enlarges the domain of application of K-T conditions to include concave functions. Still K-T conditions remain necessary and sufficient.

Remark. A NLPP in which the conditions that $f(X)$ and $g_i(X)$ are convex is not met, the K-T conditions are necessary, but not sufficient.

13.5 Quadratic Programming

Here, we shall be concerned with convex quadratic programming problems (CQPP) defined as

$$\min \quad f(X) = X^T C X + P^T X$$
$$\text{s.t.} \quad AX \leq b$$
$$X \geq 0,$$

where $X = (x_1, x_2, \ldots, x_n)^T$, C is a $n \times n$ symmetric positive definite or positive semi-definite matrix, $A = (a_{ij})_{m \times n}$ matrix, $P = (p_1, p_2, \ldots, p_n)^T$, and $b = (b_1, b_2, \ldots, b_m)^T$.

The constraints $AX \leq b$ can be written as $AX - b \leq 0$. Since $f(X)$ and all the constraints (being linear) are convex functions, it follows that CQPP is a particular case of CNLPP. Thus, K-T conditions are applicable for a CQPP. The solution of above problem was suggested by Wolfe. This is nothing but simplex method with the control on entering variable by K-T conditions. This means that while entering variable is decided the K-T conditions should not be violated.

Also, CNLPP in Example 6 can be solved by Wolfe's method discussed below provided all $g_i(X)$ are linear.

Example 7. Use Wolfe's method to solve the following nonlinear programming problem

$$\begin{aligned} \max \quad & f(X) = -2x_1^2 - 3x_2^2 - 4x_1x_2 + 6x_1 + 3x_2 \\ \text{s.t.} \quad & x_1 + x_2 \leq 1 \\ & 2x_1 + 3x_2 \leq 4 \\ & x_1, x_2 \geq 0 \end{aligned}$$

Writing the problem in appropriate format:

$$\begin{aligned} \min \quad & f(X) = 2x_1^2 + 3x_2^2 + 4x_1x_2 - 6x_1 - 3x_2 \\ \text{s.t.} \quad & x_1 + x_2 - 1 \leq 0 \\ & 2x_1 + 3x_2 - 4 \leq 0 \\ & x_1, x_2 \geq 0 \end{aligned}$$

The objective function is of the form $X^TCX + P^TX$, where X^TCX is positive definite. Any way, the objective function (being sum of convex functions) and constraints (as these are linear) are convex and hence this is a CQPP. We use Wolfe's method. The Lagrange function is

$$L = 2x_1^2 + 3x_2^2 + 4x_1x_2 - 6x_1 - 3x_2 + \lambda_1(x_1 + x_2 - 1) + \lambda_2(2x_1 + 3x_2 - 4).$$

The K-T conditions are

$$4x_1 + 4x_2 - 6 + \lambda_1 + 2\lambda_2 \geq 0, \qquad x_1(4x_1 + 4x_2 - 6 + \lambda_1 + 2\lambda_2) = 0$$

$$4x_1 + 6x_2 - 3 + \lambda_1 + 3\lambda_2 \geq 0, \qquad x_2(4x_1 + 6x_2 - 3 + \lambda_1 + 3\lambda_2) = 0$$

$$x_1 + x_2 - 1 \leq 0, \qquad \lambda_1(x_1 + x_2 - 1) = 0$$

$$2x_1 + 3x_2 - 4 \leq 0, \qquad \lambda_2(2x_1 + 3x_2 - 4) = 0$$

$$x_1, x_2, \lambda_1, \lambda_2 \geq 0.$$

Since K-T conditions are necessary and sufficient, the BFS of the above system will be the optimal solution of the problem. We find BFS by using Phase-I of two phase method.

The auxiliary LPP with left side restrictions is

$$\min \quad r_0 = R_1 + R_2$$

$$\text{s.t.} \quad 4x_1 + 4x_2 + \lambda_1 + 2\lambda_2 - s_1 + R_1 = 6, \qquad x_1 s_1 = 0$$

$$4x_1 + 6x_2 + \lambda_1 + 3\lambda_2 - s_2 + R_2 = 3, \qquad x_2 s_2 = 0$$

$$x_1 + x_2 + s_3 = 1, \qquad \lambda_1 s_3 = 0$$

$$2x_1 + 3x_2 + s_4 = 4, \qquad \lambda_2 s_4 = 0$$

$$\text{all var} \geq 0$$

Now, simply apply the simplex algorithm with the conditions that the right side conditions in the above system, i.e., $x_1 s_1 = 0$, $x_2 s_2 = 0$, $\lambda_1 s_3 = 0$ and $\lambda_2 s_4 = 0$ should not be violated at any iteration.

Table 1

B V	x_1	x_2	λ_1	λ_2	s_1	s_2	R_1	R_2	s_3	s_4	Soln
r_0	0	0	0	0	0	0	−1	−1	0	0	0
R_1	4	4	1	2	−1	0	1	0	0	0	6
R_2	4	6	1	3	0	−1	0	1	0	0	3
s_3	1	1	0	0	0	0	0	0	1	0	1
s_4	2	3	0	0	0	0	0	0	0	1	4

Table 1 is not in simplex format as there must be relative cost 0 below the basic variables R_1 and R_2. The simplex format is achieved by adding the sum of the first and second row to r_0-row resulting in Table 2.

Table 2

B V	$x_1 \downarrow$	x_2	λ_1	λ_2	s_1	s_2	R_1	R_2	s_3	s_4	Soln
r_0	8	10	2	5	-1	-1	0	0	0	0	9
R_1	4	4	1	2	-1	0	1	0	0	0	6
$\leftarrow R_2$	4	6	1	3	0	-1	0	1	0	0	3
s_3	1	1	0	0	0	0	0	0	1	0	1
s_4	2	3	0	0	0	0	0	0	0	1	4

As usual, x_2 having the most positive relative cost is a candidate for entering the basis. But here we to select the exact variable to be considered for entering into the basis by taking most positive value of $\theta_j(z_j - c_j)$ for all nonbasic variables, see rules of entering variable in Section 3.1. Compute $\theta_j(z_j - c_j)$ as

$$x_1 : \theta_1(z_1 - c_1) = \tfrac{3}{4} \times 8 = 6$$
$$x_2 : \theta_2(z_2 - c_2) = \tfrac{3}{6} \times 10 = 5$$
$$\lambda_1 : \theta_3(z_3 - c_3) = 3 \times 2 = 6$$
$$\lambda_2 : \theta_4(z_4 - c_4) = 1 \times 5 = 5$$

The most positive $\theta_j(z_j - c_j)$ is for x_1 and λ_1, i.e., there is a tie for the entering variable. Obviously, we would like to enter the decision variable x_1 first. Note that with the entry of x_1, the left-hand side conditions are not violated. The variable R_2 leaves the basis. Complete the remaining entries of the table just by row transformations. After the first iteration, we get Table 3. The next table is

Table 3

B V	x_1	x_2	λ_1	λ_2	s_1	$s_2 \downarrow$	R_1	R_2	s_3	s_4	Soln
r_0	0	-2	0	-1	-1	1	0	-2	0	0	3
R_1	0	-2	0	-1	-1	1	1	-1	0	0	3
x_1	1	3/2	1/4	3/4	0	$-1/4$	0	1/4	0	0	3/4
$\leftarrow s_3$	0	$-1/2$	$-1/4$	$-3/4$	0	$\boxed{1/4}$	0	$-1/4$	1	0	1/4
s_4	0	0	$-1/2$	$-3/2$	0	1/2	0	$-1/2$	0	1	5/2

Perform next simplex iteration in Table 3, s_2 enters and s_3 leaves without violating any of the condition in the left side of K-T conditions. After the second iteration, we get Table 4.

Table 4

B V	x_1	x_2	$\lambda_1 \downarrow$	λ_2	s_1	s_2	R_1	R_2	s_3	s_4	Soln
r_0	0	0	1	2	-1	0	0	-1	-4	0	2
$\leftarrow R_1$	0	0	$\boxed{1}$	2	-1	0	1	0	-4	0	2
x_1	1	1	0	0	0	0	0	0	1	0	1
s_2	0	-2	-1	-3	0	1	0	-1	4	0	1
s_4	0	1	0	0	0	0	0	0	-2	1	2

Table 4 indicates that λ_2 must enter, whereas R_1 must leave the basis. But this violate the K-T conditions as λ_2 and s_4 together will be in the basis. Hence, we take the next most positive variable to enter the basis and this variable is λ_1. Obviously, R_1 leaves the basis. The next table is

Table 5

B V	x_1	x_2	λ_1	λ_2	s_1	s_2	R_1	R_2	s_3	s_4	Soln
r_0	0	0	0	0	0	0	-1	-1	0	0	0
λ_1	0	0	1	2	-1	0	1	0	$-1/4$	0	2
x_1	1	1	0	0	0	1	0	0	1	0	1
s_2	0	-2	0	-1	-1	0	1	-1	0	0	3
s_4	0	1	0	0	0	0	0	0	-2	1	2

Table 5 is the optimal table, and

optimal solution: $x_1 = 1$, $x_2 = 0$, maximum value $= 4$.

Note that the maximum value is not available from the table, and hence it is to be computed by direct substitution in the objective function of the problem.

Remark. If, during simplex iterations, the right-hand side K-T conditions are ignored, then one of the BFS of the system involving left-hand side conditions is

$x_1 = 0$, $x_2 = 1$, $\lambda_1 = 0$, $\lambda_2 = 1$, $s_2 = 6$, $s_4 = 1$ maximum value $= 0$.

This is not the optimal solution of the problem as it does not satisfy $x_2 s_2 = 0$ and $\lambda_2 s_4 = 0$.

13.6   Separable Programming

Definition 3. A function $f(x_1, x_2, \ldots, x_n)$ is said to be separable if it can be expressed as

$$f(x_1, x_2, \ldots, x_n) = f_1(x_1) + f_2(x_2) + \cdots + f_n(x_n)$$

Definition 4. A mathematical programming problem in which the objective function and constraints are separable functions is called separable programming problem.

Remark. Every LPP is a separable programming problem. However, we are interested here in NLPP which is separable.

Formulation. Let us formulate separable programming problems.

$$\text{opt} \quad f(X) = f_1(x_1) + f_2(x_2) + \cdots + f_n(x_n)$$

$$\text{s.t.} \quad g_1^i(x_1) + g_2^i(x_2) + \cdots + g_n^i(x_n) \leq => \geq b_i, \quad i = 1, 2 \ldots, m$$

$$x_1, x_2, \ldots, x_n \geq 0$$

For the separable LPP, the simplex method is used. While in applying the simplex method to the separable NLPP, we shall approximate the nonlinear functions by linear functions which are very close to them.

Consider the case when $f(x)$ is a single real variable function, see Fig. 13.2.

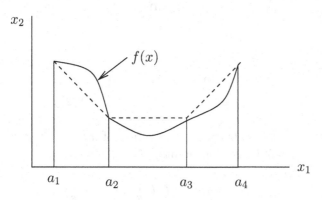

Figure 13.2

Then

$$f(\hat{x}) \simeq f(x) = \alpha_1 f(a_1) \quad + \alpha_2 f(a_2);$$

$$\text{with } \alpha_1 + \alpha_2 = 1, \alpha_1, \alpha_2 \geq 0, \forall x \in [a_1, a_2]$$

$$f(\hat{x}) \simeq f(x) = \alpha_2 f(a_2) \quad + \alpha_3 f(a_3);$$

$$\text{with } \alpha_2 + \alpha_3 = 1, \alpha_2, \alpha_3 \geq 0, \forall x \in [a_2, a_3]$$

$$f(\hat{x}) \simeq f(x) = \alpha_3 f(a_3) \quad + \alpha_4 f(a_4);$$

$$\text{with } \alpha_3 + \alpha_4 = 1, \alpha_3, \alpha_3 \geq 0, \forall x \in [a_3, a_4]$$

Hence,

$$f(\hat{x}) = \alpha_1 f(a_1) + \alpha_2 f(a_2) + \alpha_3 f(a_3) + \alpha_4 f(a_4); \quad \alpha_1 + \alpha_2 + \alpha_3 + \alpha_4 = 1,$$
$\alpha_j \geq 0$ subject to

(i) at most two α_j can be positive and the remaining are set at zero.

(ii) these two positive α's should be adjacent.

These are called the *basis restrictions*.

Remark. For available x_i, the breaking points are $a_1^i, a_2^i, \ldots, a_n^i$ and the weight associated with these points are taken $\alpha_1^i, \alpha_2^i, \ldots, \alpha_n^i$.

Example 8. Formulate the approximate LPP for the following separable programming problem and find its solution:

$$\max \quad f(X) = \sin(\pi x_1/4) + x_2$$
$$\text{s.t.} \quad x_1^2 + 3x_2 \leq 4$$
$$x_1, x_2 \geq 0.$$

Here, we have

$$f_1(x_1) = \sin(\pi x_1/4); \qquad f_2(x_2) = x_2$$
$$g_1^1(x_1) = x_1^2; \qquad\qquad g_2^1(x_2) = 3x_2$$

For $x_2 = 0$ in the constraint, the range of $x_1^2 \leq 4$ is given by $a_1^1 = 0$, $a_2^1 = 1$, $a_3^1 = 2$ (breaking points). Now, calculate

k	a_k^1	$f_1\left(a_k^1\right)$	$g_1(a_k^1)$
1	0	0	0
2	1	$1/\sqrt{2}$	1
3	2	1	4

Hence,

$$\sin \frac{\pi x_1}{4} = f_1(x_1) \simeq \alpha_1^1 \cdot 0 + \alpha_2^1/\sqrt{2} + \alpha_3^1; \ \alpha_1^1 + \alpha_2^1 + \alpha_3^1 = 1, \ \alpha_j^1 \geq 0$$

$$x_1^2 = g_1^1(x_1) \simeq \alpha_1^1 \cdot 0 + \alpha_2^1 \cdot 1 + \alpha_3^1 \cdot 4$$

The approximate LPP is written as

$$\max \quad x_0 = \alpha_2^1/\sqrt{2} + \alpha_3^1 + x_2$$
$$\text{s.t.} \quad \alpha_2^1 + 4\alpha_3^1 + 3x_2 \leq 4$$
$$\alpha_1^1 + \alpha_2^1 + \alpha_3^1 = 1$$
$$\text{all var} \geq 0$$

Let us solve the above LPP by the simplex method. First, write this LPP in its standard form

$$\max \quad x_0 = \alpha_2^1/\sqrt{2} + \alpha_3^1 + x_2$$
$$\text{s.t.} \quad \alpha_2^1 + 4\alpha_3^1 + 3x_2 + s_1 = 4$$
$$\alpha_1^1 + \alpha_2^1 + \alpha_3^1 = 1$$
$$\text{all var} \geq 0$$

Keeping in view the basis restrictions, all the computations are shown in the following table:

B V	α_2^1	α_3^1	$x_2 \downarrow$	s_1	α_1^1	Soln
x_0	$-1/\sqrt{2}$	-1	-1	0	0	0
$\leftarrow s_1$	1	4	$\boxed{3}$	1	0	4
α_1^1	1	1	0	0	1	1
x_0	$\frac{1}{3} - \frac{1}{\sqrt{2}} \downarrow$	$1/3$	0	$1/3$	0	$4/3$
x_2	$1/3$	$4/3$	1	$1/3$	0	$4/3$
$\leftarrow \alpha_1^1$	$\boxed{1}$	1	0	0	1	1
x_0	0	$1/\sqrt{2}$	0	$1/3$	$\frac{1}{\sqrt{2}} - \frac{1}{3}$	$1 + 1/\sqrt{2}$
x_2	0	1	1	$1/3$	$-1/3$	1
α_2^1	1	1	0	0	1	1

From the table, the optimal value is $1 + 1/\sqrt{2}$. The problem has been approximated well and we use the linear expression to find

$$x_1^2 \simeq \alpha_2^1 + 4\alpha_3^1 = 1 \implies x_1 = 1.$$

The approximate optimal solution is

$$x_1 = 0, \quad x_2 = 1, \quad \text{optimal value} = 1 + \frac{1}{\sqrt{2}}.$$

Example 9. Solve the following problem as separable programming problem

$$\max \quad f(X) = 5x_1 + 2x_2^2$$
$$\text{s.t.} \quad 2x_1 + x_2^2 \leq 9$$
$$x_1, x_2 \geq 0$$

The separable functions are:

$$f_1(x_1) = 5x_1, \qquad f_2(x_2) = 2x_2^2$$
$$g_1^1(x_1) = 2x_1, \qquad g_2^1(x_2) = x_2^2$$

For $x_1 = 0$ in the constraint, we have $x_2^2 \le 9$. This gives the range of x_2 as $0, 1, 2, 3$ (the braking points).

k	a_k^2	$f_2\left(a_k^2\right)$	$g_2^1(a_k^2)$
1	0	0	0
2	1	2	1
3	2	8	4
4	3	18	9

$$2x_2^2 = f_2(x_2) \simeq \alpha_1'' \cdot 0 + \alpha_2'' \cdot 2 + \alpha_3'' \cdot 8 + \alpha_4'' \cdot 18;$$
$$\text{with} \quad \alpha_1'' + \alpha_2'' + \alpha_3'' + \alpha_4'' = 1, \ \alpha_i'' \ge 0$$

$$x_2^2 = g_2(x_2) \simeq \alpha_1'' \cdot 0 + \alpha_2'' \cdot 1 + \alpha_3'' \cdot 4 + \alpha_4'' \cdot 9;$$

The approximate LPP is

$$\begin{aligned}
\max \quad & f(\tilde{X}) = 5x_1 + 2\alpha_2'' + 8\alpha_3'' + 18\alpha_4'' \\
\text{s.t.} \quad & 2x_1 + \alpha_2'' + 4\alpha_3'' + 9\alpha_4'' \le 9 \\
& \alpha_1'' + \alpha_2'' + \alpha_3'' + \alpha_4'' = 1 \\
& \text{all var} \ge 0
\end{aligned}$$

The standard form of the above LPP is

$$\begin{aligned}
\max \quad & f(\tilde{X}) = 5x_1 + 2\alpha_2'' + 8\alpha_3'' + 18\alpha_4'' \\
\text{s.t.} \quad & 2x_1 + \alpha_2'' + 4\alpha_3'' + 9\alpha_4'' + s_1 = 9 \\
& \alpha_1'' + \alpha_2'' + \alpha_3'' + \alpha_4'' = 1 \\
& \text{all var} \ge 0
\end{aligned}$$

Now, we solve the problem by the simplex method keeping in view the basis restrictions at each iteration. The optimal solution is

$$x_1 = 9/2, \ x_2 = 0, \quad \text{optimal value} = 45/2.$$

To compute x_2, we have used

$$x_2^2 \simeq \alpha_2'' + 4\alpha_3'' + 9\alpha_4'' = 0 \implies x_2 = 0.$$

Remarks. Whenever a nonlinear function is to be evaluated after the optimal table, we use the linear expression representing it. In the above case, the value of x_2 has not been taken from the optimal table, but is computed by the linear approximating expression.

2. In case a constraint is of the type

$$2x_1 + x_2^2 \leq 11$$

the breaking points are given by $x_2^2 \leq 11$, and since $\sqrt{11} > 3$, we should take breaking points as $x_2 = 0, 1, 2, 3, 4$.

Example 10. Solve the following NLPP using separable programming.

$$\max \quad f = x_1 + x_2^4$$
$$\text{s.t.} \quad 3x_1 + 2x_2^2 \leq 9$$
$$x_1, x_2 \geq 0$$

The separable functions are:

$$f_1(x_1) = x_1; \qquad f_2(x_2) = x_2^4;$$
$$g_1^1(x_1) = 3x_1; \qquad g_2^1(x_2) = 2x_2^2.$$

The function $f_1(x)$ and $g_1^1(x)$ are linear and hence are left in their existing form. Consider $f_2(x_2)$ and $g_2^1(x_2)$.

As usual, for $x_1 = 0$ in the constraint, $2x_2^2 \leq 9 \Rightarrow x_2^2 \leq 4.5 \Rightarrow x_2$ can not exceed 3 and hence there are four breaking points.

k	a_k^2	$f_1\left(a_k^2\right)$	$g_2^1(a_k^2)$
1	0	0	0
2	1	1	2
3	2	16	8
4	3	81	18

This ensures

$$x_2^4 \simeq \alpha_1^2 \cdot 0 + \alpha_2^2 \cdot 1 + \alpha_3^2 \cdot 16 + \alpha_4^2 \cdot 16 = \alpha_2^2 + 16\alpha_3^2 + 81\alpha_4^2$$
$$2x_2^2 \simeq \alpha_1^2 \cdot 0 + \alpha_2^2 \cdot 2 + \alpha_3^2 \cdot 8 + \alpha_4^2 \cdot 18 = 2\alpha_2^2 + 8\alpha_3^2 + 18\alpha_4^2$$

Then the approximation LPP turns out to be

$$\max \quad f = x_1 + \alpha_2^2 + 16\alpha_3^2 + 81\alpha_4^2$$
$$\text{s.t.} \quad 3x_1 + 2\alpha_2^2 + 8\alpha_3^2 + 18\alpha_4^2 \leq 9$$
$$\alpha_1^2 + \alpha_2^2 + \alpha_3^2 + \alpha_4^2 = 1$$
$$\text{all variables} \geq 0$$

B.V.	x_1	α_2^2	$\alpha_3^2 \downarrow$	α_4^2	s_1	α_1^2	Soln
$f(\tilde{X})$	-1	-1	-16	-81	0	0	0
s_1	3	2	8	18	1	0	9
$\leftarrow \alpha_1^2$	0	1	$\boxed{1}$	1	0	1	1
$f(\tilde{X})$	-1	15	0	$-65 \downarrow$	0	16	16
$\leftarrow s_1$	3	-6	0	$\boxed{10}$	1	-8	1
α_3^2	0	1	1	0	0	1	1
$f(\tilde{X})$	$37/2$	-24	0	0	$13/2$	-36	$45/2$
α_4^2	$3/10$	$-3/5$	0	1	$1/10$	$-4/15$	$1/10$
α_3^2	$-3/10$	$8/5$	1	0	$-1/10$	$9/5$	$9/10$

We have followed basis restrictions in simplex iterations as follows:

(i) α_4^2 being most negative should enter and s_1 should leave the basis, but this is not permitted by basis restriction (b). The next choice goes to α_3^2 to enter and s_1 to leave the basis. This is permitted and the first iteration is made complete.

(ii) In the second iteration, α_4^2 should enter and s_1 should leave. This is permitted as both α_3^2 and α_4^2 come together in the basis. Complete the second simplex iteration.

(iii) Now, only α_2^2 can enter and α_3^2 can leave the basis. But this is not permitted by basis restriction (b). The process ends at this stage, even though the table is not optimal. How much we could do to improve the value of objective function, we have done.

To find solution of the problem, note that $x_1 = 0$, $\alpha_1^2 = 0$, $\alpha_2^2 = 0$, $\alpha_3^2 = 9/10$, $\alpha_4^2 = 1/10$. Using

$$2x_2^2 \simeq 0 \cdot 0 + 0 \cdot 2 + \frac{9}{10} \cdot 8 + \frac{1}{10} \cdot 18 = 9 \implies x_2 \simeq 2.1.$$

Hence, the approximate optimal solution: $x_1 = 0, x_2 = 2.123$ with optimal value: 20.25.

13.7 Duality in Nonlinear Programming

The duality for LPP has already been discussed at length in Chapter 4. The concept of duality in nonlinear programming plays an important role for advanced studies. The aim of this section is to introduce the concept of duality for nonlinear programming problems.

The primal nonlinear programming is taken in the form

$$\begin{aligned}
\min \quad & x_0 = f(X) \\
\text{s.t.} \quad & g_i(X) \leq 0, \quad i = 1, 2 \ldots, m,
\end{aligned}$$

where $f(X)$ and $g_i(X)$ have the first order partial derivatives.

The dual of the above NLPP is written as

$$\max \quad \phi(X, \lambda) = f(X) + \lambda^T G(X) \tag{13.12}$$
$$\text{s.t.} \quad \nabla f^T(X) + \lambda^T \nabla G(X) \geq 0 \tag{13.13}$$
$$\lambda \geq 0, \tag{13.14}$$

where

$$\nabla f(X) = \left(\frac{\partial f}{\partial x_1}, \frac{\partial f}{\partial x_2}, \cdots, \frac{\partial f}{\partial x_n}\right)^T : \text{gradient vector}$$

$$\lambda = (\lambda_1, \lambda_2, \ldots, \lambda_m)^T : \text{dual variables}$$

$$G(X) = (g_1(X), g_2(X), \ldots, g_m(X))^T \text{ and}$$

$$\nabla G(X) = \text{Jacobian matrix defined as}$$

$$\begin{bmatrix} \frac{\partial g_1}{\partial x_1} & \frac{\partial g_1}{\partial x_2} & \cdots & \frac{\partial g_1}{\partial x_n} \\ \frac{\partial g_2}{\partial x_1} & \frac{\partial g_2}{\partial x_2} & \cdots & \frac{\partial g_2}{\partial x_n} \\ \vdots & \vdots & \vdots & \\ \frac{\partial g_m}{\partial x_1} & \frac{\partial g_m}{\partial x_2} & \cdots & \frac{\partial g_m}{\partial x_n} \end{bmatrix} = \begin{bmatrix} \nabla g_1^T(X) \\ \nabla g_2^T(X) \\ \vdots \\ \nabla g_m^T(X) \end{bmatrix}$$

Remark. If the nonnegative restriction $X \geq 0$ is added in the format of NLPP (amenable to have dual), the dual is written as

$$\max \quad \phi(X, \lambda) = f(X) + \lambda^T G(X) - X^T \nabla f(X) - X^T \nabla G^T(X)\lambda$$
$$\text{s.t.} \quad \nabla f^T(X) + \lambda^T \nabla G(X) \geq 0$$

$$X \geq 0, \lambda \geq 0$$

Example 11. Write the dual of the problem

$$\min \quad f(X) = -4x_1 - 2x_2 + x_1^2 + x_2^2$$
$$\text{s.t.} \quad 2x_1 - x_2 \leq 7$$
$$\quad -x_1 + x_2 \leq -2$$
$$\quad x_1, x_2 \geq 0$$

First, we write the problem in appropriate format

$$\min \quad f(X) = -4x_1 - 2x_2 + x_1^2 + x_2^2$$
$$\text{s.t.} \quad 2x_1 - x_2 - 7 \leq 0$$
$$\quad -x_1 + x_2 + 2 \leq 0$$
$$\quad x_1, x_2 \geq 0$$

Since the nonnegative restrictions are given, we use the above remark to write the dual. Now, $X = (x_1, x_2)^T, \lambda = (\lambda_1, \lambda_2)^T$,

$$G(X) = \begin{bmatrix} 2x_1 - x_2 - 7 \\ -x_1 + x_2 + 2 \end{bmatrix}, \quad \nabla f(X) = \begin{bmatrix} -4 + 2x_1 \\ -2 + 2x_2 \end{bmatrix},$$

$$\nabla G(X) = \begin{bmatrix} 2 & -1 \\ -1 & 1 \end{bmatrix}$$

Substituting these values in the expression for dual in the above remark, we have

$$\max \quad y_0 = -4x_1 - 2x_2 + x_1^2 + x_2^2 + (\lambda_1, \lambda_2) \begin{bmatrix} 2x_1 - x_2 - 7 \\ -x_1 + x_2 + 2 \end{bmatrix}$$

$$- (x_1, x_2) \begin{bmatrix} -4 + 2x_1 \\ -2 + 2x_2 \end{bmatrix} - (x_1, x_2) \begin{bmatrix} -2 & -1 \\ -1 & 1 \end{bmatrix} \begin{bmatrix} \lambda_1 \\ \lambda_2 \end{bmatrix}$$

$$\text{s.t.} \quad (-4 + 2x_1, -2 + 2x_2) + (\lambda_1, \lambda_2) \begin{bmatrix} 2 & -1 \\ -1 & 1 \end{bmatrix} \geq 0$$

After simplification, we get

$$\begin{aligned} \max \quad & y_0 = -x_1^2 - x_2^2 - 7\lambda_1 + 2\lambda_2 \\ \text{s.t.} \quad & 2x_1 + 2\lambda_1 - \lambda_2 \geq 4 \\ & 2x_2 - \lambda_1 + \lambda_2 \geq 2 \\ & \text{all var} \geq 0 \end{aligned}$$

Note. The significant difference between the dual of a LPP and NLPP is that the dual of LPP is a function of dual variable, while the dual of NLPP is a function of primal variable and dual variable both. It can be seen obviously in (13.12), (13.13) and (13.14).

For theoretical exposition of the duality theory, we define P_F and S_D as the set of feasible solutions of primal and dual, respectively.

$$P_F = \{X \in \mathbb{R}^n : g_i(X) \leq 0, \ i = 1, 2, \ldots, m\};$$
$$S_D = \{(X, \lambda) : X \in \mathbb{R}^n, \lambda \in \mathbb{R}^m, \nabla f^T(X) + \lambda^T \nabla G(X) \geq 0\}.$$

Theorem 3. Let $f(X)$ and $G(X)$ be convex differentiable functions, and let $X_1 \in P_F$ and $(X_2, \lambda) \in S_D$. Then

$$f(X_1) \geq \phi(X_2, \lambda)$$

Theorem 4. Let $f(X)$ and $G(X)$ be convex differentiable functions, and let X^* be an optimal solution of primal. Then there exists a λ^* such that (X^*, λ^*) is an optimal solution of the dual (i.e., satisfies K-T conditions) and

$$f(X^*) = \phi(X^*, \lambda^*)$$

Theorem 5. Let $f(X)$ and $G(X)$ be convex differentiable functions. Let X_0 be an optimal solution of primal. If (X^*, λ^*) is an optimal solution of dual and $\phi(X, \lambda^*)$ is strictly convex function of X. Then $X_0 = X^*$, i.e., X^* is an optimal solution of primal, and

$$f(X^*) = \phi(X^*, \lambda^*)$$

Problem Set 13

1. Test the definiteness of the following functions

   (a) $x_1^2 + 4x_2^2 + 4x_3^2 + 4x_1x_3 + 16x_2x_3$;   (b) $x_1^2 + x_2^2 + x_3^2 - 2x_1x_2$.

2. Find the range of x_1, x_2, x_3 for which

   (a) $f(X) = x_1^3 - 2x_1x_2^2 + x_1x_3^2 + x_2^2x_3$ (b) $g(X) = x_1^3 + x_2^3 + x_1x_2$

   are a convex functions.

3. Using Hessian matrix, decide the convexity of the following function
   $$f(X) = x_1^2 + x_2^2 + 3x_3^2 - 2x_1x_2 + x_1x_3 + 16.$$

4. Consider the problem

   $$\min \quad z = x_1^2 + x_2^2 + x_3^2$$
   $$\text{s.t.} \quad 4x_1 + x_2^2 + 2x_3 = 14.$$

   Solve this problem by the Lagrange multiplier method.

5. Prove that $f(X) = C^T X$ is a convex function.

6. Let $f_1, f_2, \ldots, f_n$ be convex functions defined on a convex set S. Then show that the linear combination $\alpha_1 f_1 + \alpha_2 f_2 + \cdots + \alpha_n f_n$, $\alpha_i \geq 0$, $i = 1, 2 \ldots, n$ is convex over S.

7. In a convex nonlinear programming problem (CNLPP), if the kth variable is unrestricted in sign, show that $\partial L(X, \lambda)/\partial x_k = 0$.

   Suggestion. Write $x_k = x_k^+ - x_k^-$.

8. If in a CNLPP the pth constraint is an equality constraint, show that pth Lagrange multiplier will be unrestricted in sign.

   Suggestion. Write $g_p(X) = 0$ as $g_p(X) \leq 0$ and $-g_p(X) \leq 0$.

9. Use the Kuhn-Tucker conditions to solve the following nonlinear problem

$$\max \quad z = 2x_1 - x_1^2 + x_2$$
$$\text{s.t.} \quad 2x_1 + 3x_2 \leq 6$$
$$2x_1 + x_2 \leq 4$$
$$x_1, x_2 \geq 0$$

10. Solve the following NLPP using Wolfe's method

$$\max \quad f(X) = x_2 - x_1^2$$
$$\text{s.t.} \quad x_1 + x_2 \leq 1$$
$$x_1, x_2 \geq 0$$

11. Solve the following NLPP using Wolfe's method

$$\min \quad f(X) = x_1^2 + x_2^2 + x_3^2$$
$$\text{s.t.} \quad 2x_1 + x_2 - x_3 \leq 0$$
$$1 - x_1 \leq 0$$
$$x_1, x_2 \geq 0, x_3 \leq 0$$

12. Solve the following quadratic programming problem using Wolfe's method

$$\max \quad f(X) = 2x_1 + x_2 - x_1^2$$
$$\text{s.t.} \quad 2x_1 + 3x_2 \leq 6$$
$$2x_1 + x_2 \leq 4$$
$$x_1, x_2 \geq 0$$

13. A company manufactures two products A and B. It takes 40 minutes to process one unit of product A and 20 minutes for each unit of B. The maximum machine time available is 35 hours per week. Products A and B require 3 kg and 4 kg of raw material per unit, respectively. The available quantity of raw material is envisaged to be 170 kg per week. The products A and B which have unlimited market potential sell for $220 and $600, respectively. The manufacturing costs for products A and B are equal to two and three times of square of the quantity produced, respectively. Find the optimum quantity of each product to be produced so as to maximize the profit.

14 A company sells two types of items A and B. Item A sells for $30 per unit. No quantity discount is given. The sales revenue for item B decreases as the number of its units sold increases and is given by $(40 - 0.4x_2)x_2$, where x_2 is the number of units sold of item B. The marketing department has only 1200 hours available for distributing these items in the next year. Further, the company estimates the sales time function is given by

$$\text{Sales time} = x_1 + 0.4x_1^2 + 4x_2 + 0.45x_2^2$$

The company can only procure 6000 units of items A and B for sales in the next year. Find the number of items A and B be produced by the company so as to maximize its total revenue.

15. A factory is faced with a decision regarding the number of units of a product it should produce during months of January and February respectively. At the end of January sufficient units must be on hand so as to supply regular customers with a total of 200 units. Furthermore, at the end of February, the required quantity will be 300 units. Assume that factory ceases production at the end of February. The production cost C is a simple function of output x and is given by $C = 3x^2$. In addition to production cost, units produced in January which are not sold until February incur an inventory cost of $8 per unit. Assume the initial inventory to be zero. Formulate the problem and find the minimum cost solution. The number of units produced must be equal to the number demanded and distributed.

16. A manufacturing company produces two products : Radios and TV sets. The sales-price relationship for these two products are given below

Products	Quantity Demanded	Unit Price
Radios	2000-7p	p
TV Sets	4000-15q	q

The total cost function for these two products are given by $250x+0.2x^2$ and $400y+0.2y^2$, respectively. The production takes place on two assembly lines. Radio sets are assembled on Assembly line I and TV sets are assembled on Assembly line II. Because of the limitations of the assembly line capacities, the daily production is limited to no more than 90 radio sets and 70 TV sets. The production of both types of products require electronic components. The production of each of these sets require five units and seven units of electronic equipment respectively. The electronic components are supplied by another manufacturer, and the supply is limited to 700 units per day. The company has 170 employees, i.e., the labour supply amount to 480 man-days. The production of one unit of radio set requires 1 man-day of labour, whereas 2 man-days of labour are required for a TV set. How many units of radio and TV sets should the company produce in order to maximize the total profit ? Formulate the problem and find the optimal solution.

17. Solve the following problem

$$\min \quad f = (x_1 - 2)^2 + 4(x_2 - 6)^2$$
$$\text{s.t.} \quad 2x_1 + x_2 \leq 14$$
$$x_1, x_2 \geq 0$$

Suggestion. Replace x_1 by $x_1 + 2$ and x_2 by $x_2 + 6$.

18. Consider the problem

$$\max \quad f = 6x_1 + 3x_2 - 4x_1x_2 - 2x_1^2 - 3x_2^2$$
$$\text{s.t.} \quad x_1 + x_2 \leq 1$$
$$2x_1 + 3x_2 \leq 4$$
$$x_1, x_2 \geq 0$$

Show that f is strictly concave and then find the solution.

19. Solve the following problem using separable programming technique

$$\min \quad f = x_1^2 + 2x_2^2$$
$$\text{s.t.} \quad 2x_1^2 + x_2^2 \leq 8$$
$$x_1, x_2 \geq 0$$

20. Solve the following NLPP using separable programming.

$$\max \quad f = x_1 + x_2^4$$
$$\text{s.t.} \quad 3x_1 + 2x_2^2 \leq 5$$
$$- x_1 + 2x_2 + 3x_3 \geq -4$$
$$2x_1 + 3x_2 - 4x_3 \geq 3$$
$$x_1 + x_2 + x_3 = 2$$
$$x_1, x_2 \geq 0$$

Chapter 14

Search Techniques

This chapter is devoted to unconstrained optimization of nonlinear problems. Here we discuss some search techniques which are in common practice and are applicable to optimize a function of one or several variables. In previous chapter for optimizing nonlinear functions the condition of convexity have been assumed, whenever necessary. However, in search techniques we avoid such type of strong conditions on the function to be optimized.

14.1   Unimodal function

A unimodal function is one that has only one peak in a given interval. Thus, a function of one variable is said to be unimodal on a given interval $[a, b]$ if it has either unique minimum or maximum on $[a, b]$.

For two values of the variable on the same side or enclosing the optimum point, the one nearer to this optimum gives the better approximation to the functional value. Mathematically, we can write it as follows.

Let x^* be a minimum point of the function $f(x)$ which is unimodal on $[a, b]$, if for $x_1 < x_2$, see Fig. 14.1.

$$(i)\ x_1 < x^* \Rightarrow f(x_1) > f(x^*),$$

$$(ii)\ x_2 > x^* \Rightarrow f(x_2) > f(x^*).$$

Note that in unimodal case we do not need $f(x_1) \geq f(x_2)$, instead strict inequality.

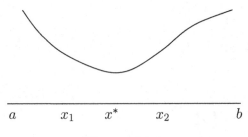

$$a \qquad x_1 \qquad x^* \qquad x_2 \qquad b$$

Figure 14.1

The initial interval being considered is called the interval of uncertainty. Here, $[a, b]$ is the interval of uncertainty. We term the evaluation of one functional value in search techniques as one experiment.

Let $f(x)$ be the unimodal function on the interval of uncertainty $[a, b]$ having a point of minimum. Take two arbitrary points (experiments) x_1 and x_2 such that $x_1 < x_2$ in this interval, then the interval of uncertainty is reduced as, see Fig. 14.2.

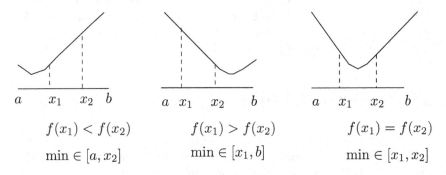

$$a \qquad x_1 \quad x_2 \quad b \qquad\qquad a \ x_1 \quad x_2 \qquad b \qquad\qquad a \qquad x_1 \quad x_2 \quad b$$

$$f(x_1) < f(x_2) \qquad\qquad f(x_1) > f(x_2) \qquad\qquad f(x_1) = f(x_2)$$

$$\min \in [a, x_2] \qquad\qquad \min \in [x_1, b] \qquad\qquad \min \in [x_1, x_2]$$

Figure 14.2

From the above figure, it is obvious that after two experiments the interval of uncertainty becomes $[a, x_2]$ or $[x_1, b]$ or $[x_1, x_2]$. It seems in the first graph of Fig. 14.2 that interval of uncertainty should be taken $[a, x_1]$. But this is not correct as the point x_1 may be on the left of the minimum point and in that case $[a, x_1]$ fails to contain the minimum point. Similar reasoning is applicable for the second graph of Fig. 14.2.

Measure of effectiveness. Suppose L_0 is the width of the interval of uncertainty. Further, suppose that L_n be the width of interval of uncertainty after n experiments. Then measure of effectiveness of any search technique is defined as

$$\alpha = \frac{L_n}{L_0} \le 1$$

Note that measure of effectiveness is used to compare the efficiency of different search methods.

14.2   Dichotomous Search Method

In this method two experiments are placed as close as possible to the centre of the interval of uncertainty. Now, based on the relative values of the objective function at the two points, almost half of the interval of uncertainty is eliminated. Let (x_L, x_R) be the initial interval of uncertainty. The two experiments will now be performed at x_1 and x_2 given by

$$x_1 = x_L + \frac{L_0 - \delta}{2},$$

$$x_2 = x_L + \frac{L_0 + \delta}{2},$$

where $L_0 = x_R - x_L$, and δ is the small number chosen so that the two experiments give significantly different results. Fig. 14.3 illustrates these calculations.

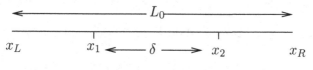

Figure 14.3

Now, based on the function values at x_1 and x_2, the new interval of uncertainty is given by $(L_0 + \delta)/2$. The next two experiments are now performed in this reduced interval of uncertainty. This procedure can be repeated until a desired accuracy is achieved.

The interval of uncertainty at the end of the different pairs of experiments can be given as

Number of experiments	2	4	6
Width of interval of uncertainty	$\frac{L_0+\delta}{2}$	$\frac{1}{2}\frac{L_0+\delta}{2} + \frac{\delta}{2}$	$\frac{1}{2}\left(\frac{L_0+\delta}{4} + \frac{\delta}{2}\right) + \frac{\delta}{2}$

In general, the final interval of uncertainty after conducting n experiments (n, being even) is given by

$$L_n = \frac{L_0}{2^{n/2}} + \delta\left(1 - \frac{1}{2^{n/2}}\right) \tag{14.1}$$

Thus,

$$\alpha = \frac{1}{2^{n/2}} + \frac{\delta}{L_0}\left(1 - \frac{1}{2^{n/2}}\right). \tag{14.2}$$

The procedure is well explained in the following example.

Example 1. Find the maximum of $f(x) = x(1.5 - x)$ in the interval $[0, 1]$ to within 10% of the exact value. Take $\delta = 0.001$.

If the middle point of the final interval of uncertainty is taken as the optimum point, then

$$\frac{1}{2}\frac{L_n}{L_0} \leq 0.1.$$

In view of (14.2),

$$\frac{1}{2^{n/2}} + \frac{\delta}{L_0}\left(1 - \frac{1}{2^{n/2}}\right) \leq \frac{1}{5}.$$

For $\delta = 0.001$ and $L_0 = 1$, we have

$$\frac{1}{2^{n/2}} + \frac{1}{1000}\left(1 - \frac{1}{2^{n/2}}\right) \leq \frac{1}{5},$$

i.e.,

$$\frac{999}{1000}\frac{1}{2^{n/2}} \leq \frac{995}{5000} \quad \text{or} \quad n \geq \frac{999}{199} \simeq 5.$$

Since n is even, the minimum admissible value is 6.

Perform starting two experiments as

$$x_1 = \frac{L_0}{2} - \frac{\delta}{2} = 0.5 - .0005 = 0.4995,$$

$$x_2 = \frac{L_0}{2} + \frac{\delta}{2} = 0.5 + .0005 = 0.5005.$$

with the function values

$$f(x_1) = 0.4995(1.0005) = 0.49975,$$

$$f(x_2) = 0.5005(0.9995) = 0.50025.$$

Since $f(x_1) < f(x_2)$, the next interval of uncertainty is $[0.4995, 1]$. Perform the second pair of experiments as

$$x_3 = 0.4995 + \frac{1 - 0.4995}{2} - 0.0005 = 0.74925,$$

$$x_4 = 0.4995 + \frac{1 - 0.4995}{2} + 0.0005 = 0.75025.$$

The corresponding function values are

$$f(x_3) = 0.74925 \times 0.75075 = 0.5624994,$$

$$f(x_4) = 0.75025 \times 0.74975 = 0.5624999.$$

Since $f(x_3) < f(x_4)$, the next interval of uncertainty is $[0.75037, 1]$. Now, perform the final pair of experiments as

$$x_5 = 0.5624994 + \frac{1 - 0.5624994}{2} - 0.0005 = 0.7807497,$$

$$x_6 = 0.562499 + \frac{1 - 0.5624999}{2} + 0.0005 = 0.78174995.$$

The function values at these points are

$$f(x_5) = 0.7807497 \times 0.625875 = 0.5615545,$$

$$f(x_6) = 0.78174995 \times 0.624875 = 0.56149194.$$

Since $f(x_5) > f(x_6)$, the new interval of uncertainty is given by

$$[x_4, x_6] = [0.75025, 0.78174995].$$

The center of this interval can be taken as the optimum point, and we have

$$x_{\max} \simeq 0.7649997, \quad f_{\max} \simeq 0.562275.$$

14.3 Fibonacci Search Method

As the name of the method suggests, this method uses the Fibonacci numbers generated by the recursive relation:

$$F_n = F_{n-1} + F_{n-2}, \quad n > 1,$$

where $F_0 = F_1 = 1$.

The Fibonacci search method gives an infinite sequence

$$F_2 \quad F_3 \quad F_4 \quad F_5 \quad F_6 \quad F_7 \quad F_8 \quad F_9 \quad F_{10} \quad F_{11} \quad \cdots$$

$$2 \quad 3 \quad 5 \quad 8 \quad 13 \quad 21 \quad 34 \quad 55 \quad 89 \quad 144 \quad \cdots$$

Obviously, any number is the sum of two previous numbers.

Let $L_0 = b - a$ be the length of the initial interval of uncertainty, and let n be the total number of experiments to be conducted.

Place the first two points x_1 and x_2 according to the formulas

$$x_1 = a + \frac{F_{n-2}}{F_n} L_0, \tag{14.3}$$

$$x_2 = b - \frac{F_{n-2}}{F_n} L_0 = a + \frac{F_{n-1}}{F_n} L_0. \tag{14.4}$$

The equivalent expression for x_2 is due to the property of the Fibonacci sequence. The simple consequence of (14.3) and (14.4) is that $b - x_2 = x_1 - a$.

This shows that the points x_1 and x_2 are symmetrically placed in reference to the end points, see Fig. 14.4.

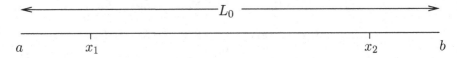

Figure 14.4

Since we are considering unimodal functions, the interval of uncertainty after two experiments x_1 and x_2 is reduced and its position depends whether $f(x_1)$ is $<$ or $> f(x_2)$. Hence,

$$L_2 = \text{Length of } [a, x_2] \text{ or length of } [x_1, b]$$

$$= x_2 - a \text{ or } b - x_1$$

$$= \frac{F_{n-1}}{F_n} L_0, \quad \text{using (14.4)}.$$

Suppose interval of uncertainty is $[a, x_2]$ (the studies for $[x_1, b]$ are similar). The next iteration is on L_2. Take two observations at x_1' and

x_2' defined by

$$x_1' = a + \frac{F_{n-3}}{F_{n-1}} L_2 \tag{14.5}$$

$$x_2' = b - \frac{F_{n-3}}{F_{n-1}} L_2 = a + \frac{F_{n-2}}{F_{n-1}} L_2. \tag{14.6}$$

Figure 14.5 depicts the next placements

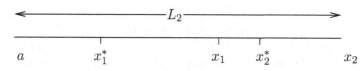

Figure 14.5

Using (14.5) and (14.6), we have

$$x_2 - x_2' = x_2 - a - \frac{F_{n-2}}{F_{n-1}} L_2$$

$$= L_2 - \frac{F_{n-2}}{F_{n-1}} L_2$$

$$= \frac{F_{n-1} - F_{n-2}}{F_{n-1}} L_2$$

$$= \frac{F_{n-3}}{F_{n-1}} L_2 = x_1' - a.$$

This ensures that x_1' and x_2' are symmetrically placed with respect to a and x_2, respectively. Next, consider

$$x_2 - x_2' = \frac{F_{n-3}}{F_{n-1}} L_2$$

$$= \frac{F_{n-3}}{F_n} L_0 = x_2 - x_1, \quad \text{using (14.3) and (14.4).}$$

The last equation implies that x_2' and x_1 coincide. At x_1, we already know $f(x_1)$ in the first iteration. Thus, we have added only one new observation in the second iteration. This is the reason for the best performance of the Fibonacci method. Suppose $f(x_1') < f(x_1)$. Then the length of interval of uncertainty after three iterations is

$$L_3 = x_1 - a = \frac{F_{n-2}}{F_n} L_0, \quad \text{using (14.3).}$$

Continuing in this manner, we get after n iterations

$$L_n = \frac{F_1}{F_n}L_0 = \frac{L_0}{F_n} \implies \alpha = \frac{1}{F_n}.$$

Note that in our analysis we have avoided the case $f(x_1) = f(x_2)$, since we are considering the interval of uncertainty at worst.

Limitations. (1) The interval of uncertainty should be known to us.

(2) The number of experiments n to be performed in the search has to be specified before hand. In case n is not given, then measure of effectiveness will be given to compute n as $\alpha \leq \alpha_0 \implies (1/F_n) \leq \alpha_0$. This determines n.

Example 2. Minimize the function $x^2 - 2x$, $0 \leq x \leq 1.5$ within the interval of uncertainty $0.25L_0$, where L_0 is the initial interval of uncertainty.

Here

$$\alpha = \frac{1}{F_n} \leq 0.25 \implies F_n \geq 4 \implies n = 4.$$

Thus, we have to take 4 observations. The first experiment is at

$$x_1 = a + \frac{F_{n-2}}{F_n}L_0 = 0 + \frac{F_2}{F_4}L_0 = \frac{2}{5} \times 1.5 = 0.6.$$

Since initial two starting observations are symmetrically placed, x_2 will be at distance 0.6 from the right end point, i.e., at 0.9. The next figure depicts this.

0   .6   .9   1.5

$$\left.\begin{array}{l} f(x_1) = f(0.6) = -0.84 \\ f(x_2) = f(0.9) = -0.99 \end{array}\right\} \implies f(0.6) > f(0.9) \implies \text{reject } [0, 0.6]$$

So, the next interval of uncertainty is $[0.6, 1.5]$ and $L_2 = 1.5 - 0.6 = 0.9$. The next iteration, by theory one point will be same as 0.9. Since 0.9 is at a distance of 0.3 from 0.6, the new point will at a distance of 0.3 from 1.5. This new point is 1.2 as shown in the next figure.

.6   .9   1.2   1.5

Since $f(x_3) = f(1.2) = -0.96 > f(0.9)$. So, the next interval of uncertainty is $[0.6, 1.2]$ and $L_3 = 1.2 - 0.6 = 0.6$

.6   .9   1.2

Now, we can not add a new point as 0.9 is equally spaced from 0.6 and 1.2. This will always happen at last iteration for all n, since $F_0/F_2 = 1/2$, i.e., last interval of uncertainty is halved.

At this juncture we find the last point x_4 very nearer to 0.9 on either side, say 0.95 and compute $f(0.95) = -0.9975$. Since $f(0.9) < f(0.95)$ thus the final interval of uncertainty is $[.9, 1.2]$ and $L_4 = 0.3$. The condition $L_4/L_0 \leq 0.3/1.5 = 0.2 \leq 0.25$ is satisfied. We take x^* as the mid point of the last interval of uncertainty, i.e., $f(1.05) = -0.9975$.

$$x^*$$

.9   1.05   1.2

The correct value is $x^* = 1$ and $f(x^*) = -1$. This is a good approximation.

14.4 Golden Section Method

This method can be thought of an extension of the Fibonacci method. The only difference is that the number of experiments to be conducted is not needed to be specified in advance. In this method, we start with the assumption that we are going to conduct a large number of experiments.

As a consequence of the Fibonacci method for n experiments, we have

$$L_k = L_{k+1} + L_{k+2}, \quad 1 \leq k \leq n - 2, \quad L_0 = L_1. \tag{14.7}$$

Suppose the intervals of uncertainty is reduced at a fixed rate γ, and let $n \to \infty$. Then

$$\frac{L_{k+1}}{L_k} = \frac{F_{n-k}}{F_{n-k+1}} = \gamma \ \forall \ k \geq 1. \tag{14.8}$$

Since $0 \leq F_{n-k}/F_{n-k+1} \leq 1$ and monotonic decreasing, the convergence is ensured for some positive number γ.

Note that the relation $L_k = L_{k+1} + L_{k+2}$, $k \geq 1$ requires that the points at which experiments are to be made must be symmetrically placed with respect to the end points of the interval of uncertainty. The distance of these points are determined as follows. Using (14.7) and (14.8), we get

$$L_k = \gamma L_k + \gamma^2 L_k, \quad \text{since } L_{k+2} = \gamma L_{k+1} = \gamma^2 L_k.$$

This gives $\gamma = (-1 + \sqrt{5})/2 = 0.618$.

Thus, the first two experiments are conducted at a distance of γL_0 from the end points of L_0. The next experiment is conducted at a distance of γL_2 from one end point of L_2 and so on.

Remark. In the Fibonacci search technique, γ is not fixed. However, Fibonacci method is slightly superior to Golden section method. This can be verified that in four experiments the interval of uncertainty is reduced to

$$L_4 = \frac{F_1}{F_4} L_0 = \frac{1}{5} L_0 = 0.2 L_0$$

for the Fibonacci method, while

$$L_4 = 0.618 L_3 = 0.618 \times 0.618 L_2$$
$$= (0.618)^3 L_0 = 0.236 L_0$$

for the Golden section method. Here $n = 4$ is predicted in advance.

The search techniques discussed in Sections 14.2-14.4 were limited to optimize functions of a single variable and that too unimodal functions. Our next studies are to optimize nonlinear functions of several variables.

14.5   Steepest Descent Method

This is also known as the gradient method. The steepest descent method can be used to find the minimum of a function of several variables. Let $f(X)$ be the given function of X, representing a n-dimensional vector. Cauchy (1847) used the negative of the gradient vector to choose the direction of the minimization.

Here, the theory and examples are discussed for the minimization problem. The method depends upon the fact that negative of grad $f(X) = \nabla f(X)$ points in the maximum decrease of the function $f(X)$.

Let us choose an initial starting point X_1, and iteratively move toward the minimum according to the following rule,

$$X_{i+1} = X_i + \lambda_i s_i$$

$$= X_i - \lambda_i \nabla f(X_i), \quad s_i = \nabla f(X_i), \quad i = 1, 2, \dots .$$

Here λ_i is the optimal step length along the search direction of $s_i = -\nabla f(X_i)$. One of the following criteria can be used to terminate the iterative process

(a) $\quad |f(X_{i+1}) - f(X_i)| \le \varepsilon_1$

(b) $\quad \left| \dfrac{f(X_{i+1}) - f(X_i)}{f(X_i)} \right| \le \varepsilon_2;$

(c) $\quad \left| \dfrac{\partial f}{\partial x_i} \right| \le \varepsilon_3, \quad i = 1, 2 \dots, n;$

(d) $\quad |X_{i+1} - X_i| \le \varepsilon_4.$

Let us work out an example to give the clear exposition of the above ideas.

Example 3. Minimize the function

$$f(x_1, x_2) = x_1^2 - x_1 x_2 + x_2^2$$

so that the error does not exceed by 0.05. The initial approximation is to be taken as $(1, 1/2)$.

Note that

$$\nabla f(X) = (2x_1 - x_2, -x_1 + 2x_2)^T$$

and the initial approximation is $X_1 = (1, 1/2)^T$.

Step 1. The first step is $f(X_1) = f(1, 1/2) = 3/4$, and

$$\nabla f(X_1) = (2x_1 - x_2, -x_1 + 2x_2)^T \text{ at } (1, 1/2)$$

$$= (3/2, 0)^T = -s_1.$$

Find the first step size λ_1 along s_1 by minimizing

$$f(X_2) = f(X_1 + \lambda_1 s_1) = f\left(\left(1, \frac{1}{2}\right) - \lambda_1 \left(\frac{3}{2}, 0\right) \right) = f\left(\frac{2 - 3\lambda_1}{2}, \frac{1}{2} \right).$$

Hence

$$f(X_2) = \frac{(2-3\lambda_1)^2}{4} - \frac{2-3\lambda_1}{4} + \frac{1}{4}.$$

This is a function of single variable, and the minimum occurs at $df/d\lambda_1 = 0$, i.e.,

$$-\frac{6}{4}(2-3\lambda_1) + \frac{3}{4} = 0 \implies \lambda_1 = \frac{1}{2}.$$

Thus, the new point and function value at this point are

$$X_2 = X_1 - \lambda_1 s_1 = (1, 1/2)^T - (1/2)(3/2, 0)^T = (1/4, 1/2)^T$$

$$f(X_2) = f\left(\frac{1}{4}, \frac{1}{2}\right) = \frac{1}{16} - \frac{1}{8} + \frac{1}{4} = \frac{3}{16}.$$

Terminate the calculations if $F(X_1) - f(X_2) < 0.05$. Obviously, the tolerance criteria is not satisfied.

Step 2. Again, start calculation replacing X_1 by X_2. So, proceed with $X_2 = (1/4, 1/2)^T$, and compute

$$\nabla f(X_2) = \left(\frac{2}{4} - \frac{1}{2}, -\frac{1}{4} + 2\frac{1}{2}\right) = \left(0, \frac{3}{4}\right) = -s_2.$$

The second step size λ_2 along s_2 is obtained by minimizing

$$f(X_3) = f(X_2 - \lambda_2 \nabla f(X_2)) = f((1/4, 1/2) - \lambda_2(0, 3/4))$$
$$= f(1/4, (2-3\lambda_2)/4).$$

Hence

$$f(X_3) = \frac{1}{16} - \frac{2-3\lambda_2}{16} + \frac{(2-3\lambda_2)^2}{16}.$$

Again, f is a function of single variable, and the minimum occurs at $df/d\lambda_2 = 0$, i.e.,

$$\frac{3}{16} - \frac{6}{16}(2-3\lambda_2) = 0 \implies \lambda_2 = 1/2.$$

Thus, the next point and function value are

$$X_3 = (1/4, 1/8)^T, \quad f(X_3) = \frac{1}{16} - \frac{1}{32} + \frac{1}{64} = \frac{3}{64}.$$

Since

$$f(X_2) - f(X_3) = 9/64 < .05,$$

the tolerance limit is satisfied, and we stop further computations.

14.6 Conjugate Gradient Method

This is considered to be the best as its rate of convergence is faster than the steepest descent method. Expand f in Taylor series about X_0 as

$$f(X) = f(X_0) + (X - X_0)^T \nabla f(X_0) + \frac{1}{2}(X - X_0)^T H(X_0)(X - X_0) + \cdots,$$

where $H(X_0)$ is the positive definite Hessian matrix computed at X_0. If X_0 is a point of minimum, then $\nabla f(X_0) = 0$, and the above expression reduces to

$$f(X) = f(X_0) + \frac{1}{2}(X - X_0)^T H(X_0)(X - X_0).$$

The second term on the right is positive, since X_0 is a minimum point. This suggests that a function can be approximated by a positive definite quadratic form in the neighbourhood of a minimum point.

Definition. Let H be $n \times n$ symmetric matrix. Two vectors s_1 and s_2 in $\mathbb{R}^n$ are said to be conjugate (orthogonal) with respect to H if

$$s_1^T H s_2 = 0.$$

For example, $(-1, 1)^T$ and $(0, 2)^T$ are conjugate vectors with respect to the matrix

$$\begin{bmatrix} 4 & 2 \\ 2 & 4 \end{bmatrix}$$

Moreover, a finite set of vectors $s_1, s_2, \ldots, s_n$ is said to be conjugate with respect to H, if

$$s_i^T H s_j = 0 \ \forall \ i \neq j, \ i, j \in \{1, 2, \ldots, n\}.$$

Note that, if $H = I$, the conjugate vectors become orthogonal in the usual sense.

Theorem 1. Every finite set $s_1, s_2, \ldots, s_n$ of conjugate vectors with respect to a symmetric and positive definite matrix is linearly independent.

The proof, being straightforward is left to the reader.

Remark. For a positive definite quadratic $X^T A X$, A is symmetric positive definite matrix and $H = 2A$.

Next, we generalize $X^T A X$. Consider

$$f(X) = X^T A X + P^T X \alpha$$

be quadratic function of n variables $x_1, x_2, \ldots, x_n$ with its Hessian matrix H as positive definite. Here P is a column vector and α is a scalar.

Theorem 2. Let $F(X)$ be a quadratic function of n variables with positive definite Hessian matrix. If the successive optimal steps are taken along $s_1, s_2, \ldots s_n$, then the point of minimum is reached in exactly n iterations.

Proof. Let X_1 be starting point. Then

$$X_{k+1} = X_k + \lambda_k s_k, \quad k = 1, 2, \ldots, n, \qquad (14.9)$$

where step size λ_k along s_k is determined by minimizing the function

$$f(X_k + \lambda_k s_k).$$

By Taylor series, we have

$$f(X_k + \lambda_k s_k) = f(X_k) + \lambda_k s_k^T \nabla f(X_k) + \frac{1}{2}\lambda_k^2 s_k^T H s_k.$$

The above series terminates at the third term, because $f(X)$ is a quadratic (H is a constant matrix, so that the third order determinant vanishes. Also,

$$\frac{d}{d\lambda_k} f(X_k + \lambda_k s_k) = s_k^T \nabla f(X_k) + \lambda_k s_k^T H S_k = 0.$$

This gives

$$\lambda_k = -\frac{s_k^T \nabla f(X_k)}{s_k^T H s_k} \qquad (14.10)$$

Let X^* be the point of minimum for $F(X)$. The conjugate directions $s_1, s_2, \ldots, s_n$ are linearly dependent because $F(X)$ is positive definite, see Theorem 1. Hence they form a basis in $\mathbb{R}^n$. Thus,

$$X^* - X_1 = \alpha_1 s_1 + \alpha_2 s_2 + \cdots + \alpha_n s_n, \quad \alpha_i \text{ are scalars.}$$

Pre-multiplying by $s_k^T H$, we get

$$\alpha_k = \frac{s_k^T H(X^* - X_1)}{s_k^T H s_k}, \quad k = 1, 2, \ldots, n. \qquad (14.11)$$

On the other hand, in view of (14.9), we can write

$$X_k = X_1 + \lambda_1 s_1 + \lambda_2 s_2 + \cdots + \lambda_{k-1} s_{k-1}, \ k > 1. \qquad (14.12)$$

For $k = n + 1$, this reduces to

$$X_{n+1} - X_1 = \lambda_1 s_1 + \lambda_2 s_2 + \cdots + \lambda_n s_n. \qquad (14.13)$$

After the first iteration, we get X_2 (X_1 is the initial approximation). So, after n iterations we determine X_{n+1}, and our objective becomes to establish that $X_{n+1} = X^*$. Thus, if we show that $\alpha_1 = \lambda_1, \alpha_2 = \lambda_2, \ldots, \alpha_n = \lambda_n$, we are done. From (14.12),

$$X_k - X_1 = \lambda_1 s_1 + \cdots + \lambda_{k-1} s_{k-1}$$

and pre-multiplying by $s_k^T H$ and noticing that $s_1, s_2, \ldots, \alpha_k$ are conjugate vectors, we obtain

$$s_k^T H(X_k - X_1) = 0, \quad k = 2, 3, \ldots, n.$$

For $k = 1$, the above result is trivially true.

Using (14.12) and the above condition, one gets

$$\alpha_k = \frac{s_k^T H(X^* - X_k + X_k - X_1)}{s_k^T H s_k} = \frac{s_k^T H(X^* - X_k)}{s_k^T H s_k}.$$

Applying Taylor series expansions to the function $\partial f / \partial x_1, \partial f / \partial x_2, \ldots, \partial f / \partial x_n$, we get

$$\nabla f(X) = \nabla f(X_k) + H(X - X_k)$$

At $X = X^*$,

$$\nabla f(X^*) = 0 = \nabla f(X_k) + H(X^* - X_k),$$

since X^* is the point of minimum. Using this and (14.9), we have

$$\alpha_k = -\frac{s_k^T \nabla f(X_k)}{s_k^T H s_k} = \lambda_k, \quad k = 1, 2, \ldots, n.$$

This proves the result.

To apply the above procedure we must have conjugate vectors $s_1, s_2, \ldots, s_n$ with respect to the H. This theorem says that if the step size is taken according to (14.9), then the minimum point X^* is obtained exactly in n-steps. For this we need n and H-conjugate directions at our choice. Thus, prominent work is to generate $s_1, s_2, \ldots, s_n$. If we have n linearly independent vectors in $\mathbb{R}^n$, then in the same manner as in Gram-Schmidt orthogonalization process, we can generate n conjugate vectors with respect to the positive definite symmetric matrix H. Take

$$s_1 = -\nabla f(X_1), \quad X_1 \quad \text{is the initial approximation}$$

$$s_{k+1} = -\nabla f(X_{k+1}) + \beta_k s_k, \quad k \geq 1,$$

where

$$\beta_k = \frac{\nabla f^T(X_{k+1}) H s_k}{s_k^T H s_k}.$$

Algorithm. Suppose $f(X)$ is quadratic function of n variable with positive symmetric definite matrix. The algorithm proceeds as

Let $\nabla f(X_k)$ be the gradient at X_1, where X_1 is the initial approximation for the minimum point. Then, compute recursively for $k = 1, 2, \ldots,$

$$s_1 = -\nabla f(X_1)$$

$$\lambda_k = \frac{-s_k^T \nabla f(X_k)}{s_k^T H s_k}$$

$$X_{k+1} = X_k + \lambda_k s_k$$

$$\beta_k = \frac{\nabla f^T(X_{k+1}) H s_k}{s_k^T H s_k}$$

$$s_{k+1} = -\nabla f(X_{k+1}) + \beta_k s_k.$$

At last X_{n+1} is the exact point of minimum.

Remarks. 1. If the function is not quadratic in n variables, the Hessian matrix H is not constant. Even then the above algorithm works nicely, provided we replace H by $H(X_k)$. But, now n iterations may not give the exact minimum, and in this case perform iterations till

$$|F(X_k) - f(X_{k+1})| < \varepsilon,$$

where $\varepsilon > 0$ is the tolerance limit.

2. There are other ways of considering the tolerance limits, mentioned in the introduction part of the steepest descent method.

3. The successive points X_{k+1} are computed as in gradient. The direction s_1 is same for both methods. The only difference is that $s_2, s_3, \ldots, s_3$ are determined in a different way so that $s_1, s_2, \ldots, s_n$ are conjugate with respect to H.

Example 4. Use conjugate gradient method to optimize the nonlinear function

$$\min x_1^2 - x_1 x_2 + 3x_2^2, \quad \text{initial approximation } X_1 = (1, 2)^T.$$

Here $n = 2$, so $X_3 = X^*$ will give the exact minimum point.

tion 1. Compute s_1, λ_1, X_2, and β_1 using relations given in the algorithm as

$$s_1 = -\nabla f(X_1) = - \begin{bmatrix} 2x_1 - x_2 \\ -x_1 + 6x_2 \end{bmatrix}_{(1,2)} = \begin{bmatrix} 0 \\ -11 \end{bmatrix}$$

$$H = \begin{bmatrix} 2 & -1 \\ -1 & 6 \end{bmatrix}$$

$$\lambda_1 = \frac{-s_1^T \nabla f(X_1)}{s_1^T H s_1} = \frac{(0, -11) \begin{bmatrix} 0 \\ -11 \end{bmatrix}}{(0, -11) \begin{bmatrix} 2 & -1 \\ -1 & 6 \end{bmatrix} \begin{bmatrix} 0 \\ -11 \end{bmatrix}}$$

$$= \frac{121}{11 \times 66} = \frac{1}{6}$$

$$X_2 = X_1 + \lambda_1 s_1 = \begin{bmatrix} 1 \\ 2 \end{bmatrix} + \frac{1}{6} \begin{bmatrix} 0 \\ -11 \end{bmatrix} = \begin{bmatrix} 1 \\ 1/6 \end{bmatrix}$$

$$\nabla f(X_2) = \begin{bmatrix} 2x_1 - x_2 \\ -x_1 + 6x_2 \end{bmatrix}_{(1,1/6)} = \begin{bmatrix} 11/6 \\ 0 \end{bmatrix}$$

$$\beta_1 = \frac{(\nabla f(X_2))^T H s_1}{s_1^T H s_1} = \frac{(11/6, 0) \begin{bmatrix} 2 & -1 \\ -1 & 6 \end{bmatrix} \begin{bmatrix} 0 \\ -11 \end{bmatrix}}{11 \times 66} = \frac{1}{36}$$

Iteration 2. Proceed to compute $\beta_1, s_2, \lambda_2, X_3$, using relations given in the algorithm

$$s_2 = -\nabla f(X_2) + \beta_1 s_1 = -\begin{bmatrix} 11/6 \\ 0 \end{bmatrix} + \frac{1}{36}\begin{bmatrix} 0 \\ -11 \end{bmatrix} = \begin{bmatrix} -11/6 \\ -11/36 \end{bmatrix}$$

$$\lambda_2 = \frac{-s_2^T \nabla f(X_2)}{s_2^T H s_2} = \frac{-(-11/6, -11/36)\begin{bmatrix} 11/6 \\ 0 \end{bmatrix}}{(-11/6, -11/36)\begin{bmatrix} 2 & -1 \\ -1 & 6 \end{bmatrix}\begin{bmatrix} -11/6 \\ -11/36 \end{bmatrix}}$$

$$= \frac{6}{11}$$

$$X_3 = X_2 + \lambda_2 s_2 = \begin{bmatrix} 1 \\ 1/6 \end{bmatrix} + \frac{6}{11}\begin{bmatrix} -11/6 \\ -11/36 \end{bmatrix} = \begin{bmatrix} 0 \\ 0 \end{bmatrix}$$

Thus, the minimum point is $X^* = (0,0)^T$. The process converges in two iterations because $f(X)$ is positive definite quadratic.

Problem Set 14

1. Show that a convex function is unimodal.

2. Find the minimum of the function $f(x) = x(x - 2.5)$, using Dichotomous search method in the interval $(0,1)$ to within 5% of the intial interval of uncertaint.

3. Let $L_0 = b - a$ be the length of the initial interval of uncertainty of $f(x)$. Partition this interval into three equal parts and make two experiments such such that $x_1 - a = x_2 - x_1 = b - x_2$. Show that after n (even) iterations the interval of uncertainty is reduced to

$$L_n = \left(\frac{2}{3}\right)^{n/2} L_0.$$

Suggestion. This is known as Bolzano search technique. Consider the interval of uncertainty at worst.

4. Find the number of experiments to be conducted to reduce the interval of uncertainty to $0.001L_0$ (L_0: initial interval of uncertainty)

   (a) Bolzano search;
   (b) Dichotomous search with $\delta = .0001$;
   (c) Fibonacci search;
   (d) Golden search.

5. Find the minimum of the function

$$f(x) = 0.8 - 0.7x \tan^{-1}\left(\frac{1}{x}\right) + \frac{0.5}{1+x^2}, \quad x \in [0.2]$$

   using the following methods:

   (a) Dichotomous search method to achieve an accuracy of within 5% of the interval of uncertainty and using a value of $\delta = 0.0001$.
   (b) Fibonacci method with $n = 6$.
   (c) Golden section method with $n = 6$.

6. Find the minimum of the function

$$f(x) = \frac{1}{1+x^2} + (1+x^2)\left(3 - \frac{2}{\sqrt{1+x^2}}\right) + x$$

   on the interval $[0, 1]$, using the following search techniques:

   (a) Fibonacci method with $n = 8$.
   (b) Golden section method with $n = 8$.

7. Prove the relation (14.7), when number of experiments for the Fibonacci are fixed at n.

8. Prove the relation (14.8), when number of experiments for the Fibonacci are fixed at n.

9. Prove Theorem 1.

10. Minimize the function

$$f(x_1, x_2) = 2x_1^2 + 2x_1x_2 + x_2^2 + x_1 - x_2$$

   starting from the initial approximate minimum point $X_1 = (0,0)^T$.

11. Using steepest method, find the minimum of the function

$$f(X) = x_3 + x_2^3 - 2x_1^2 + 3x_2^2 - 8, \quad \text{initial approximation } X_0 = (1, -1)^T$$

Is it local or global minimum? How many iterations are required to get the exact minimum point.

12. An electric power of 5×10^4 kW generated at a hydro-electric plant is to be transmitted 400 km to a step-down transformer station for distribution at 11 kilovolts. The power dissipated due the resistance of the conductor is $I^2 C^{-1}$, where I is the line current in amperes and C is the conductance in ohms. Based on the cost of delivered power, the resistance loss can be expressed as $\$0.263 I^2 C^{-1}$. The power transmitted is related to the transmission line voltage at the power plant E by the relation $K = \sqrt{3} EI$, where E is in kilovolts. The cost of conductors is given by $\$3.9 \times 10^6 C$ and the investment in equipment is is needed to to accommodate the voltage E is given by $\$10^3 E$. Find the value of E and C to minimize the cost of transmission system. Use conjugate gradient method or Newton's method to solve the problem.

Chapter 15

Geometric Programming

Geometric programming is very efficient to determine the optimal solutions of highly nonlinear programming problems. Moreover, this concept has a very wide range of applications for dealing with mechanical and industrial engineering processes. Different types of geometric programming problems have been considered in this chapter.

15.1 Introduction

The subject of geometric programming was initiated by Duffin, Peterson, Zener, and later on, it was enriched by various researchers. This technique is in frequent use for solving problems involving posynomials and polynomials. The basic tool, we generally utilize, is the arithmetic mean-geometric mean inequality (AM-GM inequality), and that is why it is called geometric programming.

Arithmetic-Geometric mean inequality.. Let us first derive the AM-GM inequality inequality:

$$\frac{1}{n}\sum_{j=1}^{n} x_j \geq \left(\prod_{j=1}^{n} x_j\right)^{1/n}.$$

To derive the above inequality, we first solve the problem

$$\min \quad f(X) = \sum_{j=1}^{n} x_j \tag{15.1}$$

$$\text{s.t.} \quad \prod_{j=1}^{n} x_j = c, \quad \text{a constant} \tag{15.2}$$

$$x_j \geq 0$$

Construct the Lagrange function

$$L(X, \lambda) = f(X) + \lambda \left(\prod_{j=1}^{n} x_j - c \right).$$

The partial derivatives are given by the relations

$$\frac{\partial L}{\partial x_i} = 1 + \lambda \prod_{\substack{j=1 \\ j \neq i}}^{n} x_j, \quad i = 1, 2, \ldots, n.$$

The necessary conditions for minimization are the solutions of the above equations, i.e.,

$$\lambda = \frac{-1}{\prod_{\substack{j=1 \\ j \neq i}}^{n} x_j}.$$

Since the above relation is true for any value of $i \in \{1, 2, \ldots, n\}$, it follows that

$$\lambda = \frac{-1}{\prod_{\substack{j=1 \\ j \neq k}}^{n} x_j}, \quad k = 1, 2, \ldots, n.$$

Comparing both values of λ, we have

$$x_1 x_2 \cdots x_{i-1} \cdots x_n = x_1 x_2 \cdots x_{k-1} \cdots x_n$$

which implies that $x_i = x_k$ for all $i, k \in \{1, 2, \ldots, n\}$.

Let these values be denoted by $x_i^o = x_k^o = a$ (constant). From (15.1),

$$a^m = c \implies a = c^{1/n} \implies x_i^o = c^{1/n},$$

and hence

$$f(X^o) = \sum_{j=1}^{n} c^{1/n} = mc^{1/n}.$$

Thus, we have proved that $\min f(X) = nc^{1/n}$. Note that

$$f(x) \geq nc^{1/n} \Rightarrow \sum_{j=1}^{n} x_j \geq nc^{1/n}.$$

Hence,

$$\frac{1}{n} \sum_{j=1}^{n} x_j \geq c^{1/n} \implies \frac{1}{n} \sum_{j=1}^{n} x_j \geq \left(\prod_{j=1}^{n} x_j \right)^{1/n}.$$

Remarks. 1. This inequality can be generalized as

$$\sum_{j=1}^{n} \lambda_j x_j \geq \prod_{j=1}^{n} x_j^{\lambda_j},$$

where $\lambda_1 + \lambda_2 + \cdots + \lambda_n = 1$ and $\lambda_j \geq 0$ for all $i = 1, 2, \ldots, n$.

2. Equality holds in the AM-GM inequality $\iff x_1 = x_2 = \cdots = x_n$.

3. The AM-GM inequality can also be derived by considering the problem

$$\max \quad g(X) = \prod_{j=1}^{n} x_j$$

$$\text{s.t.} \quad \sum_{j=1}^{n} x_j = c$$

$$x_j \geq 0$$

Determination of lower bound. From the theory developed above, if

$$f(X) = f_1(X) + f_2(X) + \cdots + f_m(X), \quad f_i(X) \geq 0$$

and, in addition

$$f_1^{1/n} f_2^{1/n} \cdots f_m^{1/m} = c \text{ (constant)},$$

then the AM-GM inequality tells

$$\frac{1}{m}\sum_{i=1}^{m} f_j \geq \prod_{i=1}^{m} f_j^{1/m} \Rightarrow f(X) \geq mc.$$

Here mc is the lower bound, and the minimum can not go below this value.

Example 1. Find the lower bound for $f(x) = x^{-4} + 4x^3 + 4x$, $x > 0$

Let

$$f = f_1 + f_2 + f_3, \quad f_1 = x_{-4}, f_2 = 4x^3, f_3 = 4x.$$

In addition,

$$f_1^{1/3} f_2^{1/3} f_3^{1/3} = 2.$$

Hence, $f(X) \geq 3 \times 2 = 6$

Example 2. Find the lower bound for $f(x) = x^4 + 4x^{-3}, x > 0$

Suppose $f = f_1 + f_2$. But $f_1^{1/2} f_2^{1/2} = 2\sqrt{x}$, not a constant. Hence, an additional condition is to be satisfied, let

$$f_1^{\lambda_1} f_2^{\lambda_2} = 4^{\lambda_2} x^{4\lambda_1 - 3\lambda_2},$$

and

$$\lambda_1 + \lambda_2 = 1 \tag{15.3}$$

For the additional condition, we assume

$$4\lambda_1 - 3\lambda_2 = 0. \tag{15.4}$$

Then $f_1^{\lambda_1} f_2^{\lambda_2} = 4^{\lambda_2}$. But $\lambda_1 f_1 + \lambda_2 f_2 \neq f(x)$. Rewrite

$$F_1 = \frac{f_1}{\lambda_1}, \quad F_2 = \frac{f_2}{\lambda_2}$$

Then $\lambda_1 F_1 + \lambda_2 F_2 = f(X)$. Now, by the AM-GM inequality,

$$\lambda_1 F_1 + \lambda_2 F_2 \geq F_1^{\lambda_1} F_2^{\lambda_2} = \lambda_1^{-\lambda_1} \lambda_2^{-\lambda_2} 4^{-\lambda_2}.$$

From (15.3) and (15.4), we have $\lambda_1 = 3/7$, $\lambda_2 = 4/7$. This ensures that

$$f(x) \geq \left(\frac{3}{7}\right)^{-3/7} \left(\frac{4}{7}\right)^{-4/7} 4^{4/7}.$$

Example 3. Find the lower bound for $f(X)$ given below

    (a) $5x^{5/2} + x^{-2/3} + x^{1/2} + 2x^{-2}$;

    (b) $x^{-2} + x^{-1} + 1 + x$.

For part (a), consider $f(x) = 5x^{5/2} + x^{-2/3} + x^{1/2} + 2x^{-2}$

$$f = f_1 + f_2 + f_3 + f_4, \quad f_1 = 5x^{5/2}, f_2 = x^{-2/3}, f_3 = x^{1/2}, f_4 = 2x^{-2}.$$

But

$$f_1^{1/4} f_2^{1/4} f_3^{1/4} f_4^{1/4} = 5^{1/4} 2^{1/4} x^{-1/24}$$

is not a constant. Let $\lambda_1, \lambda_2, \lambda_3, \lambda_4$ be such that

$$\lambda_1 + \lambda_2 + \lambda_3 + \lambda_4 = 1 \tag{15.5}$$

$$f_1^{\lambda_1} f_2^{\lambda_2} f_3^{\lambda_3} f_4^{\lambda_4} = 5^{\lambda_1} x^{5\lambda_2/3} x^{\lambda_3/2} 2^{\lambda_4} x^{-2\lambda_4}$$

Fix $\lambda_3 = \lambda_4 = 1/4$ so that the equations involving λ_1, λ_2 will give unique value (just for convenience). Then

$$f_1^{\lambda_1} f_2^{\lambda_2} f_3^{\lambda_3} f_4^{\lambda_4} = 5^{\lambda_1} 2^{1/4} x^{(60\lambda_1 - 16\lambda_2 - 9)/24}.$$

To satisfy the condition, let

$$60\lambda_1 - 16\lambda_2 - 19 = 0 \tag{15.6}$$

Since $\lambda_1 f_1 + \lambda_2 f_2 + \lambda_3 f_3 + \lambda_4 f_4 \neq f(x)$. Rewrite

$$F_1 = \frac{f_1}{\lambda_1}, \quad F_2 = \frac{f_2}{\lambda_2}, \quad F_3 = \frac{f_3}{\lambda_3}, \quad F_4 = \frac{f_4}{\lambda_4}$$

Then

$$f(x) = \lambda_1 F_1 + \lambda_2 F_2 + \lambda_3 F_3 + \lambda_4 F_4$$

$$\geq F_1^{\lambda_1} F_2^{\lambda_2} F_3^{\lambda_3} F_4^{\lambda_4}$$

$$= \lambda_1^{-\lambda_1} \lambda_2^{-\lambda_2} f_1^{\lambda_1} f_2^{\lambda_2} f_3^{\lambda_3} f_4^{\lambda_4} 4^{\lambda_3} 4^{\lambda_4}$$

$$\geq \lambda_1^{-\lambda_1} \lambda_2^{-\lambda_2} 2^{1/4} 25^{\lambda_1}$$

From (15.5) and (15.6),

$$\lambda_1 + \lambda_2 = 1/2, 60\lambda_1 - 16\lambda_2 = 19 \Rightarrow \lambda_1 = 27/16, \lambda_2 = 11/76.$$

Hence,

$$f(x) \geq \left(\frac{27}{76}\right)^{-27/76} \left(\frac{11}{76}\right)^{-11/6} 2^{5/4} 5^{27/76}.$$

For part (b), $g(X) = f_1^{1/4} f_2^{1/4} f_3^{1/4} = x^{-1/2-1/4+1/4} = 1/\sqrt{x}, x > 0$. Hence, take

$$f_1^{\lambda_1} f_2^{\lambda_2} f_3^{\lambda_3} = x^{-2\lambda_1 - \lambda_2 + \lambda_3}, \quad \lambda_1 + \lambda_2 + \lambda_3 = 1.$$

Here, assume $\lambda_3 = 1/3 \Rightarrow \lambda_1 + \lambda_2 = 2/3, -2\lambda_1 - \lambda_2 = -1/3$ (infeasible). Hence, we consider

$$\lambda_2 = 1/3 \Rightarrow \lambda_1 + \lambda_3 = 2/3, -2\lambda_1 + \lambda_3 = 1/3 \Rightarrow \lambda_1 = 1/9, \lambda_2 = 1/3.$$

Thus,

$$g(x) \geq \lambda_1^{-\lambda_1} \lambda_2^{-\lambda_2} \Rightarrow g(x) \geq \left(\frac{1}{9}\right)^{-1/9} \left(\frac{1}{3}\right)^{-1/3}$$

$$\Rightarrow f(x) \geq 1 + \left(\frac{1}{9}\right)^{-1/9} \left(\frac{1}{3}\right)^{-1/3}.$$

15.2   Unconstrained Posynomial Optimization

Here, we discuss geometric programming problem in reference to posynomial functions. Let us first define these functions.

Definition 1.   A posynomial is defined as

$$f(X) = \sum_{i=1}^{m} f_i(X),$$

where
$f_i(X) = c_i x_1^{a_{i1}} x_2^{a_{i2}} \cdots x_n^{a_{in}}, i = 1, 2, \ldots, m$ and $c_i > 0, a_{ij} \in \mathbb{R}, j = 1, 2, \ldots, n$.

For instance, $f(X) = (1/2)x_1^{1/3} x_2^{-1/3} + (4/5)x_1^{1/2} x_3^{4/3}$ is a posynomial.

Consider the unconstrained posynomial minimization problem

$$\min f(X) = \sum_{i=1}^{m} f_i(X), \quad f(X) \text{ is a posynomial} \qquad (15.7)$$

From Section 13.1, we can show that if $f_1 + f_2 + \cdots + f_m = f$ then $\forall \lambda_i > 0$ such that

$$f = f_1 + f_2 + \cdots + f_m \geq \left(\frac{f_1}{\lambda_1}\right)^{\lambda_1} \left(\frac{f_2}{\lambda_2}\right)^{\lambda_2} \cdots \left(\frac{f_m}{\lambda_m}\right)^{\lambda_m} ;$$

$$\lambda_1 + \lambda_2 + \cdots + \lambda_m = 1 \tag{15.8}$$

Applying the above condition to $f(X)$, we get

$$f \geq \left(\frac{c_1}{\lambda_1}\right)^{1/\lambda_1} \left(\frac{c_2}{\lambda_2}\right)^{1/\lambda_2} \cdots \left(\frac{c_n}{\lambda_n}\right)^{1/\lambda_n}$$
$$\times x_1^{a_{11}\lambda_1 + a_{21}\lambda_2 + \cdots + a_{m1}\lambda_m}$$
$$\times x_2^{a_{12}\lambda_1 + a_{22}\lambda_2 + \cdots + a_{m2}\lambda_m} \tag{15.9}$$
$$\vdots$$
$$\times x_n^{a_{1n}\lambda_1 + a_{2n}\lambda_2 + \cdots + a_{mn}\lambda_m},$$

where $\sum_{i=1}^{m} \lambda_i = 1$, $\forall \lambda_i > 0$.

Select $\lambda_i > 0$ such that the right side of (15.9) is constant, and for this we determine λ_i from the solution of simultaneously linear equations:

$$a_{11}\lambda_1 + a_{21}\lambda_2 + \cdots + a_{m1}\lambda_m = 0$$
$$a_{12}\lambda_1 + a_{22}\lambda_2 + \cdots + a_{m2}\lambda_m = 0$$
$$\vdots \tag{15.10}$$
$$a_{1n}\lambda_1 + a_{2n}\lambda_2 + \cdots + a_{mn}\lambda_m = 0,$$

For the time being, we digress, and assume that the system (15.10) has a solution, whether it is unique or not will be considered later in posynomial optimization.

Definition 2. The dual of $\min f(x) = \sum_{i=1}^{m} f_i(X)$ is defined as

$$\max \phi(\lambda) = \phi(\lambda_1, \lambda_2, \cdots, \lambda_m) = \left(\frac{c_1}{\lambda_1}\right)^{\lambda_1} \left(\frac{c_2}{\lambda_2}\right)^{\lambda_2} \cdots \left(\frac{c_m}{\lambda_m}\right)^{\lambda_m} \tag{15.11}$$

subject to the conditions given in (15.10).

Note that (15.11) is the right side of (15.9). In view of (15.10) x_i's term do not exist.

Theorem 1. If $X_0 = (x_1^o, x_2^o, \ldots, x_n^o)^T$ is the global minimum of the unconstrained polynomial problem: $\min f(x) = \sum_{i=1}^m f_i(X)$, then

$$\lambda_j^o = \frac{f_j(X_0)}{f(X_0)}, \quad j = 1, 2, \cdots, m$$

gives the optimal solution of the dual problem (15.11) and we have

$$f(X_0) = \phi(\lambda_0), \quad \lambda_0 = (\lambda_1^o, \lambda_2^o, \ldots, \lambda_m^o)^T.$$

Proof. We shall not prove this result here. Theorem 1 will be utilized to solve the problems. For its proof, the reader may consult advanced books on the subject, see Bazara et al.

Example 4. Minimize the following function $[m = k+1, m = 3, k = 2]$

$$f = 4x_1^2 x_2^{-3} + 5x_1^{-3} x_2 + 7x_1 x_2, \quad x_1, x_2 > 0$$

Here, $f_1 = 4x_1^2 x_2^{-3}$, $f_2 = 5x_1^{-3} x_2$, $f_3 = 7x_1 x_2$. Applying the result, if $f = f_1 + f_2 + \cdots + f_m$ then $\exists \lambda_i > 0$ such that

$$f \geq \left(\frac{f_1}{\lambda_1}\right)^{\lambda_1} \left(\frac{f_2}{\lambda_2}\right)^{\lambda_2} \cdots \left(\frac{f_m}{\lambda_m}\right)^{\lambda_m}.$$

We have

$$f \geq \left(\frac{4x_1^2 x_2^{-3}}{\lambda_1}\right)^{\lambda_1} \left(\frac{5x_1^{-3} x_2}{\lambda_2}\right)^{\lambda_2} \left(\frac{7x_1 x_2}{\lambda_3}\right)^{\lambda_3}$$

$$= \left(\frac{4}{\lambda_1}\right)^{\lambda_1} \left(\frac{5}{\lambda_2}\right)^{\lambda_2} \left(\frac{7}{\lambda_3}\right)^{\lambda_3} x_1^{2\lambda_1 - 3\lambda_2 + \lambda_3} x_2^{-3\lambda_1 + \lambda_2 + \lambda_3}$$

$$= \phi(\lambda), \text{ (say)}.$$

The dual constraints are obtained using orthogonality and normality conditions as

$$2\lambda_1 - 3\lambda_2 + \lambda_3 = 0$$
$$-3\lambda_1 + \lambda_2 + \lambda_3 = 0$$
$$\lambda_1 + \lambda_2 + \lambda_3 = 1$$
$$\lambda_i \geq 0$$

$$\begin{bmatrix} 1 & 1 & 1 & 1 \\ 2 & -3 & 1 & 0 \\ -3 & 1 & 1 & 0 \end{bmatrix} \simeq \begin{bmatrix} 1 & 1 & 1 & 1 \\ 0 & -5 & -1 & -2 \\ 0 & 4 & 4 & 3 \end{bmatrix} \simeq \begin{bmatrix} 1 & 1 & 1 & 1 \\ 0 & 1 & 1/5 & 2/5 \\ 0 & 1 & 1 & 3/4 \end{bmatrix} \simeq$$

$$\begin{bmatrix} 1 & 0 & 0 & 1/4 \\ 0 & 1 & 1/5 & 2/5 \\ 0 & 1 & 1 & 3/4 \end{bmatrix} \simeq \begin{bmatrix} 1 & 0 & 0 & 1/4 \\ 0 & 1 & 1/5 & 2/5 \\ 0 & 0 & 4/5 & 7/20 \end{bmatrix} \simeq \begin{bmatrix} 1 & 0 & 0 & 1/4 \\ 0 & 1 & 1/5 & 2/5 \\ 0 & 0 & 1 & 7/16 \end{bmatrix} \simeq$$

$$\begin{bmatrix} 1 & 0 & 0 & 1/4 \\ 0 & 1 & 0 & 5/16 \\ 0 & 0 & 1 & 7/16 \end{bmatrix}$$

So,

$$\lambda_1 = 1/4, \lambda_2 = 5/16, \lambda_3 = 7/16;$$

$$f(X_0) = \phi(\lambda_0) = (16)^{1/4}(16)^{5/16}(16)^{7/16} = 16.$$

Again, using the above theorem, we find that x_1^o and x_2^o as

$$\frac{f_1(X_0)}{\lambda_1^o} = \frac{f_2(X_0)}{\lambda_2^o} = \frac{f_3(X_0)}{\lambda_3^o} = f(X_0)$$

$$\frac{4(x_1^o)^2(x_2^o)^{-3}}{1/4} = \frac{5(x_1^o)^{-3}(x_2^o)}{5/16} = \frac{7(x_1^o)(x_2^o)}{7/16} = 16$$

This implies $(x_1^o)^2(x_2^o)^{-3} = (x_1^o)^{-3}(x_2^o) = (x_1^o)(x_2^o) = 1$. Taking logarithm, we get

$$2\ln x_1^o - 3\ln x_2^o = \ln 1 = 0$$
$$-3\ln x_1^o + \ln x_2^o = 0$$
$$\ln x_1^o + \ln x_2^o = 0$$

Assuming that $\ln x_1^o = z_1$ and $\ln x_2^o = z_2$, we have

$$2z_1 - 3z_2 = 0$$
$$-3z_1 + z_2 = 0$$
$$z_1 + z_2 = 0$$

The solution of the above system is $z_1 = 0, z_2 = 0$, i.e., $\ln x_1^o = 0$ and $\ln x_2^o = 0$ or $x_1^o = 1$ and $x_2^o = 1$. Hence the optimal solution is $x_1 = 1, x_2 = 1$, and min $= 16$

Example 5 (Oil Tank Design). Design an oil storage tank having volume V m^3 for the minimum cost. The materials for the bottom and side cost C_1 and C_2 units per sq.m.,respectively.

The problem is

$$\min f(x) = C_1LW + 2C_2LH + 2C_3WH + C_4VL^{-1}W^{-1}H^{-1}$$

Here $m=4$ and $k=3$. The pre-dual is

$$\psi(\lambda) = \left(\frac{C_1LW}{\lambda_1}\right)^{\lambda_1} \left(\frac{2C_2LH}{\lambda_2}\right)^{\lambda_2} \left(\frac{2C_3WH}{\lambda_3}\right)^{\lambda_3} \left(\frac{C_4VL^{-1}W^{-1}H^{-1}}{\lambda_4}\right)^{\lambda_4}$$

The dual problem is

$$\max \quad \phi(\lambda) = \left(\frac{C_1}{\lambda_1}\right)^{\lambda_1} \left(\frac{2C_2}{\lambda_2}\right)^{\lambda_2} \left(\frac{2C_3}{\lambda_3}\right)^{\lambda_3} \left(\frac{C_4V}{\lambda_4}\right)^{\lambda_4}$$

$$\text{s.t.} \quad \lambda_1 + \lambda_2 - \lambda_4 = 0$$
$$\lambda_1 + \lambda_3 - \lambda_4 = 0$$
$$\lambda_2 + \lambda_3 - \lambda_4 = 0$$
$$\lambda_1 + \lambda_2 + \lambda_3 + +\lambda_4 = 1$$
$$\lambda_i \geq 0, i = 1, 2, 3, 4$$

The solution of this system is $\lambda_1 = 1/5, \lambda_2 = 1/5, \lambda_3 = 1/5, \lambda_4 = 2/5$. The minimum total cost in transporting the gravel is given as

$$f(X^*) = \phi(\lambda^*) = (5C_1)^{1/5} (5C_2)^{1/5} (5C_3)^{1/5} \left(\frac{5C_4V}{2}\right)^{2/5}$$

$$= (3125C_1C_2C_3C_4^2V^2)^{1/5}$$

The corresponding optimum dimensions of the box are obtained using the relations

$$C_1LW = \frac{1}{5}\phi(\lambda^*)$$

$$C_2LH = \frac{1}{5}\phi(\lambda^*)$$

$$C_3HW = \frac{1}{5}\phi(\lambda^*)$$

$$\frac{C_4V}{LWH} = \frac{1}{5}\phi(\lambda^*)$$

Solving the above equations, we get

$$L^* = \left(\frac{C_3^3 C_4 V}{C_1^2 C_2^2}\right)^{1/5}, \quad W^* = \left(\frac{C_2^3 C_4 V}{C_1^2 C_3^2}\right)^{1/5},$$

$$H^* = \frac{1}{2}\left(\frac{C_1^3 C_4 V}{C_2^2 C_3^2}\right)^{1/5}.$$

Example 6. When $m > k + 1$. Solve the problem

$$\min f(x) = 5x_1 x_2^{-1} + 2x_1^{-1} x_2 + 5x_1 + x_2^{-1} \quad (m = 4, k = 2)$$

by geometric programming.

The orthogonality and normality conditions are given by

$$\begin{bmatrix} 1 & -1 & 1 & 0 \\ -1 & 1 & 0 & -1 \\ 1 & 1 & 1 & 1 \end{bmatrix} \begin{bmatrix} \lambda_1 \\ \lambda_2 \\ \lambda_3 \\ \lambda_4 \end{bmatrix} = \begin{bmatrix} 0 \\ 0 \\ 1 \end{bmatrix}$$

Since the number of dual variables > number of equations, we do not get the unique solution. By the row reduction method, we get

$$\lambda_1 = \frac{1 - 3\lambda_4}{2}, \quad \lambda_2 = \frac{1 - \lambda_4}{2}, \quad \lambda_3 = -\lambda_4$$

Hence, the corresponding dual function is determined by

$$\max \phi(\lambda) = \left(\frac{10}{1 - 3\lambda_4}\right)^{(1-3\lambda_4)/2} \left(\frac{4}{1 - \lambda_4}\right)^{(1-\lambda_4)/2} \left(\frac{5}{\lambda_4}\right)^{\lambda_4} \left(\frac{1}{\lambda_4}\right)^{\lambda_4}$$

This dual function becomes a problem of maximizing a function of a single variable. Usual calculus can be applied now. Taking logarithm, we get

$$
\begin{aligned}
F &= \ln \phi(\lambda) \\
&= \frac{1 - 3\lambda_4}{2}[\ln 10 - \ln(1 - 3\lambda_4)] + \frac{1 - \lambda_4}{2}[\ln 4 - \ln(1 - \lambda_4)] \\
&\quad + \lambda_4[\ln 5 - \ln \lambda_4] + \lambda + 4(\ln 1 - \ln \lambda_4)
\end{aligned}
$$

The value of λ_4 maximizing $\phi(\lambda)$ is as primal has a feasible solution

$$\frac{\partial F}{\partial \lambda_4} = \frac{-3\ln 10}{2} + \frac{3\ln(1-3\lambda_4)}{2} + \frac{3}{2} - \frac{1}{2}\ln 4 + \frac{1}{2}\ln(1-\lambda_4) + \frac{1}{2}$$
$$+ \ln 5 - \ln \lambda_4 - 1 - \ln \lambda_4 - 1$$
$$= -\frac{1}{2}\ln 5 - \frac{3}{2}\ln 2 - 2\ln \lambda_4 + \frac{3}{2}\ln(1-3\lambda_4) + \frac{1}{2}\ln(1-\lambda_4) = 0$$

This implies

$$\ln \frac{(1-3\lambda_4)^{3/2}(1-\lambda_4)^{1/2}}{\lambda_4^2} = \frac{1}{2}\ln 160$$

This gives $\lambda_4 = 0.16$. Hence, $\lambda_3 = 0.16, \lambda_2 = 0.42$, and $\lambda_1 = 0.26$. The optimal value is

$$\phi(\lambda^*) = \left(\frac{5}{0.26}\right)^{0.26}\left(\frac{2}{0.42}\right)^{0.42}\left(\frac{5}{0.16}\right)^{0.16}\left(\frac{1}{0.16}\right)^{0.16} = 9.506$$

Thus,

$$f_3 = 5x_1 = 1.52, \quad f_4 = x_2^{-1} = 1.52$$

The solution here gives $x_1^* = 0.304$ and $x_2^* = 0.66$.

Example 7 (Gravel Box Design). It is required to transport $V \; m^3$ of gravel from a mine to a manufacturing plant. A rectangular box (with open top) of length L, width W and height H is to be built for this purpose. The bottom, sides and ends of the box cost C_1, C_2 and C_3 units per m^2, respectively. The transportation of box per round trip cost C_3 units. After all the gravel has been transported, the box is to be discarded. Find the minimum total cost involved in transporting the gravel and corresponding optimum dimensions of the box.

The problem is

$$\min f(x) = C_1 LW + 2C_2 LH + 2C_3 WH + C_4 V L^{-1} W^{-1} H^{-1}$$

Here $m=4$ and $k=3$. The pre-dual is

$$\psi(\lambda) = \left(\frac{C_1 LW}{\lambda_1}\right)^{\lambda_1}\left(\frac{2C_2 LH}{\lambda_2}\right)^{\lambda_2}\left(\frac{2C_3 WH}{\lambda_3}\right)^{\lambda_3}\left(\frac{C_4 V L^{-1} W^{-1} H^{-1}}{\lambda_4}\right)^{\lambda_4}$$

The dual problem is

$$\max \quad \phi(\lambda) = \left(\frac{C_1}{\lambda_1}\right)^{\lambda_1} \left(\frac{2C_2}{\lambda_2}\right)^{\lambda_2} \left(\frac{2C_3}{\lambda_3}\right)^{\lambda_3} \left(\frac{C_4V}{\lambda_4}\right)^{\lambda_4}$$

$$\text{s.t.} \quad \lambda_1 + \lambda_2 - \lambda_4 = 0$$
$$\lambda_1 + \lambda_3 - \lambda_4 = 0$$
$$\lambda_2 + \lambda_3 - \lambda_4 = 0$$
$$\lambda_1 + \lambda_2 + \lambda_3 + +\lambda_4 = 1$$
$$\lambda_i \geq 0, i = 1, 2, 3, 4$$

The solution of the system is $\lambda_1 = 1/5$, $\lambda_2 = 1/5$, $\lambda_3 = 1/5$, $\lambda_4 = 2/5$. The minimum total cost in transporting the gravel is given as

$$f(X^*) = \phi(\lambda^*) = (5C_1)^{1/5} (5C_2)^{1/5} (5C_3)^{1/5} \left(\frac{5C_4V}{2}\right)^{2/5}$$
$$= (3125C_1C_2C_3C_4^2V^2)^{1/5}$$

The corresponding optimum dimensions of the box are obtained using the relations

$$C_1LW = \frac{1}{5}\phi(\lambda^*)$$

$$C_2LH = \frac{1}{5}\phi(\lambda^*)$$

$$C_3HW = \frac{1}{5}\phi(\lambda^*)$$

$$\frac{C_4V}{LWH} = \frac{1}{5}\phi(\lambda^*).$$

Solving the above equations, we get

$$L^* = \left(\frac{C_3^3C_4V}{C_1^2C_2^2}\right)^{1/5}, \quad W^* = \left(\frac{C_2^3C_4V}{C_1^2C_3^2}\right)^{1/5}, \quad H^* = \frac{1}{2}\left(\frac{C_1^3C_4V}{C_2^2C_3^2}\right)^{1/5}.$$

15.3　Constrained Posynomial Optimization

Consider the problem

$$\min \quad f(X) = \sum_{i=1}^{m} f_i(X) \tag{15.12}$$

$$\text{s.t.} \quad g_k(X) = \sum_{k=1}^{p} f_k(X) \leq 1 \tag{15.13}$$

$$X \geq 0 \tag{15.14}$$

where $f(X)$ and each $g_k(X)$ are posynomials.

The theory developed in earlier section depends upon using AM-GM inequality so that left hand side becomes $f(X)$ and right hand side becomes constant. To use AM-GM inequality for constraint problem we modify it. The modified form is such that it can be used even for weights whose sum is not unity. Let $\alpha_1, \alpha_2, \cdots, \alpha_m$ be positive weights such that

$$\alpha = \alpha_1 + \alpha_2 + \cdots + \alpha_m$$

Define $\lambda_i = \alpha_i/\alpha$, $i = 1, 2, \ldots, m$. Then, clearly $\lambda_1 + \lambda_2 + \cdots + \lambda_m = 1$.

Let $F_i = f_i/\alpha_i$, $i = 1, 2, \ldots, m$. Then, from AM-GM inequality

$$\lambda_1 F_1 + \lambda_2 F_2 + \cdots + \lambda_m F_m \geq F_1^{\lambda_1} F_2^{\lambda_2} \cdots F_m^{\lambda_m}$$

or

$$\frac{f_1 + f_2 + \cdots + f_m}{\alpha} \geq \left(\frac{f_1}{\alpha_1}\right)^{\alpha_1/\alpha} \left(\frac{f_2}{\alpha_2}\right)^{\alpha_2/\alpha} \cdots \left(\frac{f_m}{\alpha_m}\right)^{\alpha_m/\alpha}$$

or

$$(f_1 + f_2 + \cdots + f_m)^\alpha \geq \alpha^\alpha \left(\frac{f_1}{\alpha_1}\right)^{\alpha_1} \left(\frac{f_2}{al_2}\right)^{\alpha_2} \cdots \left(\frac{f_m}{\alpha_m}\right)^{\alpha_m} = A \text{ , say} \tag{15.15}$$

From above

$$[f(X)]^\alpha \geq A \tag{15.16}$$

Again, using (15.15) for g_1 with positive weights $\beta_1, \beta_2, \ldots, \beta_{n-m}$ such that

$$\beta_1 + \beta_2 + \cdots + \beta_{n-m} = \beta,$$

we have

$$1 \geq [g_1(X)]^\beta \geq \beta^\beta \left(\frac{f_{m+1}}{\beta_1}\right)_1^\beta \left(\frac{f_{m+2}}{\beta_2}\right)_2^\beta \cdots \left(\frac{f_n}{\beta_{n-m}}\right)_{n-m}^\beta \tag{15.17}$$

Multiplying respective sides of (15.16) and (15.17), we get

$$[f(X)]^\alpha \geq \alpha^\alpha \beta^\beta \left(\frac{f_1}{\alpha_1}\right)^{\alpha_1} \left(\frac{f_2}{\alpha_2}\right)^{\alpha_2} \cdots \left(\frac{f_m}{\alpha_m}\right)^{\alpha_m}$$
$$\times \left(\frac{f_{m+1}}{\beta_1}\right)^{\beta_1} \left(\frac{f_{m+2}}{\beta_2}\right)^{\beta_2} \cdots \left(\frac{f_n}{\beta_{n-m}}\right)^{\beta_n} \tag{15.18}$$

In above if we take $\alpha = 1, \beta = 1$, then we can not apply AM-GM inequality, since

$$\alpha_1 + \alpha_2 + \cdots + \alpha_m + \beta_1 + \beta_2 + \cdots + \beta_{n-m} \neq 1.$$

This was the reason to modify AM-GM inequality. But without loss of generality we may assume

$$\alpha = \alpha_1 + \alpha_2 + \cdots + \alpha_m = 1.$$

In this case (15.18) is reduced to

$$[f(X)]^\alpha \geq \alpha^\alpha \left(\frac{f_1}{\alpha_1}\right)^{\alpha_1} \left(\frac{f_2}{\alpha_2}\right)^{\alpha_2} \cdots \left(\frac{f_m}{\alpha_m}\right)^{\alpha_m}$$
$$\times \left(\frac{f_{m+1}}{\beta_1}\right)^{\beta_1} \left(\frac{f_{m+2}}{\beta_2}\right)^{\beta_2} \cdots \left(\frac{f_n}{\beta_{n-m}}\right)^{\beta_n} \tag{15.19}$$

where

$$\alpha_1 + \alpha_2 + \cdots + \alpha_m = 1, \quad \alpha_i > 0, \ i = 1, 2, \ldots, m$$
$$\beta_1 + \beta_2 + \cdots + \beta_{n-m} = \beta, \quad \beta_j > 0, \ j = 1, 2, \ldots, n - m.$$

Now, we can use the method developed in unconstrained posynomial optimization to transform the right side of (15.19) to a constant term.

Definition 3. The dual of the problem (15.12), (15.13) and (15.14) is defined to be the following problem:

$$\max \quad \phi(\lambda) = \left(\frac{c_1}{\lambda_1}\right)^{\lambda_1} \left(\frac{c_2}{\lambda_2}\right)^{\lambda_2} \cdots \left(\frac{c_n}{\lambda_n}\right)^{\lambda_n} (\lambda_{m+1} + \cdots + \lambda_n)$$

$$\text{s.t.} \quad a_{1j}\lambda_1 + a_{2j}\lambda_2 + \cdots + a_{nj}\lambda_n = 0, \ j = 1, 2, \ldots, n \tag{15.20}$$
$$\lambda_1 + \lambda_2 + \cdots + \lambda_m = 1 \tag{15.21}$$
$$\lambda_1, \lambda_2, \ldots, \lambda_n > 0 \tag{15.22}$$

where $\phi(\lambda) = \quad \phi(\lambda) = \phi(\lambda_1, \lambda_2, \ldots, \lambda_n)$.
The relations between the solutions of Primal and Dual are character-
ized by our next results:

Theorem 2. If the primal (15.12), (15.13) and (15.14) has a feasible
solution and there exists a $\lambda > 0$ which is dual feasible, i.e., satisfies
(15.20) to (15.22), then primal has an optimal solution.

Theorem 3. If

   (i) primal has an optimal solution, and

  (ii) there exists a X_0 such that $g_i(X_0) < 1$, $i = 1, 2 \ldots, p$ for all the
      constraint,

then

  (a) the corresponding dual problem has an optimal solution;

  (b) the constraint maximum value of dual is same as the constraint
      minimum value of primal, i.e., if X^* and λ^* are the optimal so-
      lutions of primal and dual, respectively with

$$f(X^*) = \phi(\lambda^*);$$

  (c) if λ^* is the optimal dual solution, every optimal solution X^* of
      primal problem satisfies

$$c_i x_1^{a_{i1}} x_2^{a_{i2}} \cdots x_n^{a_{in}} = \lambda_i^* \phi(\lambda^*), \quad i = 1, 2, \ldots, m$$

and

$$c_i x_1^{a_{i1}} x_2^{a_{i2}} \cdots x_n^{a_{in}} = \frac{\lambda_i^*}{\beta^*}, \quad i = m+1, m+2, \ldots, n,$$

where $\beta^* = \lambda_{m+1} + \lambda_{m+2} + \cdots + \lambda_n$, $\lambda_i > 0, i = m+1, m+2, \ldots, n$.

Example 8. Solve the following problems

(a) min $f = 2x_1^2 x_2^2 + 2x_1^{-3} x_2^{-2}$ (b) min $f = x_2 + x_2^2$

      s.t. $x_1 x_2^{-1} \leq 1/4$ s.t. $x_1 x_2 \geq 1$

           $x_1, x_2 \geq 0$ $x_1, x_2 \geq 0$

For part (a), take $g_i(X) \leq 1$ as $4x_1x_2^{-1} \leq 1$. Now, the dual of (a) is obtained from

$$\psi(\lambda) = \left(\frac{2x_1^2x_2^2}{\lambda_1}\right)^{\lambda_1} \left(\frac{2x_1^{-3}x_2^{-2}}{\lambda_2}\right)^{\lambda_2} \left(\frac{4x_1x_2^{-1}}{\lambda_3}\right)^{\lambda_3}.$$

The dual problem is

$$\text{max} \quad \phi(\lambda) = \left(\frac{2}{\lambda_1}\right)^{\lambda_1} \left(\frac{2}{\lambda_2}\right)^{\lambda_2} \left(\frac{4}{\lambda_3}\right)^{\lambda_3} \lambda_3^{\lambda_3}$$

$$\text{s.t.} \quad 2\lambda_1 - 3\lambda_2 + \lambda_3 = 0$$
$$2\lambda_2 - 2\lambda_2 - \lambda_3 = 0$$
$$\lambda_1 + \lambda_2 = 1$$

The solution of this system is $\lambda_1 = 5/9$, $\lambda_2 = 4/9$, $\lambda_3 = 2/9$. Thus, the solution is unique. Now, by inspection it can be easily verified that primal has a feasible solution which satisfies $g_1(X_0) \leq 1$. In the above case, let

$$x_1 = \frac{1}{2}, \; x_2 = \frac{1}{4}.$$

By Theorem 2, the primal has an optimal solution and thus, the dual has a unique solution with $\lambda > 0$.

Using Theorem 3(b),

$$f(X^*) = \phi(\lambda^*) = \left(\frac{18}{5}\right)^{5/9} \left(\frac{9}{2}\right)^{4/9} (18)^{2/9} \left(\frac{2}{9}\right)^{2/9} = \frac{9.2^{4/9}}{5^{5/9}}$$

Again, by Theorem 3(c),

$$2x_1^2x_2^2 = \frac{5}{9} \cdot \frac{9 \cdot 2^{4/9}}{5^{5/9}} = 5^{5/9}2^{4/9} \implies x_1^2x_2^2 = \frac{5^{4/9}}{2^{5/9}}$$

$$2x_1^{-3}x_2^{-2} = \frac{4}{9} \cdot \frac{9 \cdot 2^{4/9}}{5^{5/9}} = 2^{22/9}5^{5/9} \implies x_1^{-3}x_2^{-2} = \frac{2^{13/9}}{5^{5/9}}$$

$$4x_1x_2^{-1} = \frac{2/9}{2/9} = 1 \implies x_1x_2^{-1} = \frac{1}{4}$$

Taking logarithm on both the sides, we get

$$2\ln x_1 + 2\ln x_2 = \frac{4}{9}\ln 5 - \frac{5}{9}\ln 2$$

$$-3\ln x_1 - 2\ln x_2 = \frac{13}{9}\ln 2 - \frac{5}{9}\ln 5$$

$$\ln x_1 - \ln x_2 = -2\ln 2$$

Solving the system,

$$x_1 = \frac{5^{1/9}}{2^{8/9}}, \; x_2 = \frac{5^{1/9}}{2^{-10/9}}.$$

For part (b), take $x_1^{-1}x_2^{-1} \le 1$, and

$$\psi(\lambda) = \left(\frac{x_1^2}{\lambda_1}\right)^{\lambda_1} \left(\frac{x_2^2}{\lambda_2}\right)^{\lambda_2} \left(\frac{x_1^{-1}x_2^{-1}}{\lambda_3}\right)^{\lambda_3}$$

$$\phi(\lambda) = \left(\frac{1}{\lambda_1}\right)^{\lambda_1} \left(\frac{1}{\lambda_2}\right)^{\lambda_2} \left(\frac{1}{\lambda_3}\right)^{\lambda_3} \lambda_3^{\lambda_3}$$

such that

$$2\lambda_1 - \lambda_3 = 0$$
$$2\lambda_2 - \lambda_3 = 0$$
$$\lambda_1 + \lambda_2 = 1$$

This above system ensures that $\lambda_1 = \lambda_2 = 1/2$, $\lambda_3 = 1$. Also,

$$\phi(\lambda^*) = \sqrt{2}\sqrt{2} = 2 = f(X^*)$$

Thus,

$$f_1 = x_1^2 = \frac{1}{2} \cdot 2 = 1$$
$$f_2 = x_2^2 = \frac{1}{2} \cdot 2 = 1$$
$$f_3 = x_1^{-1}x_2^{-1} = 1$$

Example 9. A circular cylindrical log of radius R and length L is available. It is required to cut the stiffest rectangular beam from this log. The stiffness of a rectangular beam is proportional to the product of its width and the cube of its depth. Find the dimensions of the stiffest beam.

Let x_1 and x_2 be the width and the depth of the rectangular beam that is being cut from a circular cylindrical log. Then, The problem is

$$\max \; f = Kx_1x_2^3$$
$$\text{s.t.} \quad x_1^2 + x_2^2 \le 4R^2$$
$$x_1, x_2 \ge 0$$

where f = stiffness of the beam; K = proportionality constant.

Putting it in the standard form, then the problem is

$$\text{min} \quad f_1 = x_1^{-1} x_2^{-3}$$
$$\text{s.t.} \quad x_1^2 + x_2^2 \leq 4R^2$$
$$x_1, x_2 \geq 0$$

where $f_1 = 1/f$.

The pre-dual is

$$\psi(\lambda) = \left(\frac{K x_1^{-1} x_2^{-3}}{\lambda_1} \right)^{\lambda_1} \left(\frac{x_1^2}{4R^2 \lambda_2} \right)^{\lambda_2} \left(\frac{x_2^2}{4R^2 \lambda_3} \right)^{\lambda_3}.$$

The dual problem is

$$\text{max} \quad \phi(\lambda) = \left(\frac{K}{\lambda_1} \right)^{\lambda_1} \left(\frac{1}{4R^2 \lambda_2} \right)^{\lambda_2} \left(\frac{1}{4R^2 \lambda_3} \right)^{\lambda_3}$$
$$\text{s.t.} \quad -\lambda_1 + 2\lambda_2 = 0$$
$$-3\lambda_1 + 2\lambda_3 = 0$$
$$\lambda_1 = 1$$
$$\lambda_i \geq 0, i = 1, 2, 3$$

The solution of this system is $\lambda_1 = 1$, $\lambda_2 = 1/2$, $\lambda_3 = 3/2$. The minimum value of $f1$ is given as

$$f_1(X^*) = \phi(\lambda^*) = K \left(\frac{1}{2R^2} \right)^{1/2} \left(\frac{1}{6R^2} \right)^{3/2} = \frac{K}{12\sqrt{3}R^4}$$

The corresponding dimensions of the stiffest beam are obtained using the relations

$$K x_1^{-1} x_2^{-3} = 1 \cdot \phi(\lambda^*)$$
$$\frac{x_1^2}{4R^2} = \frac{\lambda_2}{\lambda_2 + \lambda_3} = \frac{1}{4}$$
$$\frac{x_2^2}{4R^2} = \frac{3}{4}$$

Solving the system, we get

$$x_1^* = R, \quad x_2^* = \sqrt{3}R.$$

Example 10. Solve the following problem

$$\min \quad f = 40x_1^{-1}x_2^{-1/2}x_3^{-1} + 20x_1x_3 + 20x_1x_2x_3$$

$$\text{s.t.} \quad \frac{1}{3}x_1^{-2}x_2^{-2} + \frac{4}{3}x_1^{-1/2}x_3^{-1} \le 1$$

$$x_1, x_2, x_3 \ge 0$$

The pre dual is

$$\psi(\lambda) = \left(\frac{40x_1^{-1}x_2^{-1/2}x_3^{-1}}{\lambda_1}\right)^{\lambda_1} \left(\frac{20x_1x_3}{\lambda_2}\right)^{\lambda_2} \left(\frac{20x_1x_2x_3}{\lambda_3}\right)^{\lambda_3} \left(\frac{x^{-1}x_2^{-1}}{3\lambda_4}\right)^{\lambda_4}$$

$$\times \left(\frac{4x_1^{-1/2}x_3^{-1}}{3\lambda_5}\right)^{\lambda_5}$$

The dual function is

$$\phi(\lambda) = \left(\frac{40}{\lambda_1}\right)^{\lambda_1} \left(\frac{20}{\lambda_2}\right)^{\lambda_2} \left(\frac{20}{\lambda_3}\right)^{\lambda_3} \left(\frac{1}{3\lambda_4}\right)^{\lambda_4} \left(\frac{4}{3\lambda_5}\right)^{\lambda_5} (\lambda_4 + \lambda_5)^{\lambda_4 + \lambda_5}$$

such that

$$-\lambda_1 + \lambda_2 + \lambda_3 - 2\lambda_4 - \frac{1}{2}\lambda_5 = 0$$

$$-\frac{1}{2}\lambda_1 + \lambda_3 - 2\lambda_4 + \lambda_5 = 0$$

$$-\lambda_1 + \lambda_2 + \lambda_3 = 0$$

$$\lambda_1 + \lambda_2 + \lambda_3 = 1$$

By writing row reduced echelon, it can be verified that no unique solution exists.

Remark. The degree of difficulty of a geometric programming problem is defined as

Number of f_i in objective functions + number of constraints

− number of variables − 1.

If this difference is < 0, no solution, = 0, a unique solution, and ≥ 1, the solution becomes difficult.

Problem Set 15

1. Minimize the following functions
   (a) $f = x_1 + 4x_2 + 2x_1^{-1}$; $x_1, x_2 > 0$;
   (b) $f = 4x_1^2 x_2^{-3} + 5x_1^{-3} x_2 + 6x_1 x_2$; $x_1, x_2 > 0$.

2. Find the optimal solution of the following constrained posynomial problems

$$\begin{aligned} \text{(a)} \quad \min \quad & f = 4x_1 x_2^2 x_3^3 + x_1^{-2} x_3^2 \\ \text{s.t.} \quad & 6x_1^{-1} x_2^{-1} x_3^{-1} + 4x_2^{-3} x_3^{-9} \le 15 \\ & x_1, x_2, x_3 \ge 0 \end{aligned}$$

$$\begin{aligned} \text{(b)} \quad \min \quad & f = 5x_1^{-1} x_2^{-1} x_3^{-1} + 5x_2 x_3 \\ \text{s.t.} \quad & 2x_1 x_3 + x_1 x_2 = 4 \\ & x_1, x_2, x_3 \ge 0 \end{aligned}$$

3. Find the dimensions of a rectangle of greatest area that can be inscribed in a semicircle of radius R.

4. The torque T in newton-meters developed by a certain internal combustion engine is expressed as

$$T = 50.8\omega^{0.6} - 4.8\omega,$$

   where ω is the rotational speed in radians per second. Determine the maximum power which the engine can deliver and the corresponding rotational speed.

5. The total cost of a rectangular building shell and the land it occupies is to be minimized for a building that must have a volume of V m^3. The following costs per square meter apply: land, $\$c_1$; roof, $\$c_2$; floor, $\$c_3$; and walls, $\$c_4$. Determine the minimum cost and the optimal dimensions of the building.

6. The director of a aircraft company has to allocate a maximum sum for $\$2.5 \times 10^6$ between two development projects, one related to fighter aircraft and the other related to transport aircraft. Since the results of either of the projects are helpful to the other, it is essential to allocate money to both the projects. The expected profit from these projects can be taken as $x_1^{5/2} x_2^2$, where x_i is the amount of money allocated to the first project more than three times the amount allocated to the second project. Find the allocation policy for maximizing the expected profit.

7. A two-bar truss is to be designed to carry a load of $L = 3 \times 10^5$ N. The truss is of equilateral triangular shape with two bars being the two sides of the triangle. The load is acting at the vertex joining the two bars and makes an angle of 30 deg with the horizontal. The bars have a tubular section with a mean diameter of d meter and a wall thickness of t meter. The material of the bars has a Young's modulus of 2.0×10^{11} N/m^2 and a yield stress of $10^9 N/m^2$. Find the values of d and t for minimum truss weight with no yielding or buckling in any of the bars. Assume the density of the material as $10^4 kg/m^3$.

8. An open cylindrical vessel is to be constructed to transport V cubic units of grain from a warehouse to a factory. The sheet metal used for the bottom and side cost $\$c_1$ and $\$c_2$ per sq units respectively. If it costs c_3 for each round trip of the box, find the dimensions of the vessel for minimizing the transportation cost. Assume that the vessel has no salvage upon completion of the operation. Solve the problem if only N trips are allowed for transporting V cubic units of grain.

9. In a certain reservoir-pump installation, the first cost of the pipe is given by $(0.75D + 0.0025D^2)$, where D is the diameter of the pipe in m. The cost of the reservoir decreases with an increase in the quantity of fluid handled and is given by $(30/Q)$ where Q is the rate at which the fluid is handled (cubic meters per second). The pumping cost is given by $(500Q^2/D^5)$. Find the optimal size of the pipe and the amount of fluid handled for minimum overall cost.

10. The total annual cost of an insulated facility is the sum of the annual cost of the insulation plus the annual cost of the energy. Specifically, cost x_0, $\$/m^2$ is given by

$$x_0 = c_1 x^n + c_2/x,$$

where, x=insulation thickness in millimeters.

If the unit energy cost increases by $k\%$, what must be the percent decrease in unit insulation cost so that the total annual cost remains constant ?

11. The treatment of a water is accomplished by chemical treatment and dilution to meet effluent code requirements. The total cost

is the sum of the treatment plant, pumping power requirements and piping cost. This cost is given by the following equation:

$$x_0 = c_1 D + c_2 Q^2 / D^5 + c_3 / Q,$$

where x_0 is in rupees, D in meter and Q in $meter^3/sec$. Find the minimum cost and the corresponding values of D and Q.

12. The work done by a three-stage compressor is given by the expression

$$W = \left(\frac{P_1 V_1}{r}\right)\left[\left(\frac{P_2}{P_1}\right)^r + \left(\frac{P_3}{P_2}\right)^r + \left(\frac{P_4}{P_3}\right)^r - 3\right],$$

where P_1 is the inlet pressure to stage 1, P_2 is the discharge pressure from stage 1 and inlet pressure to stage 2, P_3 is the discharge pressure from stage 3 and r is equal to $(k-1)/k$, where k is the ratio of specific heats, a constant.

For specific inlet and exit pressures P_1 and P_4, find the intermediate pressures P_2 and P_3 that minimize the work done.

13. Determine the optimal pipe diameter for the minimum installed plus operating costs for L meter of pipe conveying a given flow rate of water. The installed cost in rupees is $c_1 \times D$, and the lifetime pumping cost in rupees is $c_2 \times 10^5 / D^5$. The diameter D is in meters.

14. The profit function for each of the three chemical reactors operating in parallel with the same feed is given by the three equations below. Each reactor is operating with a different catalyst and conditions of temperature and pressure. The profit function for each reactor has the feed rates x_1, x_2 and x_3 as the independent variable and the parameters in the equations are determined by the catalyst and operating conditions:

$$P_1 = 0.2x_1 - 2(x_1/100)^2$$
$$P_2 = 0.2x_2 - 2(x_2/100)^2$$
$$P_3 = 0.2x_3 - 2(x_3/100)^2$$

The total profit is given by

$$P = P_1 + P_2 + P_3.$$

Determine the values of the three feed rates to the reactors so as to maximize the profit. Calculate the maximum profit.

Chapter 16

Goal Programming

The goal programming is introduced as an extension of multiple objective systems with assigned priority levels. The two techniques to solve goal programming problems, viz., the partitioning algorithm and grouping algorithm are discussed.

16.1 Introduction

The linearly programming problems discussed in preceding chapters are characterized by optimization of a single objective function under given set of conditions. However, situations arise when we optimize more than one objective function under the same set of conditions or, even we search for optimal solution of multiple objectives (may be conflicting). For example, institutes reduce faculty strength and lower down pay scales, simultaneously expect better performance in terms of grades. In such situations it may not be possible to find a unique solution that optimizes all objectives. Instead we search a compromising solution based on the relative significance of every objective. This feature generated the idea of goal programming. Each objective function is assigned some priority which facilitates the determination of a compromising solution. *Remark.* There are nonlinear goal programming problems also. However, this chapter will be devoted to the linear goal programming problems.

To make the idea well conceivable, let us first formulate a linear goal programming problem.

Example 1. Suppose that a company is considering to produce three products x_1, x_2, x_3. The minimum profits for selling one unit of each these products are \$12, \$9 and \$15, and the number of persons employed for producing one unit of each product are 5, 3 and 4, respectively. The investment of the company for one unit of each product are \$5, \$7, \$8, respectively. The manufacturing time required for one unit of each product are 2, 1 and 2 hours, respectively. The company can not afford more than 20 hours daily.

The goals of the company are

(i) Achieving a long-term profit of at least \$125 million from these products;

(ii) Maintaining the current employment level of 4 thousand employees;

(iii) Holding the current investment at most \$55 million.

Probably, it will not be possible to attain all of these goals simultaneously; hence, the company assigns penalty weights of 5 for missing the profit goal, 2 for going over the employment, 4 for going under the same goal, and 3 for exceeding the capital investment. Formulate the problem as linear goal programming problem.

The goal (resource) constraints are

$$12x_1 + 9x_2 + 15x_3 \geq 125 \qquad \text{(profit goal)}$$
$$5x_1 + 3x_2 + 4x_3 = 40 \qquad \text{(employment goal)}$$
$$5x_1 + 7x_2 + 8x_3 \leq 55 \qquad \text{(investment goal)}$$

The real (rigid) constraint is

$$2x_1 + x_2 + 2x_3 \leq 20.$$

Now, we write the standard form of the above linear goal programming problem as

$$\min \quad s = P_1(5s_1^-) + P_2(2s_2^+ + 4s_2^-) + P_3(3s_3^+)$$

$$\text{s.t.} \quad 12x_1 + 9x_2 + 15x_3 + s_1^- - s_1^+ = 125 \qquad P_1$$

$$5x_1 + 3x_2 + 4x_3 + s_2^- - s_2^+ = 40 \qquad P_2$$

$$5x_1 + 7x_2 + 8x_3 + s_3^- - s_3^+ = 55 \qquad P_3$$

$$2x_1 + x_2 + 2x_3 \leq 20$$

$$\text{all var} \geq 0,$$

where P_1, P_2 and P_3 are the priorities assigned to profit, employment and investment goals, respectively such that

$$P_1 >> P_2 >> P_3.$$

16.2 Standard form of LGPP

The linear goal programming problem (LGPP) formulation considered for n variables, m constraints and t pre-emptive priority levels is defined as

$$\min \quad S = \min \sum_{k=1}^{t} P_k \sum_{i=1}^{m} \left(w_{ik}^- s_i^- + w_{ik}^+ s_i^+ \right) \tag{16.1}$$

$$\text{s.t.} \quad \sum_{j=1}^{n} e_{qj} x_j = f_q, \quad q = 1, 2, \ldots, \ell \tag{16.2}$$

$$\sum_{j=1}^{n} a_{ij} x_j + s_i^- - s_i^+ = b_i, \quad i = 1, 2, \ldots, m, \tag{16.3}$$

where

$x_j \geq 0, \; b_i \geq 0$

$s_i^- = $ under achievement for goal i, $s_i^- \geq 0$

$s_i^+ =$ over achievement for goal i, $s_i^+ \geq 0$

$P_k = k$th ordinal factor; $P_k \gg P_{k+1}$

$w_{ik}^- =$ weight assigned to d_i^- at priority P_k

$w_{ik}^+ =$ weight assigned to d_i^+ at priority P_k

$e_{qj} =$ coefficient of x_j in qth real constraint

$f_q =$ required level for the qth real constraint

$a_{ij} =$ coefficient of x_j in ith goal constraint

$b_i =$ the target level for goal i

Note that P_k's are priority factors which offer the significance of any goal, i.e., preference of a goal over the others, and do not take numerical values in the formulation. Also, two or more goals may have the same priority.

Definition. A goal is said to be satisfied if, in the optimal solution of the problem its deviational variables turn to be zero.

Remark. From the above formulation, we infer that in a LGPP the objective function (16.1) attempts to minimize the weighted sum of the deviational variables at each priority.

Elimination Theorem. In the standard LGPP (16.1) to (16.3), if the optimal table for subproblem S_k has been found then any non-basic variable t_k, where t_k can be a decision variable or a deviational variable or slack/surplus variable, having negative relative cost can be eliminated, while dealing with subproblems $S_{k+1}, \ldots, S_k$ as it will never enter the basis after S_kth problem.

16.3   Partitioning Algorithm

The algorithm is based on the principle that goals with higher priorities are optimized before lower goals are even considered. With this procedure it becomes solving a sequence of linear programming problems each using the optimal solution of the previous problems.

Step 1. Solve the system of real constraints, if any, for basic feasible solution by phase-I method. Delete the objective function row and eliminate columns of the variables which have negative relative costs, cf., , elimination theorem.

Step 2. Add the goal constraint with priority P_1 to the table obtained in Step 1, and restore the simplex format by performing row operations. While restoring simplex format, if the feasibility is disturbed then multiply this row by -1 and interchange the role of s_1^- and s_1^+ to have the identity matrix inside body matrix.

Step 3. Induce objective function of the first subproblem S_1 in x_0-row, and make it amenable to carry out the simplex iterations. Obtain the optimal table and observe the optimal solution.

Two possibilities arise: a unique optimal solution or an alternative optimal solution.

If a unique solution is obtained, then algorithm stops and this solution is taken as the optimal solution for all goal constraints with lower goals also. Calculate s_i^- and s_i^+ for these lower order goals and find sum of the weighted deviations.

In case an alternative optimal solution exists for S_1 subproblem then go to Step 4.

Step 4. The columns corresponding to nonbasic variables which have $-$ve relative cost are deleted. Also, delete the objective function row, and add the goal constraints with priority P_2 by sensitivity analysis as in Step 2.

Step 5. Now, induce the objective function of the second subproblem S_2, and make it amenable to simplex iterations. Carry out simplex iterations and find the optimal solution.

If the unique solution exists, then the algorithm stops, and calculate s_i^-, s_i^+, $i \geq 2$ to find the optimal deviation. Otherwise, in case of alternative optimal solution, continue as in Steps 4 and 5 until a unique solution is obtained or all goals are exhausted.

Example 2. Consider the problem with weight functions at unity level for all the deviational variables and priorities mentioned against each goal constraint.

$$4x_1 + 5x_2 = 400 \qquad P_1$$
$$x_1 + 5x_2 = 334 \qquad P_2$$
$$4x_1 + 3x_2 \leq 250 \qquad P_3$$
$$2x_1 - x_2 \geq 59 \qquad P_4$$
$$x_1, x_2 \geq 0.$$

(a) Write the standard form of the Linear goal programming problem;

(b) Find its solution by the partitioning algorithm.

The standard form is

$$\min \quad S = P_1(s_1^- + s_1^+) + P_2(s_2^- + s_2^+) + P_3 s_3^+ + P_4 s_4^-$$
$$\text{s.t.} \quad 4x_1 + 5x_2 + s_1^- - s_1^+ = 400$$
$$x_1 + 5x_2 + s_2^- - s_2^+ = 334$$
$$4x_1 + 3x_2 + s_3^- - s_3^+ = 250$$
$$2x_1 - x_2 + s_4^- - s_4^+ = 59$$
$$\text{all var} \geq 0.$$

To deal with part (b), we solve the problem sequentially from highest priority level. Here Step 1 is not needed as the problem has no real constraint.

Ist subproblem:

$$\min \quad S_1 = s_1^- + s_1^+$$
$$\text{s.t.} \quad 4x_1 + 5x_2 + s_1^- - s_1^+ = 400$$
$$x_1, x_2, s_1^+, s_1^- \geq 0$$

B V	x_1	$x_2 \downarrow$	s_1^+	s_1^-	Soln
S_1	4	5	−2	0	400
	0	0	−1	−1	0
← s_1^-	4	5	−1	1	400
S_1	0	0	−1	−1	0
x_2	4/5	1	−1/5	1/5	80.

From the above table, it is obvious that the first subproblem has an alternative optimal solution, because the relative cost $(z_1 - c_1)$ of nonbasic variable x_1 is zero, and hence we continue beyond Step 3. Note that, while constructing the next table we shall not consider the nonbasic variables s_1^-, s_2^+ as their relative costs are negative.

IInd subproblem:

$$\min \quad S_2 = s_2^- + s_2^+$$
$$\text{s.t.} \quad x_1 + 5x_2 + s_2^- - s_2^+ = 334$$
$$x_1, x_2, s_2^+, s_2^- \geq 0.$$

B V	$x_1 \downarrow$	x_2	s_2^+	s_2^-	Soln
S_2	3	0	0	-2	66
	0	0	-1	-1	0
x_2	4/5	1	0	0	80
s_2^-	1	5	-1	1	334
	-3	0	-1	1	-66
$\leftarrow s_2^+$	⎡3⎤	0	1	-1	66
S_2	0	0	-1	-1	0
x_2	0	1	$-4/15$	4/15	312/5
x_1	1	0	1/3	$-1/3$	22

What we have done to solve the IInd subproblem is as follows. Insert the constraint from the optimal table of the Ist subproblem and then add the constraint of IInd subproblem. Bring the body matrix into simplex format by subtracting five times of the first row from the second row. This disturbs feasibility which is stored by interchanging the role of s_2^- and s_2^+ (no need of dual simplex method). Next, insert the objective function into S_2-row and again note that this row is not in simplex format. Bring into simplex format by making relative cost to be zero for basic variables by using elementary row operations between S_2-row and appropriate rows of the body matrix.

The optimal solution is unique. Thus, the algorithm stops, and we calculate the value of remaining deviational variables merely by substitution method in goal constraints with priorities P_3 and P_4. Now,

P_3 constraint gives

$$4\times22+3\times\frac{312}{5}+s_3^- -s_3^+ = 250 \implies s_3^- -s_3^+ = -126/5 \implies s_3^+ = 125/5,$$

P_4 constraint ensures

$$2 \times 22 - \frac{312}{5} + s_4^- - s_4^+ = 50 \implies s_4^- = \frac{342}{5}.$$

Hence, the optimal solution is $x_1 = 22, x_2 = 312/5$, and the optimal value is $468/5$, i.e., total deviation.

Example 3. Consider the linear goal programming problem

$$4x_1 + 5x_2 \approx 800 \qquad P_1$$
$$4x_1 + 3x_3 \leq 500 \qquad P_2$$
$$2x_1 + 5x_2 \leq 600$$
$$x_1, x_2 \geq 0$$

for which weight factors for all deviational variables have been kept at unity level.

The standard form of the LGPP is

$$\min \quad S = P_1(s_1^- + s_1^+) + P_2 s_2^+$$
$$\text{subject}$$
$$2x_1 + 5x_2 + s = 600$$
$$4x_1 + 5x_2 + s_1^- - s_1^+ = 800$$
$$4x_1 + 3x_3 + s_2^- - s_2^+ = 500$$
$$\text{all var} \geq 0$$

Since the standard form contains one real constraint, first we find the initial BFS of the problem

$$\min \quad r_0 = R$$
$$\text{s.t.} \quad 2x_1 + 5x_2 + s + R = 600$$
$$x_1, x_2, s, R \geq 0.$$

B V	x_1	$x_2 \downarrow$	s	R	Soln
r_0	2	5	1	0	600
	0	0	0	-1	0
$\leftarrow R$	2	$\boxed{5}$	1	1	600
r_0	0	0	0	-1	0
x_2	2/5	1	1/5	1/5	120

Ist subproblem

$$\min \quad S_1 = s_1^- + s_1^+$$
$$\text{s.t.} \quad 4x_1 + 5x_2 + s_1^- - s_1^+ = 800$$
$$x_1, x_2, s_1^-, s_1^+ \geq 0$$

B V	$x_1 \downarrow$	x_2	s	s_1^-	s_1^+	Soln
S_1	2	0	-1	0	-2	600
	0	0	0	-1	-1	0
x_2	2/5	1	1/5	0	0	120
	4	5	0	1	-1	800
$\leftarrow s_1^-$	$\boxed{2}$	0	-1	1	-1	200
S_1	0	0	0	-1	-1	0
x_2	0	1	2/5	$-1/5$	1/5	80
x_1	1	0	$-1/2$	1/2	$-1/2$	100

IInd subproblem

$$\min \quad S_2 = s_2^+$$
$$\text{s.t.} \quad 4x_1 + 3x_2 + s_2^- - s_2^+ = 500$$
$$x_1, x_2, s_2^-, s_2^+ \geq 0$$

B V	x_1	x_2	s	s_2^-	s_2^+	Soln
S_2	0	0	$-4/5$	-2	0	140
	0	0	0	-1	$-$	0
x_2	0	1	$2/5$	0	0	80
x_1	1	0	$-1/2$	0	0	100
s_2^-	4	3	0	1	-1	500
	0	0	$-4/5$	1	-1	-140
s_2^+	0	0	$4/5$	-1	1	140

Optimal solution $x_1 = 100$, $x_2 = 80$, total deviation $= 140$.

In discussion so far on linear goal programming problem, our theme was to minimize the sum of weighted deviations. However, a better approach for problems having more than one goal at the same priority level is to minimize the maximum of the weighted deviations for such type of goal constraints. This is well explained in the following example.

Example 4. Solve the following linear goal programming problem

$$5x_1 + 6x_2 \approx 75 \qquad P_1$$
$$x_1 + 2x_2 \leq 20 \qquad P_2$$
$$2x_1 + x_2 \leq 10 \qquad P_2$$
$$x_1, x_2 \geq 0$$

First, the problem is solved adopting the usual procedure as done in Examples 2 and 3.

The Ist subproblem is

$$\min \quad s_1^+ + s_1^-$$
$$\text{s.t.} \quad 5x_1 + 6x_2 + s_1^- - s_1^+ = 75$$
$$\text{all var} \geq 0$$

The optimal table of the Ist subproblem is the next Table, and alternative solution exists.

B V	x_1	x_2	s_1^-	s_1^+	Soln
	0	0	-1	-1	0
x_2	5/6	1	1/6	$-1/6$	25/2

The second subproblem is

$$\min \quad s_2^+ + s_3^+$$
$$\text{s.t.} \quad x_1 + 2x_2 + s_2^- - s_2^+ = 20$$
$$2x_1 + x_2 + s_3^- - s_3^+ = 10$$
$$\text{all var} \geq 0$$

Add both the constraints with the same priority P_1 simultaneously. Restore simplex format. Note that the optimal solution is

$$x_1 = 0, \ x_2 = 25/2; \ s_2^+ + s_3^+ = 25/2.$$

Let us solve the same problem by different approach which gives better solution. The procedure is same up to the optimal table of Ist subproblem. The objective function of the second subproblem is to minimize the $\max\{s_2^+, s_3^+\}$. Thus,

$$\min \quad \alpha = \max\{s_2^+, s_3^+\}$$
$$\text{s.t.} \quad x_1 + 2x_2 + s_2^- - s_2^+ = 20$$
$$2x_1 + x_2 + s_3^- - s_3^+ = 10$$
$$\text{all var} \geq 0$$

The above is a nonlinear programming problem (NLPP). Converting this into linear program, the resulting LPP is

$$\min \quad \alpha$$
$$\text{s.t.} \quad x_1 + 2x_2 + s_2^- - s_2^+ = 20$$
$$2x_1 + x_2 + s_3^- - s_3^+ = 10$$
$$s_2^+ - \alpha + s_4 = 0$$
$$s_3^+ - \alpha + s_5 = 0$$
$$\text{all var} \geq 0$$

The above LPP is now added to the optimal table of Ist sub-problem, and the next table is obtained by adjusting the column for x_2 to make the body matrix in simplex format. The variables $x_2, s_2^-, s_3^-, s_4, s_5$ contribute to give identity submatrix in next table.

B V	x_1	x_2	α	s_2^-	s_2^+	s_3^-	s_3^+	s_4	s_5	Soln
	0	0	-1	0	0	0	0	0	0	0
x_2	5/6	1	0	0	0	0	0	0	0	25/2
s_2^-	$-2/3$	0	0	1	-1	0	0	0	0	-5
s_3^-	7/6	0	0	0	0	1	-1	0	0	$-5/2$
s_4	0	0	-1	0	1	0	0	1	0	0
s_5	0	0	-1	0	0	0	1	0	1	0

Apply dual simplex method to have the optimal solution. For the sake of space, we write the outcomes of various iterations.

s_2^- leaves and s_2^+ enters, giving the solution

$$(x_2, s_2^+, s_3^-, s_4, s_5) = (25/2, 5, -5/2, -5, 0)$$

s_4 leaves and x_1 enters, giving the solution

$$(x_2, s_2^+, s_3^-, x_1, s_5) = (25/4, 0, -45/4, 15/2, 0)$$

s_3^- leaves and s_3^+ enters, giving the solution

$$(x_2, s_2^+, s_3^+, x_1, s_5) = (25/4, 0, 45/4, 15/2, -45/4)$$

s_5 leaves and α enters, giving the solution

$$(x_2, s_2^+, s_3^+, x_1, \alpha) = (25/22, 45/4, 45/4, 30/11, 45/4)$$

Thus, the optimal solution is $x_1 = 30/11$, $x_2 = 25/22$ with total deviation $= 45/4$. This is a remarkable achievement over the previous estimate 25/2.

16.4   Grouping Algorithm

The algorithm considers all goal and real constraints together as a one group with the objective function being the sum of all weighted deviations and solves a sequence of linear programming subproblems each using the optimal solution of the previous subproblems. The simplex method and sensitivity analysis are applied to obtain the solution of the subproblems. This algorithm we call grouping algorithm as all goal and real constraints are considered as a group. The grouping algorithm solves the LGPP in two situations. In the first case, the goals are already assigned priorities by the decision maker and the algorithm finds the optimal solution of the LGPP. In the second case, the goal constraints have not been assigned any priorities and the decision maker desires to know the priorities to be assigned to the goals so that the maximum number of goals is satisfied. We explain the algorithm for the two situations.

Algorithm for Goal Constraints with Pre-emptive Priorities

In solving LGPP (16.1)-(16.3) having goal constraints with pre-emptive priorities, the algorithm solves a sequence of linear programming subproblems 1,2,...,p, where p is the last subproblem, each using the optimal solution of the previous subproblems. The details of the algorithm are:

Step 1. All real constraints and goal constraints with the introduction of deviational variables are considered a group with the objective function being the sum of all weighted deviations in the subproblem 1 expressed as the following LGPP:

$$\min \quad S_1 = \sum_{i=1}^{m}(s_i^- + s_i^+) \qquad (16.4)$$

s.t.

$$\sum_{j=1}^{n} e_{qj}x_j = f_q, \quad q = 1, 2, \ldots, \ell \qquad (16.5)$$

$$\sum_{j=1}^{n} a_{ij}x_j + s_i^- - s_i^+ = b_i, \quad i = 1, 2, \ldots, m \qquad (16.6)$$

$$x_j, b_i, s_i^-, s_i^+ \geq 0, j = 1, 2, \ldots, n; i = 1, 2, \ldots, m$$

The subproblem 1 is now solved by the simplex method and its optimal solution is obtained. For convenience of representation, the optimal table of the subproblem 1 is denoted as Table 1. Take $h = 0$.

Step 2. Examine the optimal simplex table. If either the value of objective function in the optimal solution is zero or the relative cost of at least one nonbasic variable is zero, go to Step 5.

Step 3. (a) Delete the goal constraints having the priority $t-h$ and solve the resulting subproblem h+2 given as follows:

$$\text{min} \quad S_{h+2} = \sum_{i=1}^{m}(s_i^- + s_i^+) \tag{16.7}$$

s.t.

$$\sum_{j=1}^{n} e_{qj}x_j = f_q, \quad q = 1, 2, \ldots, \ell \tag{16.8}$$

$$\sum_{j=1}^{n} a_{ij}x_j + s_i^- - s_i^+ = b_i, \quad i = 1, 2, \ldots, m - k_{t-h} \tag{16.9}$$

$$\sum_{j=1}^{n} a_{ij}x_j + s_i^- - s_i^+ + u_i\rho_i = b_i, \quad i = m - k_{t-h} + 1,$$

$$m - k_{t-h} + 2, \ldots, m \tag{16.10}$$

$$x_j, b_i, s_i^-, s_i^+, \rho_i \geq 0, j = 1, 2, \ldots, n; i = 1, 2, \ldots, m$$

Here we assume k_{t-h} goal constraints have the lowest priority $t-h$. The u_i has a value of 1 or -1 depending upon the deviation variable is s_i^- or s_i^+ corresponding to the equation (16.10).

Apply sensitivity analysis to find out the effects of deletion of the k_{t-h} goal constraints on the optimal solution of the subproblem S_{h+1} and thereby obtain the solution of the subproblem S_{h+2}.

(b) The columns below nonbasic variables having the relative costs negative as well as the rows and columns corresponding to $\rho_i, i = m - k_t + 1, m - k_t + 2, \ldots, m$ are deleted from the simplex tables constructed later as they will never enter into the basis afterwords.

Step 4. If $h = t - 1$ go to Step 5, otherwise take $h = h + 1$ and go to Step 2.

Step 5. This very solution is the optimal solution of the problem with respect to their given priorities.

Step 6. Stop.

Remarks.

1. Here we would like to obtain the optimal solution of the LGPP (16.1)-(16.3). We first consider subproblem 1 and solve it using the simplex method with the assumption that all goal constraints have the same priorities. We get the optimal value of the objective function S_1 as given by

$$S_1^{*1} = \sum_{i=1}^{m} (s_i^{-(*1)} + s_i^{+(*1)}) \tag{16.11}$$

where '*1' denotes the optimal value of the variable in the subproblem 1.

We are in fact required to minimize the objective function S given by the LGPP (16.1) along with the conditions (16.2) and (16.3) as well as the conditions of priorities assigned to the goal constraints as

$$P_k >> P_{k+1}, k = 1, 2, \ldots, t. \tag{16.12}$$

As we do not consider this condition (16.12) while we obtain the solution of the subproblem S_1. So, the S_1^{*1} may not be the optimal solution of the given LGPP (16.1)-(16.3). Now we take into the consideration of the conditions (16.12). The deletion of the constraints from the subproblem 1, in general, leads to the decrease of the objective function S_1. This fact might be more clear later in the text. As our goal is to minimize the objective function S_1, we undertake the exercise of deletion of constraints in the algorithm. Before deleting the goal constraints, we should decide which set of goal constraints to be removed first. As the goal constraints having the lowest priority t are the least important goals for the decision maker, we delete them first. Let the k goal constraints have the lowest priority t and we delete them first. The resulting subproblem 2 is given by the equations (16.7)-(16.10) considering $h = 0$.

We can solve the subproblem 2 using the simplex method. However, in this situation we have to solve the problem as a fresh. In fact, we already have the optimal table as obtained by solving the subproblem 1. Therefore, we can directly apply sensitivity analysis with a view to find out the effects of deletion of the k goal constraints on the optimal solution of the subproblem S_1 and thereby obtain the solution of the subproblem S_2. Now the optimal value of the objection function

S_2 we get

$$S_2^{*2} = \sum_{i=1}^{m-k}(s_i^{-(*2)} + s_i^{+(*2)}) \tag{16.13}$$

From the equations (16.11) and (16.13), we get

$$S_1^{*1} \geq S_2^{*2} \tag{16.14}$$

Next, we delete the goal constraints having the next higher priority $(t-1)$ and so on till we reach the highest priority 1 or till we get the lowest value of the objective function as zero. Let this condition reaches when we solve the subproblem p. We get the following relation

$$S_1^{*1} \geq S_2^{*2} \geq \ldots \geq S_p^{*p} \tag{16.15}$$

where $p \leq t$.

Thus, after solving the subproblem p we get the optimal solution of the LGPP (16.1)-(16.3) as follows:

$$s_i^- = s_i^{-(*p)}, i = 1, 2, \ldots, m \tag{16.16}$$

$$s_i^+ = s_i^{+(*p)}, i = 1, 2, \ldots, m \tag{16.17}$$

$$S = S_p^{*p} = \sum_{i=1}^{m}(s_i^{-(*p)} + s_i^{+(*p)}) \tag{16.18}$$

2. In solving the subproblem j, j=1,2,...,p, we may get the relative cost of a nonbasic variable, say, d, in an optimal table of subproblem, say, h equals to zero. It means that the nonbasic variable d can enter into the basis and we get an alternate optimal solution. In this situation, if we add the goal constraints that we have deleted, i.e., the goal constraints in the subproblem (h-1), we may get

$$S_{h-1}^{*h-1} \geq S_h^{*h} \tag{16.19}$$

If we get $S_{h-1}^{*h-1} = S_h^{*h}$ then the optimal solution of the subproblem (h-1) is the optimal solution of the LGPP (16.1)-(16.3). In case we have $S_{h-1}^{*h-1} > S_h^{*h}$, still the solution of the subproblem (h-1) would be the solution of the LGPP (16.1)-(16.3) because the increase in the value of the objective function S_{h-1}^{*h-1} compared to S_{h-1}^{*h-1} is entirely due to the fact that some deviational variables corresponding to the the goal constraints in the subproblem (h-1) have entered into the objective function S_{h-1}^{*h-1} but they have not effected the deviational variables that were already present in subproblem h. Thus, the optimal table of the preceding subproblem h is the best suited optimal table with respect to their priorities and the solution we obtained is the optimal solution of the given LGPP.

3. The deletion of a goal constraint, say, kth goal constraint given by

$$\sum_{j=1}^{n} a_{kj}x_j + s_k^- - s_k^+ = b_k \qquad (16.20)$$

is accomplished using the following procedure outlined here:

(i) A new variable ρ_k is added or subtracted depending upon the deviation variable is s_k^- or s_k^+ to the equation (16.20) so that the variable ρ_k enters the basis and the resulting equation is given as

$$\sum_{j=1}^{n} a_{kj}x_j + s_k^- - s_k^+ + u\rho_k = b_k \qquad (16.21)$$

where u has a value of 1 or -1 depending upon the deviation variable is s_k^- or s_k^+ corresponding to the equation (16.42).

(ii) A new column below this new variable ρ_k is introduced in the optimal simplex table.

Entry below the ρ_k in the S_o row: $z_k - C_k = c_B^T B^{-1} A_k - C_k$. Column corresponding to the ρ_k, $\alpha^k = B^{-1}A_k$.

where c_B = Costs of basic variables, B = Basis matrix, A_k = Basis vector for the variable ρ_k, C_k = Cost of the variable ρ_k, and α^k = Column corresponding to the ρ_k.

If the relative cost $z_k - C_k$ is positive, the optimal criteria is disturbed and the variable ρ_k enters into the basis. It is be observed that the deletion of a constraint amounts to the addition of a variable.

4. If a deviational variable, say, s_k^{-1} or s_k^{+1} corresponding to a goal constraint, say kth goal constraint, to be deleted is in the basis of the optimal table of the subproblem, say, S_p then a new column below the new variable ρ_k in the simplex table is obtained as follows:

The column below ρ_k, α^k is given by

$$\alpha^k = B^{-1}A_k \qquad (16.22)$$

where

$$B^{-1} = \begin{bmatrix} b_{11} & b_{12} & \cdots & b_{1k} & \cdots & b_{1(l+m)} \\ b_{21} & b_{22} & \cdots & b_{2k} & \cdots & b_{2(l+m)} \\ \vdots & & & & & \\ b_{k1} & b_{k2} & \cdots & b_{kk} & \cdots & b_{k(l+m)} \\ \vdots & & & & & \\ b_{(l+m)1} & b_{(l+m)2} & \cdots & b_{(l+m)k} & \cdots & b_{(l+m)(l+m)} \end{bmatrix}$$

Here $b_{ik} = 0, i = 1, 2, \ldots, k-1, k+1, \ldots, l+m$; $b_{kk} = 1$ as s_k^{-1} or s_k^{+1} is in the basis.

$$A_k = \begin{bmatrix} a_1 \\ a_2 \\ \vdots \\ a_k \\ \vdots \\ a_{(l+m)(l+m)} \end{bmatrix}$$

Here $a_i = 0, i = 1, 2, \ldots, k-1, k+1, \ldots, l+m$; $a_k = 1$ as the variable ρ_k is only in the $k$the goal constraint.

Substituting these relations in the equation (16.22) and solving, we get

$$\alpha^k = \begin{bmatrix} 0 \\ 0 \\ \vdots \\ 1 \leftarrow \\ \vdots \\ 0 \end{bmatrix} \qquad (16.23)$$

Here 1 is at kth place.

The entry below ρ_k in the S_o row is given by

$$z_k - C_k = c_B^T B^{-1} A_k - C_k \tag{16.24}$$

where

$c_B = (c_1, c_2, \ldots, c_k, \ldots, c_{(l+m))}^T$

Here $c_k = 1$ as s_k^{-1} or s_k^{+1} is a deviational variable and $C_k = 0$

Substituting the above relations in the expression (16.24) and solving it we get

$$z_k - C_k = 1 \tag{16.25}$$

5. In the case of deletion of a kth goal constraint if the value of corresponding deviational variable s_k^- or s_k^+ in the optimal table is the same as the value of the objective function S_h in a subproblem h, then the deletion of this kth goal constraint results in the objective function S_h to be equal to zero. In this situation, there is no need of deletion of the kth goal constraint and the solution already obtained is to be considered as the optimal solution of the given LGPP (16.1)-(16.3).

6. In the case of deletion of a kth goal constraint if α^k vector is such that B^{-1} remains unchanged, then the value of the basic vector X_B (in this vector we do not consider ρ_k now as kth goal constraint has already been deleted) given by $B^{-1}b$ (where $b = ([f_q, q = 1, 2, \ldots l], [b_i, i = 1, 2, \ldots m])^T$) also remains unchanged. However, the value of objective function given by $c_B^T X_B$ will change as the c_B is different now.

7. The value of the objective function S_h in a subproblem h is equal to zero if all the elements of c_B are zero or all the elements of X_B are zero. In other words, if costs of all the deviational variables or the values of all basic variables are zero, the value of objective function is equal to zero. Thus, in the case when we delete kth goal constraint if $c_B = (0, 0, 0, 0, 0)^T$, then the objective function S_h will be equal to $c_B^T X_B = 0$. In this condition, we should not perform deletion of the kth goal constraint and consider the solution found in the subproblem (h-1) as the optimal solution of the given LGPP (16.1)-(16.3).

8. In the standard LGPP (16.1)-(16.3), if the optimal table for subproblem S_k has been found then any nonbasic variable t_k, where t_k can be a decision variable or a deviational variable or slack/surplus variable, if any having negative relative cost can be eliminated, while dealing with subproblems $S_{k+1}, \ldots, S_t$ as it will never enter the basis after S_kth problem.

We solve the following problem using the grouping algorithm.

Example 5.

$$x_1 - x_2 - x_3 \quad \leq 2 \quad P_1$$
$$x_1 + x_2 + 2x_3 \quad \leq 3 \quad P_2$$
$$4x_1 - x_2 + 6x_3 \quad = 9 \quad P_3$$
$$3x_1 - 2x_2 - 2x_3 \quad \geq 6 \quad P_4$$
$$-2x_1 - x_2 + x_3 \quad \geq 2 \quad P_5$$
$$x_1, x_2, x_3 \quad \geq 0$$

There are five goal constraints with priorities from P_1 to P_5. Assuming weight functions of all deviational variables to be at unity level, we formulate the above problem as the following LGPP:

$$\min \quad S = P_1 s_1^+ + P_2 s_2^+ + P_3(s_3^- + s_3^+) + P_4 s_4^- + P_5 s_5^-$$

s.t.

$$x_1 - x_2 - x_3 + s_1^- - s_1^+ = 2$$
$$x_1 + x_2 + 2x_3 + s_2^- - s_2^+ = 3$$
$$4x_1 - x_2 + 6x_3 + s_3^- - s_3^+ = 9$$
$$3x_1 - 2x_2 - 2x_3 + s_4^- - s_4^+ = 6$$
$$-2x_1 - x_2 + x_3 + s_5^- - s_5^+ = 2$$
$$x_j, s_i^-, s_i^+ \geq 0, j = 1, 2, 3; i = 1, 2, \ldots, 5.$$

Step 1. All five goal constraints after introducing deviational variables are considered a group with the objective function being the sum of all weighted deviations in the subproblem 1 mathematically represented as

$$\min \quad S_1 = s_1^+ + s_2^+ + s_3^- + s_3^+ + s_4^- + s_5^-$$

s.t.

$$x_1 - x_2 - x_3 + s_1^- - s_1^+ = 2$$
$$x_1 + x_2 + 2x_3 + s_2^- - s_2^+ = 3$$
$$4x_1 - x_2 + 6x_3 + s_3^- - s_3^+ = 9$$
$$3x_1 - 2x_2 - 2x_3 + s_4^- - s_4^+ = 6$$
$$-2x_1 - x_2 + x_3 + s_5^- - s_5^+ = 2$$
$$x_j, s_i^-, s_i^+ \geq 0, j = 1, 2, 3; i = 1, 2, \ldots, 5.$$

Now, the subproblem 1 is solved by the simplex method. The simplex table is as follows:

Table 1

B V	x_1	x_2	x_3	s_1^-	s_2^-	s_3^-	s_4^-	s_5^-	s_1^+	s_2^+	s_3^+	s_4^+	s_5^+	Soln
S_o	0	0	0	0	0	-1	-1	-1	-1	-1	-1	0	0	0
s_1^-	1	-1	-1	1	0	0	0	0	-1	0	0	0	0	2
s_2^-	1	1	2	0	1	0	0	0	0	-1	0	0	0	3
s_3^-	4	-1	6	0	0	1	0	0	0	0	-1	0	0	9
s_4^-	3	-2	-2	0	0	0	1	0	0	0	0	-1	0	6
s_5^-	-2	-1	1	0	0	0	0	1	0	0	0	0	-1	2
S_o	5	-4	5↓	0	0	0	0	0	-1	-1	-2	-1	-1	-17
s_1^-	1	-1	-1	1	0	0	0	0	-1	0	0	0	0	2
← s_2^-	1	1	2	0	1	0	0	0	0	-1	0	0	0	3
s_3^-	4	-1	6	0	0	1	0	0	0	0	-1	0	0	9
s_4^-	3	-2	-2	0	0	0	1	0	0	0	0	-1	0	6
s_5^-	-2	-1	1	0	0	0	0	1	0	0	0	0	-1	2

Table 1(contd.)

B V	x_1	x_2	x_3	s_1^-	s_2^-	s_3^-	s_4^-	s_5^-	s_1^+	s_2^+	s_3^+	s_4^+	s_5^+	Soln
S_o	$\frac{5}{2}\downarrow$	$\frac{-13}{2}$	0	0	$\frac{-5}{2}$	0	0	0	-1	$\frac{3}{2}$	-2	-1	-1	$\frac{19}{2}$
s_1^-	$\frac{3}{2}$	$\frac{-1}{2}$	0	1	$\frac{1}{2}$	0	0	0	-1	$\frac{-1}{2}$	0	0	0	$\frac{7}{2}$
x_3	$\frac{1}{2}$	$\frac{1}{2}$	1	0	$\frac{1}{2}$	0	0	0	0	$\frac{-1}{2}$	0	0	0	$\frac{3}{2}$
$\leftarrow s_3^-$	1	-4	0	0	-3	1	0	0	0	3	-1	0	0	0
s_4^-	4	-1	0	0	1	0	1	0	0	-1	0	-1	0	9
s_5^-	$\frac{-5}{2}$	$\frac{-3}{2}$	0	0	$\frac{-1}{2}$	0	0	-1	0	$\frac{1}{2}$	0	0	-1	$\frac{1}{2}$
S_o	0	$\frac{7}{2}$	0	0	$5\downarrow$	$\frac{-5}{2}$	0	0	-1	-6	$\frac{1}{2}$	-1	-1	$\frac{19}{2}$
s_1^-	0	$\frac{11}{2}$	0	1	5	$\frac{-3}{2}$	0	0	-1	-5	$\frac{3}{2}$	0	0	$\frac{7}{2}$
x_3	0	$\frac{5}{2}$	1	0	2	$\frac{-1}{2}$	0	0	0	-2	$\frac{1}{2}$	0	0	$\frac{3}{2}$
x_1	1	-4	0	0	-3	1	0	0	0	3	-1	0	0	0
$\leftarrow s_4^-$	0	15	0	0	13	-4	1	0	0	-13	4	-1	0	9
s_5^-	0	$\frac{-23}{2}$	0	0	-8	$\frac{5}{2}$	0	1	0	8	$\frac{-5}{2}$	0	-1	$\frac{1}{2}$
S_o	0	$\frac{-59}{26}$	0	0	0	$\frac{-25}{26}$	$\frac{-5}{13}$	0	-1	-1	$\frac{-27}{26}$	$\frac{-8}{13}$	-1	$\frac{157}{26}$
s_1^-	0	$\frac{-7}{26}$	0	1	0	$\frac{1}{26}$	$\frac{-5}{13}$	0	-1	0	$\frac{-1}{26}$	$\frac{5}{13}$	0	$\frac{1}{26}$
x_3	0	$\frac{5}{26}$	1	0	0	$\frac{3}{26}$	$\frac{-2}{13}$	0	0	0	$\frac{-3}{26}$	$\frac{2}{13}$	0	$\frac{3}{26}$
x_1	1	$\frac{-7}{13}$	0	0	0	$\frac{1}{13}$	$\frac{3}{13}$	0	0	0	$\frac{-1}{13}$	$\frac{-3}{13}$	0	$\frac{27}{13}$
s_2^-	0	$\frac{15}{13}$	0	0	1	$\frac{-4}{13}$	$\frac{1}{13}$	0	0	-1	$\frac{4}{13}$	$\frac{-1}{13}$	0	$\frac{9}{13}$
s_5^-	0	$\frac{-59}{26}$	0	0	0	$\frac{1}{26}$	$\frac{8}{13}$	1	0	0	$\frac{-1}{26}$	$\frac{-8}{13}$	-1	$\frac{157}{26}$

Step 2. The Table 1 is examined for the optimal solution. The value of the objective function S_1 as well as the relative costs of all nonbasic variables are nonzero.

Step 3. The last 5th goal constraint, which has the lowest priority P_5,

$$-2x_1 - x_2 + x_3 + s_5^- - s_5^+ = 2$$

is deleted using the procedure outlined earlier, i.e., Since the nonzero deviation variable is s_5^-, a new variable ρ_5 is added to this equation as given below.

$$-2x_1 - x_2 + x_3 + s_5^- - s_5^+ + \rho_5 = 2$$

The resulting subproblem 2 is

$$\min\ S_2 = s_1^+ + s_2^+ + s_3^- + s_3^+ + s_4^- + s_5^-$$

s.t.

$$x_1 - x_2 - x_3 + s_1^- - s_1^+ = 2$$
$$x_1 + x_2 + 2x_3 + s_2^- - s_2^+ = 3$$
$$4x_1 - x_2 + 6x_3 + s_3^- - s_3^+ = 9$$
$$3x_1 - 2x_2 - 2x_3 + s_4^- - s_4^+ = 6$$
$$-2x_1 - x_2 + x_3 + s_5^- - s_5^+ + \rho_5 = 2$$
$$x_j, s_i^-, s_i^+, \rho_5 \geq 0, j = 1, 2, 3; i = 1, 2, \ldots, 5.$$

In order to solve the subproblem 2, we apply the sensitivity analysis on the optimal table of the subproblem 1. A new column below this new variable ρ_5 in the optimal table of subproblem 1 is obtained as follows:

Column below ρ_5, $\alpha^5 = B^{-1}A_5$ as

$$\begin{bmatrix} 1 & 0 & 1/26 & -5/13 & 0 \\ 0 & 0 & 3/26 & -2/13 & 0 \\ 0 & 0 & 1/13 & 3/13 & 0 \\ 0 & 1 & -4/13 & 1/13 & 0 \\ 0 & 0 & 1/26 & 8/13 & 1 \end{bmatrix} \begin{bmatrix} 0 \\ 0 \\ 0 \\ 0 \\ 1 \end{bmatrix} = \begin{bmatrix} 0 \\ 0 \\ 0 \\ 0 \\ 1 \end{bmatrix}$$

Entry below ρ_5 in the S_o row:

$$z_5 - C_5 = c_B^T B^{-1} A_5 - C_5 = (0,0,0,0,1)(0,0,0,0,1)^T - 0 = 1$$

Insert variable ρ_5, $z_5 - C_5$ and α^5 in Table 2. As the relative cost $z_5 - C_5$ is positive, the optimal criteria is disturbed and the variable ρ_5 will enter into the the basis as shown in Table 2.

Table 2

B V	x_1	x_2	x_3	s_1^-	s_2^-	s_3^-	s_4^-	s_5^-	s_1^+	s_2^+	s_3^+	s_4^+	s_5^+	$\rho_5\downarrow$	Soln
S_2	0	$\frac{-59}{26}$	0	0	0	$\frac{-25}{26}$	$\frac{-5}{13}$	0	-1	-1	$\frac{-27}{26}$	$\frac{-8}{13}$	-1	1	$\frac{157}{26}$
s_1^-	0	$\frac{-7}{26}$	0	1	0	$\frac{1}{26}$	$\frac{-5}{13}$	0	-1	0	$\frac{-1}{26}$	$\frac{5}{13}$	0	0	$\frac{1}{26}$
x_3	0	$\frac{5}{26}$	1	0	0	$\frac{3}{26}$	$\frac{-2}{13}$	0	0	0	$\frac{-3}{26}$	$\frac{2}{13}$	0	0	$\frac{3}{26}$
x_1	1	$\frac{-7}{13}$	0	0	0	$\frac{1}{13}$	$\frac{3}{13}$	0	0	0	$\frac{-1}{13}$	$\frac{-3}{13}$	0	0	$\frac{27}{13}$
s_2^-	0	$\frac{15}{13}$	0	0	1	$\frac{-4}{13}$	$\frac{1}{13}$	0	0	-1	$\frac{4}{13}$	$\frac{-1}{13}$	0	0	$\frac{9}{13}$
$\leftarrow s_5^-$	0	$\frac{-59}{26}$	0	0	0	$\frac{1}{26}$	$\frac{8}{13}$	1	0	0	$\frac{-1}{26}$	$\frac{-8}{13}$	-1	1	$\frac{157}{26}$
S_2	0	0	0	0	0	-1	-1	-1	-1	-1	-1	0	0	0	0
s_1^-	0	$\frac{-7}{26}$	0	1	0	$\frac{1}{26}$	$\frac{-5}{13}$	0	-1	0	$\frac{-1}{26}$	$\frac{5}{13}$	0	0	$\frac{1}{26}$
x_3	0	$\frac{5}{26}$	1	0	0	$\frac{3}{26}$	$\frac{-2}{13}$	0	0	0	$\frac{-3}{26}$	$\frac{2}{13}$	0	0	$\frac{3}{26}$
x_1	1	$\frac{-7}{13}$	0	0	0	$\frac{1}{13}$	$\frac{3}{13}$	0	0	0	$\frac{-1}{13}$	$\frac{-3}{13}$	0	0	$\frac{27}{13}$
s_2^-	0	$\frac{15}{13}$	0	0	1	$\frac{-4}{13}$	$\frac{1}{13}$	0	0	-1	$\frac{4}{13}$	$\frac{-1}{13}$	0	0	$\frac{9}{13}$
ρ_5^-	0	$\frac{-59}{26}$	0	0	0	$\frac{1}{26}$	$\frac{8}{13}$	1	0	0	$\frac{-1}{26}$	$\frac{-8}{13}$	-1	1	$\frac{157}{26}$

Step 4. Examining the above simplex table, we get the value of objective function S_2 equals to zero. It is to be observed that the relative costs of nonbasic variables x_2 and s_4^+ are zero. Thus, the preceding optimal table shown in Table 1 is the optimal solution of the case problem undertaken here.

Step 7. This very solution is the optimal solution of the case problem with respect to their given priorities.

Hence, the optimal solution is $x_1 = 27/13, x_2 = 0, x_3 = 3/26$.

The number of iterations required by the algorithm to find the optimal solution of the given problem is 3. The same problem has also been solved by LINDO software using lexicographic minimization method. The lexicographic minimization is defined as a sequential minimization of each priority whilst maintaining the minimum values reached by all higher priority level minimizations. It is found that the lexicographic minimization method requires 5 iterations to find the optimal solution.

Remarks.

1. We observe that the value of the deviation variable $s_5^- = 157/26$ in the optimal table as shown in Table 1 is the same as the value of the objective function $S_o = 157/26$. Applying Remarks 5 of Section 3, the deletion of the constraint (5) results in S_o to be equal to zero and thus the solution already obtained in Table 1 is the optimal solution of the given case problem. In fact, there is no need of deletion of the 5th constraint.

2. It is to be noted that as the vector $\alpha^5 = (0, 0, 0, 0, 1)^T$, the basis variable s_5^- will go out of the basis and B^{-1} remains unchanged. And as a result, the value of the basic vector X_B (i.e., $B^{-1}b$ where, $b = (2, 3, 9, 6, 2)^T$) also remains same. With regard to the objective function, since $c_B = (0, 0, 0, 0, 0)^T$, the objective function S_o will be equal to $c_B^T X_B = 0$. Applying Remarks 6 of Section 3, the solution already obtained in Table 1 is the optimal solution of the given case problem. Thus, we should find out the value of c_B before we finally delete the 5th constraint.

3. It is to be observed that in the simplex table the α^{j-} entries for the deviational, slack or surplus variables s_j^- shall be negative of the α^{j+} entries for s_j^+.

4. The columns below nonbasic variables having the relative costs negative as well as the row and column corresponding to ρ_5 may be deleted from the simplex tables constructed later as they will never enter into the basis afterwards.

5. In the case the algorithm finds the optimal solution in Step 2 when $h = 0$, then it, in general, requires less number of iterations than the number of iterations required using the lexicographic minimization. As the size of the case problems increases the algorithm performs better in the case when more than fifty percent of the goals in the order of decreasing priorities are satisfied. In other words, the algorithm works better if a large number of goals is satisfied.

6. It is worth mentioning that the addition of a goal constraint as in the lexicographic minimization approach, first disturbs the simplex format and when the simplex format is restored, then the feasibility gets disturbed. The feasibility is restored by the dual simplex method in order to find the new optimal solution. However, such problems do not arise in the grouping algorithm where we delete a goal constraint from an LGPP.

Algorithm for Goal Constraints with Emptive Priority Factors

We take the LGPP wherein goal constraints have not been assigned any priorities, i.e., goal constraints with emptive priorities. Here the decision maker desires to know the priorities to be assigned to the goals so that the maximum number of goals is satisfied. Let N_g be the number of goal constraints that is satisfied. Then LGPP formulation considered for n variables, l real constraints, m goal constraints and t emptive priority levels corresponding to the goal constraints is given as

$$\max \quad N_g \tag{16.26}$$

$$\text{s.t.}$$

$$S = \min \sum_{k=1}^{t} P_k \sum_{i=1}^{m} \left(w_{ik}^{-} s_i^{-} + w_{ik}^{+} s_i^{+} \right) \tag{16.27}$$

$$\sum_{j=1}^{n} e_{qj} x_j = f_q, \quad q = 1, 2, \ldots, \ell \tag{16.28}$$

$$\sum_{j=1}^{n} a_{ij} x_j + s_i^{-} - s_i^{+} = b_i, \quad i = 1, 2, \ldots, m \tag{16.29}$$

$$x_j, b_i, s_i^{-}, s_i^{+} \geq 0, j = 1, 2, \ldots, n; i = 1, 2, \ldots, m$$

To solve LGPP (16.26)-(16.29), the algorithm solves a sequence of linear programming subproblems 1,2,...,p, where p is the last subproblem, each using the optimal solution of the previous subproblems. The steps involved in the algorithm are:

Step 1. All real constraints and goal constraints after introducing deviational variables are considered a group with the objective function being the sum of all weighted deviations in the subproblem

1 expressed as follows:

$$\min \quad S_1 = \sum_{i=1}^{m}(s_i^- + s_i^+) \tag{16.30}$$

s.t.

$$\sum_{j=1}^{n} e_{qj} x_j = f_q, \quad q = 1, 2, \ldots, \ell \tag{16.31}$$

$$\sum_{j=1}^{n} a_{ij} x_j + s_i^- - s_i^+ = b_i, \quad i = 1, 2, \ldots, m \tag{16.32}$$

$$x_j, b_i, s_i^-, s_i^+ \geq 0, j = 1, 2, \ldots, n; i = 1, 2, \ldots, m$$

Here, it is assumed that all deviational variables are having unit weight functions. We are indeed required to maximize the number of goal constraints, i.e., N_g, to solve LGPP (16.26)-(16.29). We do this exercise by first solving the subproblem 1 by the simplex method and then we obtain the corresponding maximum number of goal constraints $N_{g_1}^{*1}$ that are satisfied. Take $h = 0$.

Step 2. The simplex table is examined for the optimal solution. If the objective function S_{h+1} has zero value, go to Step 9.

Step 3. The optimal simplex table $h + 1$ is marked for three observations for all m goal constraints:

(i) all deviational variables, slack and surplus variables are at zero level;

(ii) the slack or surplus variables are zero but the deviational variables are nonzero;

(iii) the deviational variables are zero but the slack or surplus variables are nonzero.

For all the goal constraints satisfying observation (i) or (iii), go to Step 8.

Step 4. Delete all the goal constraints in the subproblem (h+1) satisfying the observation (ii) and assign them the next higher priority $(t - h)$. Let k_{t-h} goal constraints satisfy the observation (ii).

The resulting subproblem (h+2) is given as follows:

$$\min \quad S_{h+2} = \sum_{i=1}^{m}(s_i^- + s_i^+) \tag{16.33}$$

s.t.

$$\sum_{j=1}^{n} e_{qj}x_j = f_q, \quad q = 1, 2, \ldots, \ell \tag{16.34}$$

$$\sum_{j=1}^{n} a_{ij}x_j + s_i^- - s_i^+ = b_i, \quad i = 1, 2, \ldots, m - k_{t-h} \tag{16.35}$$

$$\sum_{j=1}^{n} a_{ij}x_j + s_i^- - s_i^+ + u_i\rho_i = b_i, \quad i = m - k_{t-h} + 1,$$

$$m - k_{t-h} + 2, \ldots, m \tag{16.36}$$

$$x_j, b_i, s_i^-, s_i^+, \rho_i \geq 0, j = 1, 2, \ldots, n; i = 1, 2, \ldots, m$$

here, the u_i has a value of 1 or -1 depending upon the deviation variable is s_i^- or s_i^+ corresponding to the equation (16.36).

Apply sensitivity analysis to find out the effects of deletion of the k_{t-h} goal constraints on the optimal solution of the subproblem (h+1) and thereby obtain the solution of the subproblem (h+2).

Step 5. If the deletion of all goal constraints from the set may cause the objective function S_{h+2} to be zero and $(z_j - C_j)$ entries below nonbasics are zero, then go to Step 7. In the deletion of the goal constraints, the next possibility may result in the objective function S_{h+2} to be zero but $(z_j - C_j)$ entries below some nonbasic variables are nonzero, then go to Step 8. Furthermore, it is also possible that after deletion of the goal constraints, the objective function S_{h+2} is nonzero, then go to Step 6.

Step 6. The optimal simplex table $h + 2$ is marked for three observations for all $m - k_{t-h}$ goal constraints:

(i) all deviational variables, slack and surplus variables are at zero level;

(ii) the slack or surplus variables are zero but the deviational variables are nonzero;

(iii) the deviational variables are zero but the slack or surplus variables are nonzero.

For all the goal constraints satisfying observation (i) or (iii), go to Step 8. For all the goal constraints satisfying observation (ii), $h = h + 1$ and go to Step 4.

Step 7. The preceding table $h + 1$ gives the optimal solution of the given LGPP (16.26)-(16.29). Go to Step 9.

Step 8. This very table gives the optimal solution and the remaining constraints may be assigned priorities at the disposal of the decision maker. Go to Step 10.

Step 9. The optimal solution is obtained and all the goal constraints are assigned their respective priorities so that the maximum number of goals are satisfied.

Step 10. Stop.

Remarks.

1. If a kth goal constraints satisfies the observation (i) or (iii), then the deletion of it may have no effect on the objective function since deviational variables corresponding to the kth goal constraint are zero. Therefore, the kth goal constraint can be assigned any priorities. In other words, the observation (i) or (iii) is a nonbinding on the optimal solution. So, we write for all the goal constraints satisfying observations (i) or (iii), go to Step 8.

2. If a kth goal constraint satisfies the observation (ii), then the deletion of it may have impact on the objective function since deviational variables corresponding to the kth goal constraint are nonzero. Therefore, the kth goal constraint can not be assigned any priorities. In other words, the observation (ii) is a binding on the optimal solution.

3. If the optimal value of objective function S_1^{*1} equals to zero, it means that all m goal constraints may be kept at same priority levels.

4. In order to find the optimal solution of the the LGPP (16.26)-(16.29) the algorithm solves a sequence of linear programming subproblems 1,2,...,p, where p is the last subproblem, each using the optimal solution of the previous subproblems. In fact, we are interested to maximize the number of goal constraints N_g that is satisfied and thereby assign priorities to the goal constraints. In other words, we would like to have the value of objective function zero by deleting

the minimum number of goal constraints. We first solve the subproblem 1 and obtain its optimal solution using the simplex method. The optimal value of the objective function S_1 obtained is given by

$$S_1^{*1} = \sum_{i=1}^{m}(s_i^{-(*1)} + s_i^{+(*1)}) \tag{16.37}$$

Let $N_{g_1}^{*1}$ be the maximum number of goal constraints satisfied while we solve subproblem 1. It means that deviations corresponding to $N_{g_1}^{*1}$ goal constraints are zero. If $S_1^{*1} = 0$, then

$$N_{g_1}^{*1} = m \tag{16.38}$$

where m is the total number of goal constraints in the LGPP (16.26)-(16.29).

In case $S_1^{*1} \neq 0$, we apply the sensitivity analysis to find all those goal constraints in Step 3 that may have effect on the objective function S_1. Let k_t be the number of goal constraints that have impact on the value of objective function S_1. Then we assign t priorities to these k_t goal constraints and delete all of them. We solve the resulting subproblem 2 using sensitivity analysis and get the optimal value of the objective function S_2 as given below.

$$S_2^{*2} = \sum_{i=1}^{m-k_t}(s_i^{-(*2)} + s_i^{+(*2)}) \tag{16.39}$$

The corresponding maximum number of goal constraints satisfied is, say, $N_{g_2}^{*2}$. From the equations (16.38) and (16.39), we get

$$S_1^{*1} \geq S_2^{*2} \tag{16.40}$$

Using the relation (16.40), we can write

$$N_{g_1}^{*1} \leq N_{g_2}^{*2} \tag{16.41}$$

Next, we delete the goal constraints that have effect on the value of the objective function and assign them the next higher priority $(t-1)$ and so on till we assign all the priorities or till we get the lowest value of the objective function as zero. Let this condition reaches when we solve the subproblem p. We get the following relation

$$N_{g_1}^{*1} \geq N_{g_2}^{*2} \geq \ldots \geq N_{g_p}^{*p} \tag{16.42}$$

where $p \leq t$.

Thus, after solving the subproblem p we get the optimal solution of the LGPP (16.26)-(16.29) as follows:

$$s_i^- = s_i^{-(*p)}, i = 1, 2, \ldots, m \tag{16.43}$$

$$s_i^+ = s_i^{+(*p)}, i = 1, 2, \ldots, m \tag{16.44}$$

$$S = S_p^{*p} = \sum_{i=1}^{m}(s_i^{-(*p)} + s_i^{+(*p)}) \tag{16.45}$$

$$N_g = N_{g_p}^{*p} \tag{16.46}$$

Now, the algorithm is applied to the problems as follows:

Example 6.

$$
\begin{aligned}
x_1 + x_2 &\leq 10 \\
x_1 &\geq 7 \\
2x_1 + x_2 &\leq 12 \\
x_1 + 4x_2 &\leq 4 \\
x_1, x_2 &\geq 0.
\end{aligned}
$$

Here, there are two variables ($n = 2$) and four goal constraints ($m = 4$). We are required to assign priorities P_1 to P_4 to the goal constraints so that the maximum goals are satisfied.

Step 1. All four goal constraints after introducing deviational variables are considered a group with the objective function being the sum of all weighted deviations in the subproblem 1 as follows:

$$\min \ S_o = s_1^+ + s_2^- + s_3^+ + s_4^+$$

s.t.

$$
\begin{aligned}
x_1 + x_2 + s_1^- - s_1^+ &= 10 \\
x_1 + s_2^- - s_2^+ &= 7 \\
2x_1 + x_2 + s_3^- - s_3^+ &= 12 \\
x_1 + 4x_2 + s_4^- - s_4^+ &= 4 \\
x_j, s_i^-, s_i^+ &\geq 0, j = 1, 2; i = 1, 2, 3, 4.
\end{aligned}
$$

Here, we assume weight functions of all deviational variables to be at unity level. The problem is put on the simplex algorithm. The simplex table is as follows:

Table 3

B V	x_1	x_2	s_1^-	s_2^-	s_3^-	s_4^-	s_1^+	s_2^+	s_3^+	s_4^+	Soln
S_1	0	0	0	-1	0	0	-1	0	-1	-1	0
s_1^-	1	1	1	0	0	0	-1	0	0	0	10
s_2^-	1	0	0	1	0	0	0	-1	0	0	7
s_3^-	2	1	0	0	1	0	0	0	-1	0	12
s_4^-	1	4	0	0	0	1	0	0	0	-1	4
S_1	1 ↓	0	0	0	0	0	-1	-1	-1	-1	7
s_1^-	1	1	1	0	0	0	-1	0	0	0	10
s_2^-	1	0	0	1	0	0	0	-1	0	0	7
s_3^-	2	1	0	0	1	0	0	0	-1	0	12
← s_4^-	1	4	0	0	0	1	0	0	0	-1	4
S_o	0	-4	0	0	0	-1	-1	-1	-1	0	3
s_1^-	0	-3	1	0	0	-1	-1	0	0	1	6
s_2^-	0	-4	0	1	0	-1	0	-1	0	1	3
s_3^-	0	-7	0	0	1	-2	0	0	-1	2	4
x_1	1	4	0	0	0	1	0	0	0	-1	4

Step 2. The Table 3 is examined for the optimal solution. The objective function S_1 is nonzero.

Step 3. The optimal simplex tableau in Table 3 is marked for three observations for all goal constraints:

(i) All deviational variables, slack and surplus variables are at zero level: constraint (4).

(ii) The slack or surplus variables are zero but the deviational variables are nonzero: constraint (2).

(iii) The deviational variables are zero but the slack or surplus variables are nonzero: constraints (1) and (3).

Step 4. The constraint (2) satisfying observations (ii)

$$x_1 + s_2^- - s_2^+ = 7$$

is deleted and assigned the lowest priority P_4. As the deviation variable is s_2^-, a new variable ρ_2 is added to this equation as follows:

$$x_1 + s_2^- - s_2^+ + \rho_2 = 7$$

The resulting subproblem 2 is

$$\min S_2 = s_1^+ + s_2^- + s_3^+ + s_4^+$$

s.t.

$$x_1 + x_2 + s_1^- - s_1^+ = 10$$
$$x_1 + s_2^- - s_2^+ + \rho_2 = 7$$
$$2x_1 + x_2 + s_3^- - s_3^+ = 12$$
$$x_1 + 4x_2 + s_4^- - s_4^+ = 4$$
$$x_j, s_i^-, s_i^+, \rho_2 \geq 0, j = 1, 2; i = 1, 2, 3, 4.$$

For solving the subproblem 2, we apply the sensitivity analysis on the optimal table of the subproblem 1. A new column below this new variable ρ_2 is obtained using the procedure explained in Section 3 and then it is introduced in the optimal simplex table as shown below.

Table 4

B V	x_1	x_2	s_1^-	s_2^-	s_3^-	s_4^-	s_1^+	s_2^+	s_3^+	s_4^+	$\rho_2 \downarrow$	Soln
S_2	0	-4	0	0	0	-1	-1	-1	-1	0	1	3
s_1^-	0	-3	1	0	0	-1	-1	0	0	1	0	6
$\leftarrow s_2^-$	0	-4	0	1	0	-1	0	-1	0	1	1	3
s_3^-	0	-7	0	0	1	-2	0	0	-1	2	0	4
x_1	1	4	0	0	0	1	0	0	0	-1	0	4
S_2	0	0	0	-1	0	0	-1	-1	-1	-1	0	0
s_1^-	0	-3	1	0	0	-1	-1	0	0	1	0	6
ρ_2	0	-4	0	1	0	-1	0	-1	0	1	1	3
s_3^-	0	-7	0	0	1	-2	0	0	-1	2	0	4
x_1	1	4	0	0	0	1	0	0	0	-1	0	4

Step 5. The deletion of the constraint (2) causes the value of objective function S_2 to be zero but $(z_j - C_j)$ entries below some nonbasic variables are nonzero.

Step 7. This very table (Table 4) gives the optimal solution and the remaining constraints may be assigned any priorities.

Step 9. The optimal solution is $x_1 = 4, x_2 = 0, N_g = 3$.

The goal constraints with the respective priorities being assigned are as follows:

$$x_1 + x_2 \quad \le 10$$
$$x_1 \quad\quad\quad \ge 7 \quad P_4$$
$$2x_1 + x_2 \quad \le 12$$
$$x_1 + 4x_2 \quad \le 4$$
$$x_1, x_2 \ge 0$$

Here, the goal constraints (1), (3) and (4) may have any priorities from P_1 to P_3.

Remarks.

1. The columns below nonbasic variables having the relative costs negative as well as the row and column corresponding to ρ_2 are deleted from the simplex tables constructed later. The incorporation of these facts in the algorithm makes the method easy and convenient for solving the problems as well as it results in faster computation.

2. In general, the solution obtained by the algorithm gives the varieties of choices to the decision maker to assign priorities to the goals. As in the Example problem 6, the decision maker may assign any priorities to the goals (1), (3) and (4) from P_1 to P_3. The algorithm is thus useful for classifying many goal constraints required for some problem into nested subclasses $m_1, m_2, \ldots, m_t$ with different priority levels. Sometimes, it becomes essential for the decision maker to change the priorities of goal constraints for the successful completion of the project at hand. In such situations, the algorithm can find the solution with ease.

Problem Set 16

1. Solve the formulated LGPP in example of Section 14.1.

2. Consider the following LGPP

$$2x_1 + x_2 \simeq 11 \qquad P_1$$

$$x_1 + x_2 \simeq 9 \qquad P_2$$

$$x_1 + 4x_2 \geq 4$$

$$x_1 \geq 7$$

The project manager decides that the third and fourth constraint should be satisfied as real constraint. Write the standard form of the LGPP, and then find its optimal solution.

3. Show that any LGPP does not have infeasible or unbounded solution.

4. Solve the following linear goal programming problem by the grouping algorithm.

$$(a) \qquad x_1 + x_2 \leq 10 \qquad P_1$$
$$2x_1 + x_2 \leq 15 \qquad P_2$$
$$x_1 + 4x_2 \leq 20 \qquad P_3$$
$$x_1, x_2 \geq 0$$

$$(b) \qquad 2x_1 + 5x_2 + 4x_3 = 600$$
$$3x_1 + 7x_2 + 5x_3 = 500$$
$$2x_1 + 5x_2 + 6x_3 = 450 \qquad P_1$$
$$4x_1 + 3x_2 + 3x_3 \leq 500 \qquad P_2$$
$$2x_1 + 3x_2 + 4x_3 \leq 600 \qquad P_3$$
$$3x_1 + 3x_2 + 5x_3 \geq 500 \qquad P_4$$
$$3x_1 + 4x_2 + 7x_3 \leq 900 \qquad P_5$$
$$5x_1 + 3x_2 + 3x_3 \leq 800 \qquad P_6$$
$$3x_1 + 6x_2 + 4x_3 \geq 400 \qquad P_7$$
$$3x_1 + 2x_2 + 2x_3 \leq 500 \qquad P_8$$
$$x_j \geq 0, j = 1, 2, 3$$

Also find the number of iterations required to obtain the optimal solution as compared to the lexicographic minimization method using the LINDO software.

5. Can you prove that in the lexicographic minimization approach, the addition of a goal constraint (provided it affects the optimal solution) to an LGPP always worsens the current optimal value of the objective function whereas, in the grouping algorithm, the deletion of a goal constraint (provided it affects the optimal solution) from an LGPP always improves the current optimal value of the objective function?

6. The decision maker wish to assign the priorities to the goals

$$x_1 - x_2 - x_3 \leq 2$$
$$x_1 + x_2 + 2x_3 \leq 3$$
$$4x_1 - x_2 + 6x_3 = 9$$
$$3x_1 - 2x_2 - 2x_3 \geq 6$$
$$-2x_1 - x_2 + x_3 \geq 2$$
$$x_1, x_2, x_3 \geq 0$$

with an objective that the maximum goals are satisfied. Find the optimal solution of the problem along with the priorities to be assigned to the goals.

7. A company manufactures three products A, B and C. Each product has to pass through three operations. The time each product takes in each operation is given below. Also, given are the maximum time for each an operation can work and profit per item. Find the number of units of each product to be manufactured so that the following goals in order of priorities are satisfied.

(a) Meet a profit goal of $\$1,500$;

(b) Take care of the avoidance of ideal time in operation 1 and overtime in operations 2;

(c) It is twice as significance the operation 2 as compared to operation 3.

Operations Products Time required
A B C

	A B C	
1	1 0 1	480
2	0 3 2	5000
3	2 4 0	360
Profit/item	4 3 6	

Chapter 17

Games Theory

The concept of games theory is introduced. In the end, the linear programming technique to solve two person zero sum game has been included.

17.1 Introduction

There is competition in every walk of life. Many a decision is taken in a competitive situation in which the outcome depends not on that decision alone but rather on the interaction between the decision maker and that of competitors. The term "game" now includes not only plausible activities of this kind, but also more earnest competitive situations of war and peace, love and hate, and die and survive. Such situations arise in business, politics, military operations, etc.

In the year 1928, Von Neumann who is also the father of games theory developed the theory of games that is based on the minimax and maximin principle which implies that competitors will act to minimize his maximum loss or maximize the minimum gain. There are finite number of players and each player has finite number of possible courses of action, called strategies. All of the strategies must be known to each other but must not know which of these will be chosen. A play is said to played when each of the player chooses a single course of action. After all players have chosen a course of action, their respective gains are finite.

Let us define some terms frequently used in our studies on games theory.

Pay-off. This is the outcome of the game or the gain the strategy to a player for any given counter strategy of the competitor.

Pay-off matrix. This is the matrix whose entries the payoff of different strategies of the game.

Optimal strategy. This is the course of action or plan which put the player in the most preferred position. Any deviation from this strategy results in decrease of pay for the player.

Pure Strategies. If chance does not determine any move and both players choose their strategies deterministically, then such type of strategies are called pure strategies. The objective is to maximize the gain or to minimize the loss.

Mixed Strategies. When some of the moves are determined by chance the situation is nondeterministic and the objective is to maximize the expected gain.

In this chapter we are mainly concerned about zero-sum two person games.

17.2   Two Person Zero Sum Game (Pure Strategies)

It is a game between two person in which losses of one player are equal to the gains of other player so that the sum of net gain is zero.

Strategy. The alternative which one person A has to move for each possible moves of other person B is called strategy.

Note. Similarly A may have different strategies for the same move B.

Let us write the pay-off matrix of the player A.

Pay-off matrix of A

B's strategies

	B_1	B_2		B_J		B_n
A_1	a_{11}	a_{12}	$\cdots$	a_{1j}	$\cdots$	a_{1n}
A_2	a_{21}	a_{22}	$\cdots$	a_{2j}	$\cdots$	a_{2n}
$\vdots$						
A_i	a_{i1}	a_{i2}	$\cdots$	a_{ij}	$\cdots$	a_{in}
$\vdots$						
A_m	a_{m1}	a_{m2}	$\cdots$	a_{mj}	$\cdots$	a_{mn}

A's strategies

If player A chooses the ith strategy and B chooses the jth strategy, then a_{ij} is pay-off to A from B. Now

$$a_{ij} > 0 \text{ pay-off to A from B}$$
$$a_{ij} < 0 \text{ pay-off to B from A}$$

The matrix $(a_{ij})_{m \times n}$ is pay-off matrix of A.

Minimum (Maximum) criterion. Consider the pay-off matrix of A. If A chooses ith strategy, then he is sure of getting

$$\min_j \{a_{ij}\}, \quad j \text{ varies over the strategies of player B.}$$

Then A will choose the strategy given by

$$\max_i \min_j \{a_{ij}\} = \underline{a}, \quad i = 1, 2, \ldots, m, \quad j = 1, 2, \ldots, n$$

Conversely, if B chooses jth strategy, then he is sure that A does not get more than

$$\max_i \{a_{ij}\}, \quad i \text{ varies over the strategies of player A.}$$

B will choose naturally the strategy given by

$$\min_j \max_i \{a_{ij}\} = \bar{a}, \quad i = 1, 2, \ldots, m, \quad j = 1, 2, \ldots, n.$$

Now,

$$\underline{a} \le \text{value of game} \le \overline{a}.$$

If $\underline{a} = \overline{a} = $ game value, then the game is said to have saddle point. In this case an optimal solution of the game exists. Let

$$a_{st} = \max_{i} \min_{j}\{a_{ij}\} = \min_{j} \max_{i}\{a_{ij}\} \quad \text{such that} \quad a_{st} = \underline{a} = \overline{a}.$$

Hence, the optimal strategy of player A is A_s and optimal strategy of player B is B_t. Game value $= a_{st}$.

For example, below is given the pay-off matrix of A and its solution

-2	1	0	1	-2
4	2	2	3	2
1	3	1	3	1
4	3	2	3	

A_2 : Best strategy for A

B_3 : Best strategy for B

Game value $= 2$

Again, we encounter with payoff matrix of A in which alternative solution exists.

4	0	-2	3	-2
3	11	4	1	1
4	5	8	4	4
2	13	17	1	1
4	11	17	4	

The optimal solution is

A_3 : Best strategy for A

B_1 : Best strategy for B

Game value $= 4$

The alternative optimal solution is

$$A_3 : \text{Best strategy for } A$$
$$B_4 : \text{Best strategy for } B$$
$$\text{Game value } = 4$$

17.3 Two Person Zero Sum Game (Mixed Strategies)

The concept of mixed strategies is utilized to solve game problems where saddle point does not exist, i.e., $\underline{a} \neq \bar{a}$. Let $p_1, p_2, \ldots, p_m$ and $q_1, q_2, \ldots, q_n$ be the probabilities of the events $A_1, A_2, \ldots, A_m$ and $B_1, B_2, \ldots, B_n$, respectively. Obviously,

$$\sum_{i=1}^{m} p_i = 1, \ p_i \geq 0, \qquad \sum_{j=1}^{n} q_j = 1, \ q_j \geq 0.$$

The pay-off matrix of A is as follows:

		B_1	B_2		B_J		B_n
		q_1	q_2		q_J		q_n
A_1	p_1	a_{11}	a_{12}	$\cdots$	a_{1j}	$\cdots$	a_{1n}
A_2	p_2	a_{21}	a_{22}	$\cdots$	a_{2j}	$\cdots$	a_{2n}
$\vdots$							
A_i	p_i	a_{i1}	a_{i2}	$\cdots$	a_{ij}	$\cdots$	a_{in}
$\vdots$							
A_m	p_m	a_{m1}	a_{m2}	$\cdots$	a_{mj}	$\cdots$	a_{mn}

Our objective is to find p_i ($i = 1$ to m), and q_j ($j = 1$ to n) when a_{ij}'s are given.

Thus, A's problem is max{*gains*} and B's problem is min{*losses*}. Let us first, decide the problem of A.

For deciding A's problem, let B select the jth strategy. Then, A's expected pay-off will be

$$p_1 a_{1j} + p_2 a_{2j} + \cdots + p_m a_{mj} = \sum_{i=1}^{m} p_i a_{ij}.$$

Player B moves all strategies and he pays to A

$$\sum_{i=1}^{m} a_{i1}, \sum_{i=1}^{m} a_{i2}, \ldots, \sum_{i=1}^{m} a_{in}.$$

Now, B's intention will be to pay minimum of the above pay-offs, i.e.,

$$\min \left\{ \sum_{i=1}^{m} a_{i1}, \ \sum_{i=1}^{m} a_{i2}, \ \ldots, \ \sum_{i=1}^{m} a_{in} \right\}.$$

But A's interest is to maximize these pay-offs, i.e.,

$$A\text{'s problem} \max \left\{ \min \left\{ \sum_{i=1}^{m} a_{i1}, \ \sum_{i=1}^{m} a_{i2}, \ \ldots, \ \sum_{i=1}^{m} a_{in} \right\} \right\} \quad (17.1)$$

$$\text{s.t.} \, p_1 + p_2 + \cdots + p_m = 1, \ p_i \geq 0 \quad\quad\quad (17.2)$$

For deciding B's problem, let A select the ith strategy. Then, B will loose the following expected pay-offs to A (as A's pay off matrix is given)

$$q_1 a_{i1} + q_2 a_{i2} + \cdots + q_n a_{in} = \sum_{j=1}^{n} q_i a_{ij}.$$

A moves all strategies with the thinking that he gives maximum loss to B, i.e.,

$$\max \left\{ \sum_{j=1}^{n} a_{1j}, \ \sum_{j=1}^{n} a_{2j}, \ \ldots, \ \sum_{j=1}^{n} a_{mj} \right\}.$$

Now, B's problem is to minimize these losses. Hence,

$$B\text{'s problem} \min \left\{ \max \left\{ \sum_{j=1}^{n} a_{1j}, \ \sum_{j=1}^{n} a_{2j}, \ \ldots, \ \sum_{j=1}^{n} a_{mj} \right\} \right\} (17.3)$$

$$\text{s.t.} \, q_1 + q_2 + \cdots + q_n = 1, \ q_i \geq 0 \quad\quad\quad (17.4)$$

17.4   Games Theory vs Linear Programming

Games theory and linear programming are closely related. In fact every two person zero sum can be represented by a linear programming problem. The concept of duality will also play significant role in games theory.

Convert (17.1) and (17.2) into LPP as follows. Let

$$g = \min \left\{ \sum_{i=1}^{m} p_i a_{i1}, \ \sum_{i=1}^{m} p_i a_{i2}, \ \ldots, \ \sum_{i=1}^{m} p_i a_{in} \right\}$$

Then, A's problem is

$$\max \quad z_0 = g$$
$$\text{s.t.} \quad \sum_{i=1}^{m} p_i a_{ij} - g \geq 0, \ j = 1, 2, \ldots, n$$
$$\sum_{i=1}^{m} p_i = 1, \ p_i \geq 0$$

If $g \geq 0$, nothing is to be done, otherwise for $g < 0$, a positive constant is added so that new g becomes positive. Then, this new g is used. The optimal value of the objective function is obtained by subtracting the constant c. Assume

$$x_i = \frac{p_i}{g}, \ i = 1, 2, \ldots, m$$

Then the above LPP reduces to

$$\max \quad z_0 = g$$
$$\text{s.t.} \quad \sum_{i=1}^{m} a_{ij} \geq 1$$
$$\sum_{i=1}^{m} x_i = \frac{1}{g}$$

Since

$$\max \ g = \min \ \frac{1}{g} = \min x_1 + x_2 + \cdots + x_m,$$

the above LPP is finally converted into the form given below which represents the LPP corresponding to A

$$\min \quad x_0 = x_1 + x_2 + \cdots + x_m$$
$$\text{s.t.} \quad a_{1j}x_1 + a_{2j}x_2 + \cdots + a_{mj}x_m \geq 1$$
$$x_j \geq 0, \ j = 1, 2, \ldots, n.$$

Similarly, using (17.3) and (17.4) the LPP associated with B's problem can be derived.

Example 1. Consider the game between two players A and B. The pay-off matrix of A is given below.

-1	1	2
2	-3	3

First, we check whether saddle point exits. The maximin and minimax are computed as

-1	1	2	-1
2	-3	3	-3
2	1	4	

Since $\underline{a} = -1$ and $\bar{a} = 1$, it follows that $\underline{a} \neq \bar{a}$. This suggests that we have to utilize the concept of mixed strategies.

If maximin is negative, then add a constant $c \geq -$maximin$+1$ such that each entry of the pay-off matrix turns to be nonnegative. Thus, add $c \geq 2$. Let us take $c = 2$ and add each entry of the pay-off matrix.

The maximin and minimax for the revised matrix are shown in the following matrix with the decision variables and probabilities.

			y_1	y_2	y_3	
			q_1	q_2	q_3	
			B_1	B_2	B_3	
x_1	p_1	A_1	1	3	4	1
x_2	p_2	A_2	4	-1	3	5
			4	5	4	

The LPP giving A's problem

$$\min \quad x_0 = x_1 + x_2$$
$$\text{s.t.} \quad x_1 + 4x_2 \geq 1$$
$$3x_1 - x_2 \geq 1$$
$$4x_1 + 5x_2 \geq 1$$
$$x_1, x_2 \geq 0$$

The LPP giving B's problem

$$\max \quad y_0 = y_1 + y_2 + y_3$$
$$\text{s.t.} \quad y_1 + 3y_2 + 4y_3 \leq 1$$
$$4y_1 - y_2 + 5y_3 \leq 1$$
$$y_1, y_2 \geq 0$$

As both the problems are dual to each other, solve any one which is convenient and find the optimal solution of the other one using $C_B^T B^{-1}$. Since B's problem has smaller number of constraints, we prefer to solve B's LPP.

The simplex iterations are

B V	y_1	$y_2 \downarrow$	y_3	s_1	s_2	Soln
y_0	-1	-1	-1	0	0	0
$\leftarrow s_1$	1	$\boxed{3}$	4	1	0	1
s_2	4	-1	5	0	1	1
y_0	$-2/3 \downarrow$	0	$1/3$	$1/3$	0	$1/3$
y_2	$1/3$	1	$4/3$	$1/3$	0	$1/3$
$\leftarrow s_2$	$\boxed{13/3}$	0	$19/3$	$1/3$	1	$4/3$
y_0	0	0	$24/3$	$5/13$	$2/13$	$7/13$
y_2	0	1	$11/3$	$4/13$	$-1/3$	$3/13$
y_1	1	0	$19/3$	$1/13$	$3/13$	$4/13$

The optimal solution of B's problem: $y_1 = 4/13$, $y_2 = 3/13$,

$$y_3 = 0; \quad y_0 = 7/13.$$

The optimal value of $q_i = y_i/y_0$: $q_1 = 4/7$, $q_2 = 3/7$, $q_3 = 0$.

From the optimal table of B's problem, we have

The optimal solution of A's problem: $x_1 = 5/13$, $x_2 = 2/13$,

$$x_0 = 7/13.$$

The optimal value of $p_i = x_i/x_0$: $p_1 = 5/7$, $p_2 = 2/7$.

The value of game $= 1/y_0 - c = 1/x_0 - c = -1/7$.

Remarks. 1. Normally pay-off matrix of A is given. In case pay-off matrix of B is given then multiply all entries by -1 and solve the problem as pay-off matrix of A.

2. Theoretically, it is better to add

$$c = -\min \{\text{negative entries of the matrix}\} + 1$$

.

3. If strategies of A and B are given corresponding to columns and rows and it is given that it is A's pay-off matrix then take transpose of the matrix and solve as usual with the thinking A's pay-off matrix is given.

   4. The dual or primal may have alternative optimal solution. Hence the game problem may have alternate solution. Find alternate optimal solution also.

5. Always use $\theta_j(z_j - c_j)$ for entering variable in case of the game problem.

Example 2. Prove that

$$a_{pq} = \max_i \min_j a_{ij} \leq \min_j \max_i a_{ij} = a_{rs} \text{ or } \underline{a} \leq \overline{a}$$

Proof. $\max_i a_{ij}$ for any i and fixed j. $\min_j a_{ij} \leq a_{ij}$ for any j and fixed i. Let

$$\max_i a_{ij} = a_{rj} \text{ for fixed } j$$

$$\min_j a_{ij} = a_{iq} \text{ for fixed } i$$

Then

$$a_{rj} \geq a_{ij} \geq a_{iq} \ \forall \ i,j.$$

Hence

$$\min_j a_{rj} \geq a_{ij} \max_i a_{iq}.$$

or

$$a_{rs} \geq a_{ij} \geq a_{pq}.$$

Remark. This chapter is concentrated only on two-person zero sum game with a finite number of strategies. It does not mean that game theory is limited to such type of game. In fact extensive research has been done on n-person game, where more than two players may participate in the game. This is often the case, for example, in competition among business firms, in international diplomacy, and so forth. However, the existing theory for such games is less satisfactory than for two-person games.

Another generalization is the nonzero-sum game, where the sum of the payoffs to the players need not be zero (or any other fixed constant). This reflects the fact that many competitive situations include noncompetitive aspects that contribute to the mutual advantage or mutual disadvantage of the players. For example the, the advertising strategies of competing companies can affect not only how they will split the market but also the total size of the market for their competing products.

Still another extension is to the class of infinite games, where the players have an infinite number of pure strategies available to them. These games are designed for the kind of situation where the strategy to be selected can be represented by a continuous decision variable.

Problem Set 17

1. Find probabilities of optimal strategies for players A and B involved in a two person zero-sum game given that A's pay-off matrix is given in the table

2	3	4	3
3	4	5	2

   Also, write the value of the game for optimal solution.

2. Prove that the value of the game with a skew symmetric pay-off matrix is zero.

3. A and B play a game in which each has three coins: 5 paise, 10 paise and 20 paise. Each selects a coin with the knowledge of others choice. If the sum of coins is odd, A wins B's coin, and if the sum is even, B wins A's coin. Find the best strategy for each player and the value of the game.

   Suggestion. The pay-off matrix of A is

B

	5 cents	10 cents	20 cents
5 cents	-5	10	20
A 10 cents	5	-10	-10
20 cents	5	-20	-20

4. Players A and B take out one or two matches from a match box and guess how many opponent has taken. If one of the players guess exactly then the loser has to pay him as many rupees as the sum of the number of matches held by both players. Otherwise the payment is zero. Write the pay-off matrix and obtain the optimal strategies.

5. Two players P and Q play a game where each of them has to choose one of the three colours white (W), black (B) and red (R) independently of the other. Thereafter the colours are compared. If both P and Q have chosen white (W,W) neither wins any thing. If player P selects W and Q selects B, player P loses \$2. In this way all choices are considered and we get the following pay-off matrix:

Colour chosen by Q

		W	B	R
	W	0	-2	7
Colour chosen by P	B	2	5	6
	R	3	-3	8

Find the optimal strategies of both the players and also find the value of the game.

6. Find the range of values for p and q which will render the entry $(2,2)$ a saddle point

B

$$A \begin{bmatrix} 1 & q & 3 \\ p & 5 & 10 \\ 6 & 2 & 3 \end{bmatrix}$$

7. An investor is willing to invest \$100,000 among three possible schemes: Deposits, Bonds and Stocks. The rate on these investments are given in the following table

Actions↓	Nature of growth (in percentage)			
	Low growth	Medium growth	High growth	No growth
Deposit	8	8	8	8
Bonds	12	8	10	8
Stocks	-3	14	9	7

(a) What action or combination of actions are to be taken by the investor in order to have maximum rates of return?

(b) In case of Bonds, if the medium growth is 9% then how should the investor invest for optimal policy.

(c) If the subjective probability of nature of growth is estimated to be 0.2 for Low growth, 0.4 for Medium growth, 0.2 for High growth and 0.3 for No growth, then what should be the optimal policy.

8. The Australian cricket has three alternative lineups: A_1, A_2 and A_3. The West Indies team has also three alternative lineups: w_1, y_2 and y_3. The probability of an Australian team winning the different lineups of the West Indies team, which are approximately known from the previous experience, are given as follows:

	West Indies team lineups		
	W_1	W_2	W_3
A_1	0.2	0.8	0.6
Australian team lineups A_2	0.4	0.1	0.7
A_3	0.5	0.6	0.3

(a) Find with what the teams should use each of the lineups in matches against each other to score the largest number of victories

(b) In case (3, 2) cell of the table is changed to 0.7, then what would be the solution?

Chapter 18

Special Topics

This chapter contains various concepts on advanced level of the subject. The topics include a new technique to find initial BFS of transportation problem, generalized transportation problem, generalized assignment problem, and multiobjective transportation problem.

18.1 Extremum Difference Method

The method concerning initial BFS of a transportation problem has been conceived by Kasana and Kumar. This is simpler than VAM. The algorithm is based on the principle: if an allocation is not made in the lowest cost cell of a row or column having the largest extremum difference, then the cost penalty per unit cost will be higher for any other choices of rows or columns with other extreme differences. This causes increase in the objective function value.

The algorithm proceeds as follows:

Step 1. For each row calculate the difference of the lowest and the highest costs, and write these in the right-hand side in front of each row. These numbers are called the row penalties. Similarly, calculate the column penalties and write these in the bottom of the cost matrix below each column.

Step 2. Choose the row or column which has the largest penalty. For this row or column search the lowest cost cell and let it be (i, j)th cell. Then allocate $\min(a_i, \ b_j)$. Ignore the row and column for further consideration which has been satisfied.

Step 3. Now, calculate the row and column penalties for the remaining submatrix and allocate in the manner described in Step 2. Continue the process till all the rows and columns are satisfied.

Rules for ties. 1. In case of tie for the largest penalty among rows only, choose the lowest cost cell in the tied rows. Similarly for tied columns (no row has a tie with these columns), choose the lowest cost cell of columns.

2. Again, if there is a tie for the lowest cost in Rule 1, then consider the second level entries which are just next higher to the lowest cost of that row or column. Find the transportation cost for each second level entry with the assumption that allocation is made ignoring first level entries. Allocate in the smallest cost cell of tied rows and columns for which second level transportation cost turns out to be the highest.

3. This situation arises when there is tie for the largest cost among rows and columns. Look for the saddle points (entries at intersection of tied rows and columns). Make allocation at the smallest saddle point.

Consider the transportation problem of Section 7.1.

Table 1

5	2	4	3	30 20	3	2	1	1
	20	10					
6	4	9	5	40	6	6	6	–
		40					
2	3	8	1	55 40 20	7	6	1	1
15	20	20					

15	20	40	50
		20	10
5	2	5	4
–	2	5	4
–	–	5	4
–	–	4	4

From Table 1, we write

BFS: $x_{13} = 20$, $x_{14} = 10$, $x_{24} = 40$, $x_{31} = 15$, $x_{32} = 20$, $x_{33} = 20$;

Cost of transportation: 445.

Remark. 1. In case a tie exists for the highest second level transportation costs, we should proceed to third level entries and so on to decide the exact cell for allocation. When only one row or column remains unsatisfied in the last iteration, we use LCM for allocation.

18.2   Generalized Transportation Problem

A generalized transportation problem (GTP) is a LPP of specific structure. As in a transportation problem (TP), let S_i, $i = 1, 2, \ldots, m$ and D_j, $j = 1, 2, \ldots, n$ be m sources and n destinations, respectively.

We define the following quantities:

a_i = the quantity of material available at source S_i, $i = 1, 2, \ldots, m$.

b_j = the quantity of material required at destination D_j, $j = 1, 2, \ldots, n$.

c_{ij} = unit cost of transportation from source S_i to destination D_j.

The objective is to find how much material should be transported from each source S_i to each destination D_j so that the cost of transportation is minimized.

Let x_{ij} be the number of units of the material to be transported from source S_i to destination D_j. Then the generalized transportation problem is formulated as

$$\min \quad x_0 = \sum_{i=1}^{m} \sum_{j=1}^{n} c_{ij} x_{ij} \qquad (18.1)$$

$$\text{s.t.} \quad \sum_{j=1}^{n} d_{ij} x_{ij} \pm s_i = a_i, \quad i = 1, 2, \cdots, m \qquad (18.2)$$

$$\sum_{i=1}^{m} x_{ij} = b_j, \quad j = 1, 2, \cdots, n \qquad (18.3)$$

$$x_{ij} \geq 0, \ s_i \geq 0, \text{ and } d_{ij} \geq 0, \qquad (18.4)$$

where s_i is the slack or surplus variable.

Every generalized transportation problem can be represented by a matrix of order m by n, called the cost matrix or effectiveness matrix. For $m = 3$, $n = 4$, the structure of cost matrix is

	D_1	D_2	D_3	D_4	Slack/ Surplus	
S_1	d_{11} $\quad c_{11}$ $\quad\quad x_{11}$	d_{12} $\quad c_{12}$ $\quad\quad x_{12}$	d_{13} $\quad c_{13}$ $\quad\quad x_{13}$	d_{14} $\quad c_{14}$ $\quad\quad x_{14}$	± 1 $\quad\quad 0$ $\quad s_1$	a_1
S_2	d_{21} $\quad c_{21}$ $\quad\quad x_{21}$	d_{22} $\quad c_{22}$ $\quad\quad x_{22}$	d_{23} $\quad c_{23}$ $\quad\quad x_{23}$	d_{24} $\quad c_{24}$ $\quad\quad x_{24}$	± 1 $\quad\quad 0$ $\quad s_2$	a_2
S_3	d_{31} $\quad c_{31}$ $\quad\quad x_{31}$	d_{32} $\quad c_{32}$ $\quad\quad x_{32}$	d_{33} $\quad c_{33}$ $\quad\quad x_{33}$	d_{34} $\quad c_{34}$ $\quad\quad x_{34}$	± 1 $\quad\quad 0$ $\quad s_3$	a_3
	b_1	b_2	b_3	b_4		

Here S_1, S_2, S_3 are the sources, and D_1, D_2, D_3, D_4 are the destinations. The last but one column corresponds to the slack or surplus variables s_1, s_2, s_3. The entries a_i, b_j, d_{ij} and c_{ij} of the cost matrix are given. We have to determine x_{ij} such that the product $\sum c_{ij} x_{ij}$ is minimum with the restrictions that (18.2) and (18.3) must be satisfied.

Remarks. 1. The rank of coefficient matrix A is $m + n$ in GTP, while in TP it is $m + n - 1$.

2. In a GTP, x_{ij} may not be necessarily nonnegative integers while all a_i and b_j are nonnegative integers.

3. $\sum a_i = \sum b_j$ need not be true.

This problem differs from a transportation problem due to the presence of d_{ij}. These problems arise in many applications, for example, the problem of machine assignment, etc. The above problem can be solved by the simplex method. But the $u - v$ method by Charne's and Cooper after suitable modification works efficiently. In generalized transportation problem, we have

$$A_{ij} = d_{ij} e_i + e_{m+j},$$

where A_{ij} is the column of x_{ij} in the coefficient matrix A of (18.2) and

(18.3), while the column of s_i in A is $\pm e_i$ (unit n-dimensional vector having 1 at ith position)

The coordinates α_{ij}^m for the generalized problem are not necessarily 0, +1, −1. Consequently, x_{ij} may not be integers even when all a_i and b_j are integers.

The dual of the GTP (18.1) to (18.4) is written as

$$\max \quad y_0 = \sum_{i=1}^{m} a_i u_i + \sum_{j=1}^{n} b_j v_j \qquad (18.5)$$

$$\text{s.t.} \quad d_{ij} u_i + v_j \le c_{ij}, \quad i = 1, 2, \cdots, m \; j = 1, 2, \ldots, n \qquad (18.6)$$

$$\pm u_i \le 0, \quad i = 1, 2, \cdots, m \qquad (18.7)$$

$$u_i \text{ and } v_j \text{ are unrestricted}$$

where $+$ and $-$ sign in (18.7) stand for slack and surplus variables, respectively.

How to find BFS. The basic feasible solution of a generalized transportation problem can be found by the usual methods, viz., N-W corner rule, LCM, VAM, EDM.

While applying these methods the only difference is how to allocate in a cell. Suppose we have to allocate in $(i, j)$the cell. Then the maximum allocation in this cell is computed as

$$\min \left\{ \frac{a_i}{d_{ij}}, \; b_j \right\} = \alpha.$$

If $\alpha = b_j$, then the demand b_j is satisfied and the new a_i is $a_i - d_{ij}\alpha$, otherwise a_i is satisfied and the new b_j is $b_j - \alpha$.

Let us workout a problem which will make this discussion clear.

Example 1. Find the basic feasible solutions of the following generalized transportation problem using (a) N-W rule; (b) VAM.

(a) For finding the BFS by N-W corner rule, choose the x_{11} cell. The maximum possible allocation that can be made is

$$\min \left\{ \frac{80}{0.5}, \; 60 \right\} = 60.$$

Hence the demand is satisfied and the new a_1 becomes $80 - 0.5 \times 60 = 50$. Thus, the the first step is over. Now, ignore the first column and

.5	1	3	5	4	6	1	0	
								80
1	3	1	2	.4	3	1	0	
								60
3	2	.5	6	2	1	1	0	
								30

| 60 | 40 | 40 |

the north-west corner in the remaining cost matrix is x_{12} cell. Again, compute

$$\min\left\{\frac{50}{3},\ 40\right\} = \frac{50}{3}.$$

Thus, the supply at first row is $50 - 3 \times 50/3 = 0$, i.e., the first row is satisfied and new b_2 is $40 - 50/3 = 70/3$. Continue in this manner till all the demands are satisfies. The unsatisfied supplies are filled up in the corresponding slack variable column, see Table 2.

Table 2

.5	1	3	5	4	6	1	0		
60		50/3						8̶0̶ 5̶0̶	
1	3	1	2	.4	3	1	0		
		70/3		40		62/3		6̶0̶ 1̶1̶0̶/3 6̶2̶/3	
3	2	.5	6	2	1	1	0		
						30		3̶0̶	

6̶0̶	4̶0̶	4̶0̶
	7̶0̶/3	

BFS: $x_{11} = 60,\ x_{12} = 50/3,\ x_{22} = 70/3,\ x_{23} = 40,\ s_2 = 62/3,\ s_3 = 30$;

Transportation cost: $60 + \dfrac{250}{3} + \dfrac{140}{3} + 120 = 310$.

(b) Compute row and column penalties as in Chapter 7 and allocate as per rules of the generalized transportation problem. Note that the

Table 3

.5	1	3	5	4	6	1	0					
60						50		$\cancel{80}$	$\cancel{50}$	4	1	1
1	3	1	2	.4	3	1	0					
		40		25		10		$\cancel{60}$	$\cancel{20}$ $\cancel{10}$	1	1	1
3	2	.5	6	2	1	1	0					
				15				$\cancel{30}$		1	5	–

$\cancel{60}$	$\cancel{40}$	$\cancel{40}$
		$\cancel{25}$

1	3	2
–	3	2
–	3	3
–	–	–

slack column is adjusted in the last from the unsatisfied supplies, see Table 3.

BFS: $x_{11} = 60$, $x_{22} = 40$, $x_{23} = 25$, $x_{33} = 15$, $s_1 = 50$, $s_2 = 10$;

Transportation cost: 230.

Remarks. 1. If some of the b_j's remain unsatisfied in a GTP, then the problem has an infeasible solution.

2. Even if all b_j are satisfied and the ith row has surplus variable and a_i is satisfied (before the last iteration), then the GTP has also an infeasible solution.

3. There is no connection of $\sum a_i$ and $\sum b_j$ telling infeasibility.

4. The degenerate BFS may also exist as in a GTP.

Example 2. Solve the following GTP by VAM to find the basic feasible solution.

.5	1	3	5	4	6	−1	0	
								90
1	3	1	2	.4	3	−1	0	
								10
3	2	.5	6	2	1	−1	0	
								20

| 60 | 50 | 50 |

Its solution is given by the same techniques as in Example 1. The significant point is that when supply is exhausted and demands remain unsatisfied, the additional demand at destinations are are allocated to cells of the lowest cost in that column. This will increase the supplies. Hence the surplus column is adjusted in the last, see Table 4.

Table 4

.5	1	3	5	4	6	−1	0					
60		20						9̶0̶ 6̶0̶	4	1	1	1
1	3	1	2	.4	3	−1	0					
		10+20				20		1̶0̶	1	1	1	−
3	2	.5	6	2	1	−1	0					
				10+40		80		2̶0̶	1	5	−	−

6̶0̶	5̶0̶	5̶0̶
	4̶0̶	4̶0̶
	2̶0̶	

1	3	2
−	3	2
−	3	3
−	0	0

BFS: $x_{11} = 60$, $x_{12} = 20$, $x_{22} = 30$, $s_2 = 20$, $x_{33} = 50$, $s_3 = 80$

Transportation cost: 270

Note. If, in a surplus problem, all b_j's are satisfied and some a_j's remain unsatisfied, then the problem has infeasible solution. If, in

Example 2, we take $a_1 = 105$, $a_2 = 30$, $a_3 = 40$ and $b_1 = 60$, $b_2 = 35$, $b_3 = 35$, then it can be verified that the GTP becomes infeasible. Further, if in case the data 105 is replaced by 95, then the GTP becomes feasible.

18.3 Generalized Assignment Problem

Let us consider the problem associated with the assignment of individuals to jobs wherein more than one individual can be assigned a single job or, alternatively more than one job may be assigned to a single individual. Here we this problem is discussed by using EDM. The generalized assignment problem (GAP) is stated as

$$\min \quad x_0 = \sum_{i=1}^{m} \sum_{j=1}^{n} c_{ij} x_{ij}$$

$$\text{s.t.} \quad \sum_{j=1}^{n} x_{ij} = 1, \quad i = 1, 2, \ldots, m$$

$$\sum_{i=1}^{m} c_{ij} x_{ij} \le a_j, \quad j = 1, 2, \ldots, n,$$

where

$$x_{ij} = \begin{cases} 1 & \text{if } i\text{th job is assigned to worker } j \\ 0 & \text{otherwise} \end{cases}$$

c_{ij} = time required to perform job i by worker j

a_j = total time that worker j can be assigned

Algorithm. The algorithm for solving this problem is very much similar to EDM. The row penalties are determined by taking difference of the lowest and the highest entry of a row. Assignments are made one at a time to the job having the largest penalty provided such an assignment will not exceed the limit a_j, available to the worker j. Specifically, the steps are as follows. Establish the problem in matrix format as shown in Tab. 4. Now proceed as

Step 1. Compare the available time $\{a_j\}$ with the smallest time requirement $\{c_{ij}\}$ in columns j. If $\min_{i}\{c_{ij}\} > a_j$ strike out the column j. Repeat this for $j = 1, 2, \ldots, n$.

Step 2. Calculate the penalties and make an assignment to the minimum cost element in the row with the largest penalty, wherein such an assignment does not exceed the time a_j. In case a tie is observed for the largest penalty, then take maximum of all minimum cost entries for tied rows. Again if there is a tie for minimum cost, then choose cost element arbitrarily keeping in view that total time available a_j should not be surpassed.

Step 3. Strike out the row associated with the assignment of Step 2 and reduce the associated available time.

Step 4. Repeat steps 1 through 3 until all jobs are either assigned or all columns (workers) have been marked off.

Example 3. In the table, given below the data for an example problem that is used to illustrate the algorithm is summarized. In this problem, there are seven jobs and four workers. The time required to perform a job is dependent on both match of the job requirements to workers skills and the individual difficulty of each job. These times are given in the interior cost cells of the table. At the bottom of the table the maximum amount of time that each worker can devote his set of jobs (as assigned). The problem is how to assign seven jobs to four workers so as to minimize the total amount of time required for all jobs while taking care not to assign any worker to a set of jobs such that time requirement would exceed the time limits indicated. The solutions by conventional method and EDM are shown in Tables 5 and 6, respectively.

Workers Jobs	A	B	C	D
1	4	5	4	4
2	14	13	16	6
3	6	4	2	3
4	11	18	20	5
5	3	5	2	5
6	5	9	13	9
7	5	9	4	13
Available time (a_j)	14	17	13	7

Table 5

Workers Jobs	A	B	C	D
1	(3)	5	4	4
2	14	13	16	(6)
3	6	4	(2)	3
4	(11)	18	20	5
5	3	5	(2)	5
6	5	(9)	13	9
7	5	9	(4)	13
Available time (a_j)	14	17	13	7

Solution by Conventional Method

Total time required for all jobs = 37 units.

Table 6

Workers Jobs	A	B	C	D
1	(3)	5	4	4
2	14	(13)	16	6
3	6	4	(2)	3
4	11	18	20	(5)
5	3	5	(2)	5
6	(5)	9	13	9
7	5	9	(4)	13
Available time (a_j)	14	17	13	7

Solution by EDM

Total time required for all jobs = 34 units.

Remarks. 1. There are number of refinements possible with the GAP that are of interest. First, consider that we wish to maximize rather than minimize. We again suggest to come out of the conventional methods of converting the maximization problem into minimization problem by the methods available in literature. We directly deal with the maximization problem with suitable modification in Steps 2 and 3 of minimization case.

2. Again, the solution of GAP has established the fact that the EDM is more efficient than the conventional methods.

18.4   Multiobjective Transportation Problem

Here we introduce a new algorithm to solve p objective systems (p: any positive integer) for a transportation problem. The algorithm takes advantage of using optimal solutions of any objective function as the basic feasible solutions (BFSs) for the succeeding objective and reducing the set of solutions to most compromising solution to the combined objective function.

Formulation. Let us consider m origins and n destinations and also the quantities available at each origin and the quantities to be transported to each destination. The total quantities required at the destinations may differ from the total quantities available at the origins. For such situations, the problem is balanced by introducing fictitious origin or destination, whichever is needed in order to get precisely the same total quantities at the origins and the destinations. We consider specifically a balanced transportation problem as it amounts to no loss of generality.

Let x_{ij} be the quantity to be transported from origin i to destination j and for each fixed $k : k = 0, 1, \ldots, (p-1)$, α_{ij}^k, $i = 1, 2, \ldots, m$, $j = 1, 2, \ldots, n$, be the units of the parameter required for transporting one unit of the quantity from origin i to destination j. What is to be determined is the routing from the origin i to the destination j satisfying p objectives. The starting objective is termed as primary and the others are classified as secondary.

The primary objective is to minimize

$$Z_0 = \sum_{i=1}^{j} \sum_{j=1}^{n} \alpha_{ij}^o x_{ij} \tag{18.8}$$

and for $k = 1, 2, \ldots, (p-1)$, also to minimize

$$x_k = \max\{\alpha_{ij}^k : x_{ij} \geq 0, \ i = 1, 2, \ldots, m, \ j = 1, 2, \ldots, n\} \tag{18.9}$$

in order of the priorities to be assigned under the constraints

$$\sum_{j=1}^{n} x_{ij} = a_i, \quad i = 1, 2, \ldots, m \tag{18.10}$$

$$\sum_{i=1}^{m} x_{ij} = b_j, \quad j = 1, 2, \ldots, n. \tag{18.11}$$

The problem formulated on this pattern has p objective functions given by Eqs. (18.8) and (18.9).

For each arbitrarily chosen $k(k = 1, 2, \ldots, p-1)$, the set of parameters $\alpha_{ij}^k, i = 1, 2, \ldots, m, j = 1, 2, \ldots, n$ is decomposed into mutually disjoint subsets. The partitioning is done in such a way that the subsets L_{ω_k}, $\omega_k = 1, 2, \ldots, q(k); 1 \leq q(k) \leq mn$ contain all α_{ij}^k which have the same numerical value. Next, these subsets are arranged in descending order corresponding to these numerical values such that numerical values of α_{ij}^k in L_{s-1} be greater than numerical value of α_{ij}^k in L_s with $s = 2, 3, \ldots, q(k)$.

The final problem is to minimize

$$Z = \sum_{i=1}^{m}\sum_{j=1}^{n} \alpha_{ij}^o x_{ij} + M_1 \sum_{\omega_1=1}^{q(1)} (n_{\omega_1} \sum_{L_{\omega_1}} x_{ij}) + M_2 \sum_{\omega_2=1}^{q(2)} (n_{\omega_2} \sum_{L_{\omega_2}} x_{ij})$$

$$\ldots + M_{p-1} \sum_{\omega_{p-1}=1}^{q(p-1)} (n_{\omega_{p-1}} \sum_{L_{\omega_{p-1}}} x_{ij}), \tag{18.12}$$

where n_{ω_k} is internal priority factor for the kth objective with the condition that $n_1 \gg n_2 \gg \ldots \gg n_k$ and $M_1, M_2, \ldots, M_{p-1}$ are external priority factors assigned to the secondary objective functions such that

$$M_k \gg M_{k+1}, \quad k = 1, 2, \ldots, p-2. \tag{18.13}$$

The primary objective function is linear and the secondary objective functions are nonlinear functions which cause nonlinearity to the combined objective. Now we elaborate a very simple procedure to solve this multiobjective system.

Algorithm

The algorithm consists of the following steps:

Step 1. Determine the optimal solution of the primary objective function Z_0, using $u - v$ method, see Section 6.2, Chapter 6.

Step 2. If a unique solution for Z_0 exits, then go to Step 7; otherwise go to Step 3.

Step 3. Determine the set of all different alternative optimal solutions S_0 of Z_0.

Step 4. Construct a subset S_1 of S_0 so that each element of S_1 minimizes the first secondary objective function Z_1.

Step 5. Do the optimality test for each element of S_1 for Z_1. In case a single element of S_1 yields optimal for Z_1, then go to Step 7; otherwise go to Step 6.

Step 6. Repeat Step 4 and Step 5, using S_2, S_1 and Z_2 in place of S_1, S_0 and Z_0, respectively, and so on, till optimality is not disturbed. If optimality is disturbed, or all the secondary objective functions are optimized, then go to step 7.

Step 7. The current solution obtained is the solution of the combined objective problem.

Remarks 1. If optimality is not disturbed up to pth secondary objective, then from the construction of the sets $S_1, S_2, \ldots, S_{p-1}$, it is obvious that $S_{p-1} \subseteq \ldots \subseteq S_2 \subseteq S_1$. Hence the subset S_{p-1} is the optimal solution of combined objective problem (with the condition that optimality is not disturbed up to and including the last objective function Z_{p-1}).

2. In case optimality is disturbed at the intermediary stages, $S_t : 1 < t < p - 1$ (say), then elements in S_t are the optimal solutions for $Z_1, Z_2, \ldots, Z_{t-1}$ secondary objective functions and the most compromising solutions for the remaining $(p - t)$ secondary objective functions.

Example 4. Now the above algorithm is tested for $p = 2$. We consider the dead mileage system.[1] The dead mileage refers to the total

---

[1] Sharma, V. and Prakash, S. (1986), Optimizing dead mileage in urban bus routes, *Journal of Transportation Engineering*, ASCE, **112**(1), 121-129.

distances travelled by all buses in the morning from the garages to the starting point of their route. In the dead mileage system there is no earning, however the fuel is consumed. To make the system economical, the primary objective is to minimize the cumulative distance travelled by all buses from the garages to the starting points of their routes, and the secondary objective is to minimize the maximum distance among the distances travelled by individual buses from the garages to the starting point of their respective routes. The secondary objective obviously reduces the operation time for which garage supervisors are employed.

The equivalent balanced transportation problem is shown in Table 7. The cells in Table 7 correspond to the variables $x_{ij}(i = 1, 2, 3, j = 1, 2, \ldots, 8)$, where x_{ij} is the number of buses required from garage i to the starting point of their routes. Garages are denoted by g_1, g_2, g_3 and the starting points by $r_1, r_2, \ldots, r_8$. The north-west entries of all the (i, j) cells contain α_{ij}^o and α_{ij}^k. Further $a_1 = 35$, $a_2 = 120$, $a_3 = 100$ are the number of buses which can be parked over night in garages g_1, g_2, g_3, respectively and $b_1 = 40$, $b_2 = 25$, $b_3 = 15$, $b_4 = 30$, $b_5 = 35$, $b_6 = 45$, $b_7 = 20$ are the number of buses required at initial point of their routes $r_1, r_2, r_3, r_4, r_5, r_6, r_7$, respectively. Here, α_{ij}^o $(i = 1, 2, 3; j = 1, 2, \ldots, 8)$ is the unit of distance from garage i to the starting point of route j.

Table 7

Item	Starting point of								Maximum buses that can be parked overnight
	r_1	r_2	r_3	r_4	r_5	r_6	r_7	r_8	
g_1	15	12	3	10	2	18	4	0	35
g_2	7	18	2	12	3	6	16	0	120
g_3	15	5	18	12	4	10	15	0	100
Buses Required	40	25	15	30	35	45	20	45	

The two-objective functions of the equivalent balanced problem associated with the numerical problem are

$$Z_0 = \sum_{i=1}^{3} \sum_{j=1}^{8} \alpha_{ij}^o x_{ij},$$

$$Z_1 = \max\{\alpha_{ij}^k : x_{ij} > 0,\ i = 1, 2, 3,\ j = 1, 2, \ldots, 8\},$$

where $\alpha_{ij}^o = \alpha_{ij}^k \equiv d_{ij}$.

The combined objective function of the above problem is

$$Z = \sum_{i=1}^{3} \sum_{j=1}^{8} \alpha_{ij}^o x_{ij} + M_1 \sum_{\omega_1=1}^{q(1)} \left(n_{\omega_1} \sum_{L_{\omega_1}} x_{ij} \right).$$

Step 1. The optimal solution for the primary objective function Z_0 has been found out using stepping-stone method as shown in Table 8, and the optimal value of Z_o is 1235 units. The entries in the parentheses are basic cells giving optimal solution, and the entries without parentheses are the relative cost coefficients $(\alpha_{ij}^o - u_i - v_j)$. The values of the dual variables u_i and v_j, taking $u_2 = 0$ are as follows:

$u_1 = -1,\ u_2 = 0,\ u_3 = -1,$

$v_1 = 7,\ v_2 = 4,\ v_3 = 2,\ v_4 = 11,\ v_5 = 3,\ v_6 = 6,\ v_7 = 5,$
$v_8 = -1.$

Table 8

Item	\multicolumn{8}{c	}{Starting point of}	Max. buses parked overnight
	r_1	r_2	r_3	r_4	r_5	r_6	r_7	r_8	
g_1	9	9	2	(15)	0	13	(20)	2	35
g_2	(40)	14	(15)	1	(20)	(45)	11	1	120
g_3	7	(25)	15	(15)	(15)	3	9	(45)	100
Buses Req.	40	25	15	30	35	45	20	45	

Step 2. Since there is a nonbasic cell $(1, 4)$ with zero relative cost coefficient as shown in Table 8, a unique solution for Z_o does not exist. Go to Step 3.

Step 3. The set of all different alternative optimal BFSs S_o of Z_o has been determined by bringing the nonbasic cells having zero relative cost coefficients into the basic vector. This results in two alternative optimal BFSs besides the primary optimal BFS. Thus, three alternative optimal BFSs are as follows:

(a) $x_{14} = 15$, $x_{17} = 20$, $x_{21} = 40$, $x_{22} = 15$, $x_{25} = 20$, $x_{26} = 45$, $x_{32} = 25$, $x_{34} = 15$, $x_{35} = 15$, $x_{38} = 45$.

(b) $x_{15} = 15$, $x_{17} = 20$, $x_{21} = 40$, $x_{23} = 15$, $x_{25} = 20$, $x_{26} = 45$, $x_{32} = 25$, $x_{34} = 30$, $x_{35} = 0$, $x_{38} = 45$.

(c) $x_{14} = 0$, $x_{15} = 15$, $x_{17} = 20$, $x_{21} = 40$, $x_{23} = 15$, $x_{25} = 20$, $x_{26} = 45$, $x_{32} = 25$, $x_{34} = 30$, $x_{38} = 45$.

The set $S_0 = \{(a), (b), (c)\}$.

Step 4. The first secondary objective cost coefficients of (i, j) cells are shown in Table 9. The numeric values of the internal priority factors $n_{\omega_1}; \omega_1 = 1, 2, \ldots, 12$ are given by

$n_1, n_2, n_3, n_4, n_5, n_6, n_7, n_8, n_9, n_{10}, n_{11}, n_{12}$.

The values of the first secondary objective function Z_1 for the solutions (a), (b) and (c) are as follows:

For (a): $15 * n_5 + 20 * n_9 + 40 * n_6 + 15 * n_{11} + 20 * n_{10} + 45 * n_7 + 25 * n_8 + 15 * n_4 + 15 * n_9 + 45 * n_{12}$.

For (b): $15 * n_{11} + 20 * n_9 + 40 * n_6 + 15 * n_{11} + 20 * n_{10} + 45 * n_7 + 25 * n_8 + 30 * n_4 + 0 * n_9 + 45 * n_{12}$.

For (c): $0 * n_5 + 15 * n_{11} + 20 * n_9 + 40 * n_6 + 15 * n_{11} + 20 * n_{10} + 45 * n_7 + 25 * n_8 + 30 * n_4 + 45 * n_{12}$.

The minimum of the first secondary objective function Z_1 occurs for (a); hence by our construction the subset $S_1 = \{(a)\}$.

Step 5. Table 9 shows that optimality is not disturbed for the function Z_1 when we used S_1 as the basic feasible solution. The $+$ entries in parentheses depict the value of $(d_{ij} - u_i - v_j)$. The values of the dual variables u_i and v_j, taking $u_2=0$ are as follows:

$u_1 = n_5 + n_8 - n_4 - n_{10}$, $u_2 = 0$, $u_3 = n_9 - n_{10}$, $v_1 = n_6$, $v_2 = n_8 + n_{10} - n_9$, $v_3 = n_{11}$, $v_4 = n_4 + n_{10} - n_8$, $v_5 = n_{10}$, $v_6 = n_7$, $v_7 = n_9 + n_4 + n_{10} - n_5 - n_8$, $v_8 = n_{10} + n_{12} - n_9$.

The relative cost coefficients of the nonbasic cells can be found as

$(1,1)$ cell: $n_3 + n_4 + n_{10} - n_5 - n_6 - n_8$;
$(1,2)$ cell: $2*n_4 + n_9 - n_5 - 2*n_8$;
$(1,3)$ cell: $2*n_{10} + n_4 - n_5 - n_8 - n_{11}$;
$(1,5)$ cell: $n_4 + n_{11} - n_5 - n_8$;
$(1,6)$ cell: $n_1 + n_4 + n_{10} - n_5 - n_7 - n_8$;
$(1,8)$ cell: $n_4 + n_9 - n_5 - n_8 - n_{12}$;
$(2,2)$ cell: $n_1 + n_9 - n_8 - n_{10}$;
$(2,4)$ cell: $n_8 - n_{10}$;
$(2,7)$ cell: $n_2 + n_5 + n_8 - n_4 - n_9 - n_{10}$;
$(2,8)$ cell: $n_9 - n_{10}$;
$(3,1)$ cell: $n_3 + n_{10} - n_6 - n_9$;
$(3,3)$ cell: $n_1 + n_{10} - n_9 - n_{11}$;
$(3,6)$ cell: $n_5 + n_{10} - n_7 - n_9$;
$(3,7)$ cell: $n_3 + n_5 + n_8 - n_4$.

Table 9

Item	Starting point of								Max. buses parked overnight
	r_1	r_2	r_3	r_4	r_5	r_6	r_7	r_8	
g_1	n_3 (+)	n_4 (+)	n_{10} (+)	n_5 15	n_{11} (+)	n_1 (+)	n_9 20	n_{12} (+)	35
g_2	n_6 40	n_1 (+)	n_{11} 15	n_4 (+)	n_{10} 20	n_7 45	n_2 (+)	n_{12} (+)	120
g_3	n_3 (+)	n_8 25	n_1 (+)	n_4 15	n_9 15	n_5 (+)	n_3 (+)	n_{12} 45	100
Buses Required	40	25	15	30	35	45	20	45	

It is observed that optimality for primary optimal basic feasible solutions (b) and (c) is disturbed with respect to secondary objective function. In particular, for (b), the relative cost coefficients of cell $(1,4)$ turn out to be $n_5 + n_9 - n_4 - n_{11}$ and this is certainly negative in view of the assumption $n_1 \gg n_2 \gg \cdots \gg n_{12}$. Similarly, it can be verified that optimality of the secondary objective function is disturbed for the solution (c).

Step 6. Since there are no other secondary objectives left to be optimized, go to Step 7.

Step 7. Thus, in view of the proposed algorithm, the solution (a) happens to be the optimal solution of the combined problem with $Z_o = 1235$ units and $Z_1 = 12$ units.

Remarks. 1. The algorithm proposed here is simple and efficient in solving transportation problems with more than two objectives, because it accounts for additional secondary objectives in partitioning and ordering the coefficients before solving.

2. The algorithm can be applied to unbalanced transportation problems as well. Such problems can be balanced by introducing fictitious row or column whichever is desired.

Problem Set 18

1. Solve the following multiobjective transportation problem $(p = 4)$ described as

Availability: $a_1 = 30$, $a_2 = 40$, $a_3 = 30$,

Requirement: $b_1 = 20$, $b_2 = 20$, $b_3 = 25$, $b_4 = 30$, $b_5 = 5$.

Primary Objective Cost Coefficients:

$\alpha_{11}^o = 57$, $\alpha_{12}^o = 60$, $\alpha_{13}^o = 64$, $\alpha_{14}^o = 58$, $\alpha_{15}^o = 55$,

$\alpha_{21}^o = 58$, $\alpha_{22}^o = 62$, $\alpha_{23}^o = 63$, $\alpha_{24}^o = 57$, $\alpha_{25}^o = 57$,

$\alpha_{31}^o = 57$, $\alpha_{32}^o = 60$, $\alpha_{33}^o = 67$, $\alpha_{34}^o = 61$, $\alpha_{35}^o = 56$.

First Secondary Objective Cost Coefficient:

$\alpha_{11}^1 = 7(n_8)$, $\alpha_{12}^1 = 10(n_5)$, $\alpha_{13}^1 = 14(n_2)$, $\alpha_{14}^1 = 14(n_2)$, $\alpha_{15}^1 = 0(n_{10})$;

$\alpha_{21}^1 = 7(n_8)$, $\alpha_{22}^1 = 11(n_4)$, $\alpha_{23}^1 = 12(n_3)$, $\alpha_{24}^1 = 12(n_3)$, $\alpha_{25}^1 = 0(n_{10})$;

$\alpha_{31}^1 = 5(n_9)$, $\alpha_{32}^1 = 8(n_7)$, $\alpha_{33}^1 = 15(n_1)$, $\alpha_{34}^1 = 14(n_2)$, $\alpha_{35}^1 = 0(n_{10})$.

Second Secondary Objective Cost Coefficient:

$\alpha_{11}^2 = 15(n_5)$, $\alpha_{12}^2 = 20(n_3)$, $\alpha_{13}^2 = 15(n_5)$, $\alpha_{14}^2 = 10(n_7)$,
$\alpha_{15}^2 = 10(n_7)$;

$\alpha_{21}^2 = 21(n_2)$, $\alpha_{22}^2 = 26(n_1)$, $\alpha_{23}^2 = 20(n_3)$, $\alpha_{24}^2 = 5(n_8)$,
$\alpha_{25}^2 = 20(n_3)$;

$\alpha_{31}^2 = 10(n_7)$, $\alpha_{32}^2 = 15(n_5)$, $\alpha_{33}^2 = 12(n_6)$, $\alpha_{34}^2 = 5(n_8)$,
$\alpha_{35}^2 = 10(n_7)$.

Third Secondary Objective Cost Coefficient:

$\alpha_{11}^3 = 21(n_3)$, $\alpha_{12}^3 = 25(n_2)$, $\alpha_{13}^3 = 10(n_8)$, $\alpha_{14}^3 = 10(n_8)$,
$\alpha_{15}^3 = 15(n_6)$;

$\alpha_{21}^3 = 25(n_2)$, $\alpha_{22}^3 = 26(n_1)$, $\alpha_{23}^3 = 10(n_8)$, $\alpha_{24}^3 = 7(n_9)$,
$\alpha_{25}^3 = 20(n_4)$;

$\alpha_{31}^3 = 5(n_{10})$, $\alpha_{32}^3 = 20(n_5)$, $\alpha_{33}^3 = 7(n_9)$, $\alpha_{34}^3 = 5(n_{10})$,
$\alpha_{35}^3 = 12(n_7)$.

The internal priority factors are shown above within the parentheses.

Appendix:
Objective Type Questions

1. Let $S_1 = \{(x_1, x_2) : 2x_1 + 3x_2 = 5\}$, $S_2 = \{(1, 1)\}$ be two subsets of $\mathbb{R}^2$. Then $S_1 \cap S_2$ is

   (a) a convex set   (b) not a convex set

2. Let S_1 and S_2 be two convex subsets of $\mathbb{R}^n$. If S_1' and S_2' represent the compliments of S_1 and S_2, respectively. Then

   (a) $S_1 + S_2$ (b) $S_1 \cup S_2$ (c) $S_1' \cap S_2'$ (d) $S_1' \cup S_2$

   is always a convex set

3. Let S_1 and S_2 be two convex subsets of $\mathbb{R}^n$. If S_1' and S_2' represent the compliments of S_1 and S_2, respectively. Then

   (a) $S_1 - S_2$ (b) $S_1' \cup S_2'$ (c) $S_1' \cap S_2$ (d) $S_1 \cap S_2'$

   is always a convex set

4. Consider the set $S = \{(x_1, x_2) : x_2^2 \le x_1\}$. Then S has

   (a) no vertex   (b) finite number of vertices   (c) infinite number of vertices

5. The number of extreme point(s) that a hyper-plane has

   (a) infinite   (b) finite   (c) none of these

6. The set $S = \{(x_1, x_2) : x_1 + x_2 = 1\}$ has no vertex because it is

   (a) not convex   (b) not bounded
   (c) not closed   (d) none of these

7. Consider the unit simplex $S = \{(x_1, x_2, x_3) : x_1 + x_2 + x_3 = 1, \ x_1, x_2, x_3 \ge 0\}$. Then number of vertices S has

   (a) 2   (b) 4   (c) 5   (d) none of these

8. Let $S = \{X \in \mathbb{R}^2 : |X| < 1\}$. Then S has no vertex because it is

    (a) not closed   (b) empty

    (c) unbounded   (d) not convex.

9. Let $S = \{(x_1, x_2) : x_1^2 + x_2^2 \leq 1\}$. Then S has

    (a) no vertex   (b) finite number of vertices   (c) infinite number of vertices

10. Consider the set $S = \{(x_1, x_2) : x_1 + x_2 \geq -1, \ x_1 \leq 0, \ x_2 \leq 1\}$. Then S has

    (a) no vertex   (b) infinite number of vertices

    (c) only two vertices   (d) none of these

11 The vertex of the set $S = \{X : X = (1 - \alpha)X_1 + \alpha X_2, \ \alpha \geq 0, \ X_1, X_2 \in \mathbb{R}^2\}$ is

    $\ldots\ldots$

12. The set $P_F \setminus A$, where A is the set of all vertices of P_F is

    (a) convex set   (b) not a convex set

    (c) may or may not be convex   (d) none of these

13. The system of equations

$$x_1 - x_2 + x_3 = 4$$

$$2x_1 + x_2 - 5x_3 = 3$$

is equivalent to the following system with inequalities

    (a) $x_1 - x_2 + x_3 \leq 4, \ 2x_1 + x_2 - 5x_3 \leq 3, \ -x_1 + 2x_2 + 6x_3 \geq 7$

    (b) $x_1 - x_2 + x_3 \leq 4, \ 2x_1 + x_2 - 5x_3 \leq 3, \ -x_1 + 2x_2 + 6x_3 \leq 1$

    (c) $x_1 - x_2 + x_3 \leq 4, \ 2x_1 + x_2 - 5x_3 \leq 3, \ 2x_1 - 4x_3 \leq 1$

    (d) $x_1 - x_2 + x_3 \leq 4, \ 2x_1 + x_2 - 5x_3 \leq 3, \ 3x_1 - 4x_3 \geq 7$

14. The vertex of the set $S = \{X : X = (1 - \lambda)X_1 + \lambda X_2, \ 0 < \lambda \leq 1, \ X_1, X_2 \in \mathbb{R}^2\}$ is

    $\ldots\ldots$

15. For $x_1, x_2 \geq 0$, consider the system

$$x_1 + x_2 - x_3 - 2x_4 + 5x_5 = 2$$
$$x_2 + x_3 + 5x_4 + 5x_5 = 2$$

Its solution $x_1 = x_3 = x_4 = 0$, $x_2 = 7, x_5 = -1$ is

(a) a basic solution (b) a basic feasible solution (c) not a basic solution (d) feasible solution

16. Let the optimal of a LP occur at vertices X_1 and X_2. Then we know that it also occurs at each

$$X = (1 - \alpha)X_1 + \alpha X_2, \quad 0 < \alpha < 1$$

(a) X is a basic solution (b) X is not a BFS (c) X is not a basic solution (d) none of these

17. For $x_1, x_2 \geq 0$, consider the system

$$x_1 + 2x_2 - x_3 - 2x_4 - 3x_5 = -1$$
$$2x_2 + x_3 + 5x_4 - 3x_5 = -1$$

Its solution $x_1 = 0$, $x_2 = 1$, $x_3 = 0$, $x_4 = 0$, $x_5 = 1$ is

(a) a basic solution   (b) a basic feasible solution

(c) feasible solution   (d) none of these

18. In a simplex table, there is a tie for the leaving variable, then the next BFS

(a) will be nondegenerate (b) will be degenerate

(c) may be degenerate   (d) does not exist

     or nondegenerate

19. Two vertices of P_F are $(x_1, x_2, x_3, x_4) = (0, 0, 1, 2)$ and $(3, 0, 0, 1)$. Then a point of P_F which can not be the vertices

(a) $(1, 2, 0, 0)$ (b) $(0, 1, 3, 0)$ (c) $(0, 1, 2, 0)$ (d) $(1, 2, 3, 0)$

20. A LPP in standard form has m constraints and n variables. The number of basic feasible solutions will be

(a) $\binom{n}{m}$ (b) $\leq \binom{n}{m}$ (c) $\geq \binom{n}{m}$ (d) none of these

21. A LPP in standard form has m constraints and n variables. Then number of adjacent vertices corresponding to a vertex are

    (a) $n - m$ (b) $\leq n - m$ (c) $n!/m!(n-m)!$ (d) none of these

22. In Problem 19, if $m = 5$ and $n = 8$, and X is basic feasible solution with 3 components at positive level. Then, the number of bases which correspond to X due to degeneracy are

    (a) 5   (b) 10   (c) 15   (d) 20

23. In an LPP, let $p =$ number of vertices and $q =$ number of BFS. Then

    (a) $p \leq q$ (b) $p = q$ (c) $p \geq q$ (d) none of these

24. In a simplex iteration, if the leaving variable rule is violated, then the next table will

    (a) not give basic solution   (b) give a basic solution

    which is not feasible

    (c) give a nonbasic solution   (d) nothing can be said

25. For max $p_1 = -3x_1 + x_2$, subject to $3x_1 - x_2 \leq 6$; $x_2 \leq 3$; $x_1, x_2 \geq 0$, the optimal table is

B V	x_1	x_2	s_1	s_2	Soln
p_1	3	0	1	0	3
s_1	0	1	1	0	9
x_2	3	0	1	1	3

    If max $p_2 = x_1 - x_2$, then optimal solution $(x_1, x_2) = (0,3)$ remains optimal for the weighted LP: $p = $ max $\alpha_1 p_1 + \alpha_2 p_2$, $0 \leq \alpha_1 \leq \alpha_2$, then α_2 is

    (a) 1/2   (b) 3/4   (c) 1   (d) none of these

26. If in any simplex iteration the minimum ratio rule fails, then the LPP has

    (a) nondegenerate BFS   (b) degenerate BFS

    (c) unbounded solution   (d) infeasible solution

27. In a max LPP with bounded solution space, a variable having positive relative cost is permitted to enter and the minimum ratio rule is properly followed, then

(a) the next solution will not be BFS

(b) the objective function value decreases

(c) the objective function increases

(d) none of these

28. If x_j is a basic variable in some simplex table, then relative cost of x_j is

(a) positive   (b) negative   (c) infinite   (d) 0

29. In some simplex table of a maximization LPP, the column of x_j is $(3; -2, -1, -3)^T$. Then this shows that

(a) P_F is bounded   (b) solution is unbounded

(c) P_F is unbounded in x_j direction   (d) solution is infeasible

30. In phase-I of the two phase method an artificial variable turns out to be at positive level in the optimal table of Phase-I, then the LPP has

(a) no feasible solution

(b) unbounded solution

(c) optimal solution

(d) none of these

31. In a maximization problem, a basic variable corresponding to minimum ratio leaves the basis, this ensures

(a) largest increase in objective function

(b) the next solution will be a BFS

(c) decrease in objective function

(d) none of these

32. In a maximization problem, a nonbasic variable with most negative relative cost enters the basis ensures

(a) largest increase in   (b) the next solution will be

objective function   a BFS

(c) decrease in objective (d) none of these

function

33. Let $B = (A_1, A_3, A_5)$ be a basis for a LPP such that $A_4 = \alpha A_1 + \beta A_2 + \gamma A_3$. Suppose any one column of B is replaced by A_4 to have a new basis. Then

(a) $\alpha, \beta, \gamma > 0$ (b) $\alpha, \beta, \gamma \leq 0$ (c) $\alpha, \beta, \gamma \neq 0$ (d)
no such relationship required

34. The optimum of a LPP occurs at $X = (1, 0, 0, 2)$ and $Y = (0, 1, 0, 3)$. Then optimum also occurs at

(a) $(2, 0, 3, 0)$ (b) $(1/2, 1/2, 0, 5/2)$
(c) $(0, 1, 5, 0)$ (d) none of these

35. If in a simplex table the relative cost $z_j - c_j$ is zero for a non-basic variable, then there exists an alternate optimal solution, provided

(a) it is starting simplex table   (b) it is optimal simplex table

(c) it can be any simplex table   (d) none of these

36. A LPP amenable to solution by simplex method has third and fourth constraint as $x_1 + x_2 + x_3 \leq 3$ and $2x_1 + x_2 + 3x_8 \leq 8$. These constraints can be represented by a single constraint

(a) $3x_1 + 2x_2 + 4x_3 \leq 11$ (b) $x_1 + 2x_3 \leq 5$ (c) $3x_1 + x_2 + 3x_3 \leq 11$
(d) none of these

37. In canonical form of a LPP, the availability vector b

(a) is restricted to ≥ 0 (b) is restricted to ≤ 0 (c) any component
may be ≤ 0 or ≥ 0

38. Suppose, in some simplex iteration x_j enters the basis. Then, at later stage in some simplex iteration

(a) x_j can leave the basis (b) x_j can not leave the basis (c) both
(a) and (b) are possible

Suggestion. If the rule $\theta_j(z_j - c_j)$ is not followed then answer is
(c), otherwise it is (b).

39. Suppose, in some simplex iteration x_j leaves the basis. Then, in just next iteration

(a) x_j can enter the basis (b) x_j can not enter the basis (c) both (a) and (b) are possible

40. By inspecting the dual of the following LPP find the optimal value of its objective function

$$\text{max} \quad z = 2x_1 + x_2 + 3x_3$$
$$\text{s.t.} \quad x_1 - x_2 + x_3 \geq 5$$
$$x_1, x_2, x_3 \geq 0$$

41. The following LPP has

$$\text{min} \quad x_0 = -2x_1 + 10x_2$$
$$\text{s.t.} \quad x_1 - x_2 \geq 0$$
$$- x_1 + 5x_2 \geq 5$$
$$x_1, x_2 \geq 0$$

(a) alternative solution (b) unique solution (c) unbounded solution (d) none of these

Suggestion. This is an interesting problem in which the LPP has alternate optimal solution. Every point on the line $-x_1 + 5x_2 = 5$ gives optimal solution with optimal value 10. From the optimal table, it is not possible to find alternate optimal solution, since the solution space is unbounded. Except $x_1 = 5/4$, $x_2 = 5/4$, all other optimal solutions are nonbasic.

42. Let $\min f(X) = C^T X$, $AX \geq b$, $X \geq 0$ be a primal LPP. Suppose X_0 and Y_0 are the primal and dual feasible. Then

(a) $C^T X_0 \leq b^T Y_0$ (b) $C^T X_0 \geq b^T Y_0$ (c) $C^T X_0 = b^T Y_0$ (d) none of these

43. Let $\max f(X) = C^T X$, $AX \leq b$, $X \geq 0$ be a primal LPP. Suppose X_0 and Y_0 are the primal and dual feasible. Then

(a) $C^T X_0 \leq b^T Y_0$ (b) $C^T X_0 \geq b^T Y_0$ (c) $C^T X_0 = b^T Y_0$ (d) none of these

44. The dual simplex method is applicable provided

(a) optimality remains satisfied   (b) feasibility remains satisfied

(c) both remain unsatisfied   (d) optimality is satisfied but feasibility is disturbed

45. Application of dual simplex method requires that availability vector b must satisfy

(a) $b \geq 0$ (b) $b \leq 0$ (c) no restriction of (a) and (b) type

46. In dual simplex table x_j is the only variable with negative value in solution column, but all other entries in x_j-row are ≥ 0. Then LPP has

(a) unbounded solution   (b) infeasible solution

(c) alternate optimal solution   (d) none of these

47. If the primal has degenerate optimal solution, the dual has

(a) alternate optimal solution   (b) degenerate optimal solution
(c) no feasible solution

48. If a variable x_j is unrestricted in sign in a primal LPP, then the corresponding dual jth constraint in the dual will be

(a) $\leq$ (b) $\geq$ (c) equality constraint   (d) none of these

49. If the jth constraint in the primal is an equality, then the corresponding dual variable is

(a) unrestricted in sign   (b) restricted to ≥ 0 (c) restricted to ≤ 0

50. The primal LPP is

$$\begin{aligned}
\max \quad & x_0 = x_1 - 2x_2 \\
\text{s.t.} \quad & x_1 - 3x_2 \leq -3 \\
& -x_1 + 2x_2 = -2 \\
& x_1, x_2 \geq 0
\end{aligned}$$

The $C_B^T B^{-1}$ in optimal table of above LPP is $(0,1)^T$. Then the the optimal solution of the dual is

(a) $(0,1)^T$ (b) $(0,-1)^T$ (c) $(0,2)^T$ (d) none of these

51. Consider the LPP

$$\begin{aligned} \max \quad & z = x_1 + 5x_2 + 3x_3 \\ \text{s.t.} \quad & x_1 + 2x_2 + x_3 = 3 \\ & 2x_1 - x_2 = 4 \\ & x_1, x_2, x_3 \geq 0 \end{aligned}$$

Given that in optimal table of this problem x_1 and x_3 are basic variables, then the optimal solution of the dual problem is

$$\ldots\ldots\ldots$$

Suggestion. First and third dual constraints are satisfied as equality constraints.

52. Consider the LPP

$$\begin{aligned} \max \quad & z = x_1 + 5x_2 + 2x_3 \\ \text{s.t.} \quad & x_1 + 2x_2 + x_3 = 15 \\ & 2x_1 - x_2 = 10 \\ & x_1, x_2, x_3 \geq 0 \end{aligned}$$

Given that in optimal table of this problem x_1 is a basic variable and x_3 is a nonbasic variable with relative cost $1/5$, then the optimal solution of the dual problem is

$$\ldots\ldots\ldots$$

53. Let primal be a min LP and let a feasible solution which is not optimal of primal causes objective function value to 25. Then which of the following can be the value of dual objective function

(a) 25   (b) 24.5   (c) 26   (d) none of these

54. If a slack or surplus variable s_i is positive in optimal BFS of primal, then in optimal dual solution

(a) the dual variable y_i is 0   (b) the dual variable $y_i > 0$

(c) the slack or surplus of   (d) none of these

     ith dual constraint is 0

55. If the primal LPP has an unbounded solution, then the dual problem has

(a) an optimal solution (b) infeasible solution (c) an unbounded solution (d) none of these

56. If the dual LPP has an unbounded solution, then the primal problem has

(a) optimal solution (b) infeasible solution (c) unbounded solution (d) none of these

57. A primal LPP has nondegenerate optimal solution, then the optimal solution of the dual

(a) is nondegenerate   (b) is degenerate

(c) may be nondegenerate or degenerate   (d) none of these

58. If the primal LPP has infeasible solution, then the solution of the dual problem is

(a) unbounded (b) infeasible (c) either unbounded or infeasible (d) none of these

59. If in dual simplex method, the rule for entering variable is not followed, then

(a) the feasibility will further   (b) the optimality will be

deteriorate   disturbed

(c) there will be no change   (d) none of these

60. Consider the LPP:

$$\text{max} \quad z = 5x_1 + 2x_2$$
$$\text{s.t.} \quad x_1 + x_2 \leq 3$$
$$2x_1 + 3x_2 \geq 5$$
$$x_1, x_2 \geq 0$$

Which of the following primal-dual solutions are optimal:

(a) $x_1 = 3, x_2 = 1$;   (b) $x_1 = 4, x_2 = 1$;

$y_1 = 4, y_2 = 1$ $y_1 = 1, y_2 = 0$

(c) $x_1 = 3, x_2 = 0$;   (d) $x_1 = 2, x_2 = 5$;

$y_1 = 5, y_2 = 0$ $y_1 = 1, y_2 = 5$

61 In dual simplex method, let the variable x_i leave the basis and the variable x_j enter. Let $x_j > 0$. Then later on

(a) x_j can become negative (b) x_j will remain positive (c) none of these

62. For every maximization LPP with m equality constraints and n variables $(m < n)$, the number of unrestricted dual variables will always be

(a) $\leq m$ (b) m (c) $\leq n$ (d) n

Suggestion. Consult the LPP max $x_1 + 5x_2 + 3x_3$, subject to $x_1 + 2x_2 + x_3 = 3$, $2x_1 - x_2 = 4$, $x_1, x_2, x_3 \geq 0$.

63. Suppose primal is infeasible and dual is feasible. Then dual will have

(a) unbounded solution (b) finite solution (c) alternate optimal solution (d) none of these

64. The jth constraint in dual of a LPP is satisfied as strict inequality by the optimal solution. Then the jth variable of the primal will assume a value

(a) $\neq 0$ (b) ≤ 0 (c) ≥ 0 (d) 0

65. Let P_F be the feasible set of a LPP which is bounded and nonempty. If a constraint is deleted then the feasible set of the new LPP

(a) may be unbounded (b) may be empty (c) will always be bounded (d) none of these

66. A LPP is given in canonical form with $b \geq 0$, and its optimal table is

B. V.	x_1	x_2	s_1	s_2	Soln
x_0	0	0	2	1	34
x_2	1	0	2	-3	2
x_1	0	1	-3	5	1

Then, for $b' = (1, 1)^T$, $C_B^T B^{-1} b'$ is

. . . .

67. Consider the optimal table of Problem 65. The objective function of the LPP is

. . . .

68. Consider the optimal table of Problem 65. The right hand vector of the LPP is

. . . .

69. Consider the optimal table of Problem 65. The column A_1 associated with x_1 in constraint matrix of the LPP is

. . . .

Suggestion. For Problems 67 to 69, compute B from B^{-1}.

70. The optimal table of a LPP in which s_1 is a surplus variable and s_2 is a slack variable in standard form of the LPP is

B V	x_1	x_2	s_1	s_2	Soln
z	0	0	-1	-2	2
x_1	1	0	-3	-2	1
x_2	0	1	1	1	1

The right hand side vector is assigned $b' = (4,5)^T$. Then the new optimal solution is

. . . .

71. If the variable x_2 is deleted from the LPP whose optimal table is in the preceding problem, then the optimal solution of the changed problem is

. . . .

72. Following is the optimal table of a LPP (s_1, s_2, $s_3 \geq 0$ are the slack variables when LPP is written in standard form)

B V	x_1	x_2	s_1	s_2	s_3	Soln
x_0	0	0	7/6	13/6	0	218/5
s_3	0	0	3/2	$-25/2$	1	5
x_1	1	0	1/3	$-2/3$	0	16/5
x_2	0	1	$-1/6$	5/6	0	10/3

If the third constraint is deleted, then optimal solution of the revised LPP

(a) $x_1 = 14/5$, $x_2 = 11/3$ (b) $x_1 = 3$, $x_2 = 4$
(c) $x_1 = 16/5$, $x_2 = 10/3$ (d) none of these

73. Change in all the coefficients of a particular variable in a LPP

(a) disturbs feasibility   (b) disturbs optimality

(c) may disturb both feasibility (d) none of these
    and optimality

74. Change in some column of the coefficients of a LPP

(a) disturbs feasibility   (b) disturbs optimality

(c) may disturb both feasibility (d) none of these
    and optimality

75. Addition of variable and deletion of a constraint simultaneously to a LPP

(a) disturbs feasibility   (b) disturbs optimality

(c) may disturb both feasibility (d) none of these
    and optimality

76. In a max LPP, if a constraint is added then the objective function value

(a) will decrease   (b) will decrease or remains same

(c) will increase   (d) nothing can be said

77. In a LPP the costs are changed and simultaneously a constraint is deleted, then in the optimal table

(a) only feasibility may be   (b) only optimality is
    disturbed   disturbed

(c) both feasibility and optimality (d) nothing can be said
    may be disturbed

78. Write true and false for each of the following statements

(a) Dual of dual is primal

(b) To solve a LP problem with some constraints of the type $\leq$ and some of the type $\geq$, it is must to use big M-method

(c) In standard form of a LPP all constraints must be of the type $\leq$

(d) A LPP with two constraints and three variables can be solved by the graphical method

(e) A LPP may have two optimal solutions one nondegenerate and other one degenerate

(f) In optimal table of a LPP, the relative cost for a nonbasic variable is the indication of alternate optimal solution. It is always possible to find alternate optimal basic feasible solution by permitting to enter this nonbasic variable into the basis

(g) In general, the dual of a LPP with m equality constraints contains m unrestricted variables. It is possible to have a dual which has less than m unrestricted variables

(h) The number of vertices of any closed bounded set can not be infinite

(i) A LPP has 7 variables and 5 constraints. It is possible to find the optimal solution of this LPP by the graphical method

(j) It is possible to construct examples in which primal and dual are both unbounded

(k) A LPP has an optimal solution. It is possible to get unbounded solution by changing the right hand side vector arbitrarily

(l) If a linear program is infeasible, then its dual must be unbounded.

(m) If a minimization LPP problem has a feasible vector, then its dual can never go to unbounded maximum value.

Suggestion. For (f), see Problem 42.

79. In a balanced transportation problem with m sources and n destinations, the number of basic variables is

(a) $m + n$ (b) $m + n + 1$ (c) $m + n - 1$ (d) none of these

80. If some constant is added to each cost c_{ij} of a row or column in a transportation matrix, then the the optimal value
    (a) decreases   (b) increases   (c) may increase or decrease
    (d) remains same

81. In a balanced transportation problem with 3 sources and 3 destinations, the number of basic feasible solutions possible are
    (a) 100   (b) 120   (c) 124   (d) 126

82. In a balanced transportation model, if u_i's and v_j's are the dual variables associated with rows and columns, then
    (a) $u_i \geq 0$ (b) $v_j \geq 0$ (c) $u_i, v_j \geq 0$ (d) u_i, v_j unrestricted

83. In a balanced transportation problem with m sources and n destinations, the number of dual constraints will be
    (a) $m + n$ (b) $m + n + 1$ (c) $m + n - 1$ (d) mn

84. In a balanced transportation problem with m sources and n destinations, the number of nonbasic basic variables will be
    (a) mn (b) $(m-1)(n-1)$ (c) $m(n+1)$ (d) $(n+1)m$

85. In a balanced transportation problem with m sources and n destinations, the number of dual variables will be
    (a) $m + n$ (b) $m + n + 1$ (c) $m + n - 1$ (d) none of these

86. In a transportation problem, one dual variable can be assigned arbitrary
    (a) any real value   (b) only zero value
    (c) only values ≥ 0 (d) only values ≤ 0

87. For a balanced transportation problem with 2 sources and 3 destinations, the optimal solutions are $x_{11} = 10$, $x_{13} = 8$, $x_{22} = 4$, $x_{23} = 4$ and $x_{11} = 6$, $x_{13} = 12$, $x_{21} = 4$, $x_{22} = 4$, then $x_{11} = 8$, $x_{13} = 10$, $x_{21} = 2$, $x_{22} = 4$, $x_{23} = 2$
    (a) is a solution but not   (b) may or may not be optimal
        optimal   solution
    (c) is an optimal basic   (d) is a nonbasic optimal
        solution   solution

88. In TP one of the dual variable is assigned arbitrary value because

(a) solution is available (b) one of the constraint
   immediately   in TP is redundant

(c) construction of loop (d) none of these.
   becomes simple

89. In balanced TP, the dual variables are unrestricted in sign because

(a) TP is a minimization (b) TP is with equality
   problem   constraint

(c) all decision variables (d) none of these.
   are integers

90. In a balanced TP with two sources and three destinations and availabilities 30 at each source and demand 20 at each destination, the dual variables in the optimal table corresponding to sources and destinations are -1, 2 and 1, 2, 3, respectively. Then the optimal value is

(a) 90   (b) 110   (c) 80   (d) 150

91 In a transportation problem the value of the dual variables are not unique. If one dual variable is assigned two different values and remaining are computed as usual, then

(a) $z_{ij} - c_{ij}$ will also change (b) $z_{ij} - c_{ij}$ are unique (c) may or may not change

92. In an assignment problem with m jobs and m machines, the number of basic variables at zero level in a BFS is

(a) m (b) $m - 1$ (c) $m + 1$ (d) none of these

93. If some constant is added to each cost c_{ij} of the assignment matrix then the

(a) optimal solution changes   (b) optimal solution remains same.

94. Which one of the following is not a deterministic model

(a) Linear programming problem (b) Transportation problem
(c) CPM (d) PERT

95 In a minimization ILPP with x_1 and x_2 as the terminal nodes of Branch and bound algorithm at a particular stage are shown below

$x_0 = 15$	$x_0 = 13$	$x_0 = 10$	$x_0 = 16$	$x_0 = 15$	$x_0 = 13$
1, 2	1.2, 3	1, 2.3	2, 5	2, 7.2	2, 3.5
$N(1)$	$N(2)$	$N(3)$	$N(4)$	$N(5)$	$N(6)$

The next branching must be done from the node

(a) N(5)　　(b) N(3)　　(c) N(2)　　(d) none of these

96. In above problem the next branching corresponds to

(a) $x_1 \leq 1,\ x_2 \geq 2$　　(b) $x_1 \leq 2,\ x_2 \geq 3$
(c) $x_1 \leq 7,\ x_2 \geq 7$　　(d) none of these

97. State true or false

(a) The sum of two convex functions is convex

(b) The product of two convex functions is convex

(c) If in TP, two dual variables are assigned arbitrary value then the method will yield correct solution

(d) Assignment problem is not a linear model

(e) Assignment problem can be solved by using $u-v$ method

(f) A cyclic solution of an assignment problem with m machines and m jobs is also a solution of traveling salesman problem with m cities

(g) In traveling salesman problem we reduce the cost matrix first by row-wise and then by column-wise. If the process is reversed, then amount of reduction is always same

(h) In traveling salesman problem with n cities the number of possible tours are $(n-1)!$

(i) A maximum flow problem has always a unique solution

(j) The sum of two unimodal functions is unimodal

(k) A convex function is always unimodal

(l) The minimal spanning tree always gives unique solution

98. In an n-node square matrix the completion of Floyd's algorithm to find shortest route between any two nodes requires the number of comparisons:

(a) n^2　　(b) $n^2(n-1)$　　(c) $n(n-1)^2$　　(d) $n(n+1)$

99. The following table shows the machine time (in days) for 5 jobs to be processed on two different machines M_1 and M_2 in the order $M_1 M_2$:

$$\text{Job} \quad 1 \ 2 \ 3 \ 4 \ 5$$

$$M_1 \quad 3 \ 7 \ 4 \ 5 \ 7$$

$$M_2 \quad 6 \ 2 \ 7 \ 3 \ 4$$

The optimal sequence of jobs to be processed on theses machines to minimize the total elapsed time is

(a) $1 \to 2 \to 3 \to 4 \to 5$ (b) $1 \to 4 \to 5 \to 2 \to 3$

(c) $1 \to 3 \to 5 \to 4 \to 2$ (d) $1 \to 5 \to 3 \to 4 \to 2$

100. The number of all possible optimal sequences to minimize the total elapsed time required to complete the following tasks is (each job is processed in the order $M_1 M_3 M_2 M_4$).

Job	1	2	3	4
Processing time on M_1	20	17	21	25
Processing time on M_2	10	7	8	5
Processing time on M_3	9	15	10	9
Processing time on M_4	25	5	9	25

(a) 2   (b) 4   (c) 16   (d) 24

101. The function $f(X) = 3x_1^2 - 2x_2^2 + x_3^2$ is

(a) positive definite   (b) positive semi-definite   (c) negative definite   (d) indefinite

102. The function $f(X) = x_1^2 + x_2^2 + x_3^2 - 2x_1 x_2$ is

(a) convex   (b) strictly convex   (c) concave   (d) strictly concave

103. In dichotomous search technique with $\delta = 0.1$ and $L_0 = 1$, width of initial interval of uncertainty, after four experiments the width of interval of uncertainty is reduced to

(a) 0.325   (b) 0.375   (c) 0.425   (d) 0.475

104. In Fibonacci search technique with $n = 5$ and $L_0 = 1$, the measure of effectiveness is

(a) 0.001   (b) 0.01   (c) 0.1   (d) 0.125

105. Let L_4 and L_4' be the length of interval of uncertainty after four experiments in Fibonacci and Golden section search, respectively. Then

(a) $L_4 < L_4'$ (b) $L_4' < L_4$ (c) $L_4 = L_4'$ (d) none of these

106. The number of experiments in Fibonacci search that reduce the interval of uncertainty to $\approx 0.001 L_0$ (L_0: initial interval of uncertainty) is

(a) 13   (b) 14   (c) 15   (d) 16

107. The number of experiments in Golden search that reduce the interval of uncertainty to $\approx 0.05571 L_0$ (L_0: initial interval of uncertainty) is

(a) 7   (b) 8   (c) 9   (d) 10

108. For positive definite quadratic $X^T A X$ in $n(> 3)$ variables, let X_1 be the initial approximation for minimum point when the conjugate gradient method is applied. Then the exact minimum occurs at

(a) X_{n-1} (b) X_n (c) X_{n+1} (d) none of these

109. Consider the following pay-off matrix

$$\begin{bmatrix} 1 & 1 & 2 \\ 2 & -2 & 2 \end{bmatrix}$$

If solved as game of pure strategies then the game value is

(a) -2 (b) 1   (c) 2   (d) none of these

Bibliography

M. S. Bazara, J. J. Jevis and H. D. Sherali, *Linear Programming and Network Flows*, 2nd edition, John Wiley & Sons, New York, 1990.

E. M. Beal, *Numerical methods in Nonlinear Programming*, North Holland, Amsterdam, 1967.

R. Bellman, *Dynamic Programming*, Princeton University Press, 1957.

K. H. Borgwardt, *The Simplex Method : A Probabilistic Analysis*, Springer-Verlag, Berlin 1987.

Kim Chaiho, *Introduction to Linear Programming*, Holt, Rinehart and Winston, 1971.

V. Chavtal, *Linear Programming*, Freeman, San Francisco, 1983.

G. B. Dantzig, *Linear Programming and Extensions*, Princeton University Press, 1963.

I. S. Duff, A. M. Erisman and J. K. Reid, *Direct Methods for Sparse Matrices*, Claredon Press Oxford, 1986.

R. Duffin, E. Peterson and C. Zener, *Geometric Programming*, John Wiley and Sons, 1967.

R. J. Fraser, *Applied Linear Programming*, Prentice Hall, Englewood Cliffs, New Jersey, 1968.

S. I. Gauss, *Linear Programming : Methods and Applications*, 2nd edition, Mc-Graw Hill, New York, 1964.

P. E. Gill, W. Murray and F. H. Wright, *Numerical Linear Algebra and Optimization*, Addison Wesley, California 1991.

G. H. Golub and C. F. Van Loan, *Matrix Computations*, Johns Hopkins Press, Baltimore, 1983.

M. Grostschel, L. Lovasz and A. Schrijverr, *The Ellipsoid Method and Combinatorial Optimization*, Springer Verlag, Heidelberg 1988.

G. Hadley, *Linear Algebra*, Addison Wesley, 1961.

G. Hadley, *Linear Programming*, Addison Wesley Publishing Co., 1962.

G. Hadley, *Nonlinear and Dynamic Programming*, Addison Wesley Publishing Co., 1969.

J. Kelley, *The Critical Path Method: Resources Planning and Scheduling (An Industrial Scheduling)*, Muth and Thompson, Prentice Hall, 1963.

D. G. Luenbeger, *Introduction to Linear and Nonlinear Programming*, 2nd edition, Addition Wesley, Reading MA, 1973.

Claude Mcmillan Jr., *Mathematical Programming*, John Wiley and Sons, New York, 1975.

P. L. Meyer, *Introductory Probability and Statistical Applications*, Addison Wesley, 1970.

K. G. Murty, *Linear Programming*, John Wiley, New York, 1983.

J. Nazareth, *Computer Solution of Linear Programs*, Oxford University Press, Oxford and New York, 1987.

G. D. Nemhauser, *Introduction to Dynamic Programming*, John Wiley and Sons, 1970.

R. T. Rockfeller, *Convex Analysis*, Princeton University Press, New Jersey, 1970.

D. M. Simmons, *Nonlinear Programming for Operations Research*, Prentice Hall, 1975.

H. A. Taha, *Operations Research : An Introduction*, Prentice Hall of India, New Delhi, 1997.

F. A. Volantine, *Convex Sets*, McGraw-Hill, Co, 1964.

P. P. Varaiya, *Notes on Optimization*, Von Nonstrand Reinhold Co, 1972.

G. R. Wash, *Methods of Optimization*, John Wiley and Sons, New York, 1975.

D. J. White, *Introduction to Dynamic Programming*, John Wiley and Sons, 1969.

M. Zeleny, *Linear Multiobjective Programming*, Springer-Verlag, Berlin, 1974.

Answers

Problem Set 1

1. opt $z = 2x_1 + x_2' - x_3^+ + x_3^- + 1$

s.t. $2x_1 + x_2' - x_3^+ + x_3^- + s_1 = 4$

$3x_1 - 2x_2' - 3x_3^+ + 3x_3^- + s_2 = 5$

$x_1 - 3x_2' + 4x_3^+ - 4x_3^- - s_3 = 5$

$x_1 + x_2' + x_3^+ - x_3^- = 3$

$x_1, x_2', x_3^+, x_3^-, s_1, s_2, s_3 \geq 0$

2. opt $z = 2x_1 - x_2' + x_3$

s.t. $x_1 + x_2' + 2x_3 - s_1 = 2$

$u + v + s_2 = 4$

$2x_1 - x_2' - x_3 - u + v = 0$

$3x_1 + 2x_2' - 7x_3 + s_4 = 3$

$x_1, x_2', x_3, u, v, s_1, s_2, s_4 \geq 0$

3. opt $z = x_1 + 2x_2' - x_3^+ + x_3^- + 2p$

s.t. $x_1 + x_2' - x_3^+ + x_3^- + s_1 = 5 - p$

$x_1 - 2x_2' - 3x_3^+ + 3x_3^- + s_2 = 4 + 2p$

$2x_1 + 3x_2' - 4x_3^+ + 4x_3^- - s_3 = 3 - 3p$

$x_1 + x_2' + x_3^+ - x_3^- = 2 - p$

$x_1, x_2', x_3^+, x_3^-, s_1, s_2, s_3 \geq 0.$

The range of p is $-2 \leq p \leq 1$.

4. (a) This is not an LPP (b) min $z = x_1^+ + x_1^- + 2x_2^+ + 2x_2^- - x_3$

   s.t. $x_1^+ - x_1^- + x_2^+ - x_2^- - x_3 \leq 9$

   $x_1^+ - x_1^- - 2x_2^+ + 2x_2^- + 3x_3 = 11$

   $x_1^+, x_1^-, x_2^+, x_2^-, x_3 \geq 0$

5. Assume $y = |\min\{x_1, x_2\}|$. The LPP is min $z = y_1 + y_2 - 2y$

   s.t. $y_1 + y_2 - 2y + s_1 = 6$

   $y_1 - 2y_2 + y - s_2 = 3$

   $y_1, y_2, y, s_1, s_2 \geq 0$

6. Define $r = \dfrac{1}{6 + 3x_1 - x_2} > 0$. This implies $3x_1 r - x_2 r = 1 - 6r$.

   Let $rx_j = y_j$, $j = 1, 2, 3$. The required LP model is

   max $2y_1 + 5y_2 - 5y_3 - 3r$,

   s.t. $3y_1 - y_2 = 1 - 6r$

   $y_1 - y_2 \geq 0$

   $7y_1 + 9y_2 + 10y_3 \leq 30r$

   $y_1 \geq 0, y_2 \geq r, y_3 \geq 0$

7. x_j = number of units required of jth food, $j = 1, 2, \ldots, n$

   min $z = \sum_{j=1}^{n} c_j x_j$

   s.t. $\sum_{j=1}^{n} a_{ij} x_j \geq b_i$, $i = 1, 2, \ldots, m$

   $x_j \geq 0$, $j = 1, 2, \ldots, n$.

8. x_i = number of units manufactured of ith product, $i = A, B$

   max $z = 5x_A + 10x_B$

   s.t. $x_A + 5x_B \leq 10000$

   $3x_A + x_B \leq 7000$

   $x_A, x_B \geq 0$ and are integers

9. x_1 = number of belts of type A, x_2 = number of belts of type B,

   max $z = 3x_1 + 2x_2$

s.t. $2x_1 + x_2 \leq 1500$

$\quad x_1 + x_2 \leq 1000$

$\quad x_1 \leq 500$

$\quad x_2 \leq 800$

$\quad x_1, x_2 \geq 0$ and integers

10. m_1 = units of milk produced in Plant-I per day, m_2 = units of milk produced in Plant-II per day, b_1 = units of butter produced in Plant-I per day, b_2 = units of milk produced in Plant-II per day. Assume that one unit milk $\equiv 1000$ liters and one unit butter $\equiv 100$ kgs.

min $z = (15m_1 + 28b_1 + 18m_2 + 26b_2)1000$

s.t. $m_1 + m_2 \geq 10$

$\quad b_1 + b_2 \geq 8$

$\quad 3m_1 + 2b_1 \leq 16$

$\quad b_1 + 1.5b_2 \leq 16$

$\quad m_1, m_2, b_1, b_2 \geq 0$

11. max $z = 75x_1 + 50x_2$

s.t. $25x_1 + 40x_2 \leq 4400$

$\quad 30x_1 + 15x_2 \leq 3300$

$\quad \frac{6}{19} \leq \frac{x_2}{x_1} \leq \frac{17}{8}$

$\quad x_1 \geq 0, x_2 \geq 0$ and are integers

12. x_i = number of master rolls cut on the pattern p_i, $i = 1, 2, \ldots, 8$

min $z = x_1 + x_2 + \cdots + x_8$

s.t. $5x_1 + 3x_2 + 3x_3 + 2x_4 + x_5 \geq 200$

$\quad x_2 + 2x_5 + x_6 + x_7 \geq 90$

$\quad x_3 + x_6 + 2x_8 \geq 350$

$\quad x_4 + x_7 \geq 850$

$\quad x_1, x_2, \ldots, x_8 \geq 0$ and integers

13. x_i = Number of parent metallic sheets cut on the pattern p_i, $i = 1, 2, 3$.

$$\min z = x_1 + x_2 + x_3$$
$$\text{s.t. } 10x_1 + 6x_2 + 2x_3 \geq 2500$$
$$x_2 + 2x_3 \geq 1500$$
$$x_1, x_2 \geq 0 \text{ and are integers}$$

14. x_1 = number of tables and x_2 = number of chairs to be manufactured

$$\max z = 9x_1 + 6x_2$$
$$\text{s.t. } 30x_1 + 20x_2 \leq 381$$
$$10x_1 + 5x_2 \leq 117$$
$$x_1, x_2 \geq 0 \text{ and integers}$$

15. $\max z = 15x_1 + 25x_2$
$$\text{s.t. } 3x_1 + 4x_2 = 100$$
$$2x_1 + 3x_2 \leq 70$$
$$x_1 + 2x_2 \leq 30$$
$$x_1 \geq 0, x_2 \geq 3$$

16. Let x_1, x_2, x_3 be the number of models I, II, III, respectively to be manufactured.

$$\max z = 30x_1 + 20x_2 + 60x_3$$
$$\text{s.t. } 6x_1 + 3x_2 + 2x_3 \leq 4200$$
$$2x_1 + 3x_2 + 5x_3 \leq 2000$$
$$4x_1 + 2x_2 + 7x_3 \leq 3000$$
$$x_1 \geq 200, \ x_2 \geq 200, \ x_3 \geq 150 \text{ and integers}$$

17. x_{ij} = number of units of product i processed on machine j, $i = 1, 2, \ldots, m; \ j = 1, 2, \ldots, n$

$$\min z = \sum_{i=1}^{m} \sum_{j=1}^{n} c_{ij} x_{ij}$$
$$\text{s.t. } a_{i1}x_{11} + a_{i2}x_{12} + \cdots + a_{in}x_{1n} \leq b_i$$
$$x_{1j} + x_{2j} + \cdots + x_{mj} \geq d_j$$
$$x_{ij} \geq 0; \ i = 1, 2, \ldots, m, \ j = 1, 2, \ldots, n.$$

18. x_{ij} = the amount (in tons) of the ith commodity given the placement at jth position, $i = A, B, C$, $j = 1$ (for forward), 2 (for centre), 3 (for after)

$$\max z = 60(x_{A1} + x_{A2} + x_{A3}) + 80(x_{B1} + x_{B2} + x_{B3})$$
$$+ 50(x_{C1} + x_{C2} + x_{C3})$$

s.t. $x_{A1} + x_{A2} + x_{A3} \leq 6000$

$x_{B1} + x_{B2} + x_{B3} \leq 4000$

$x_{C1} + x_{C2} + x_{C3} \leq 2000$

$x_{A1} + x_{B1} + x_{C1} \leq 2000$

$x_{A2} + x_{B2} + x_{C2} \leq 3000$

$x_{A3} + x_{B3} + x_{C3} \leq 1500$

$60x_{A1} + 50x_{B1} + 25x_{C1} \leq 100000$

$60x_{A2} + 50x_{B2} + 25x_{C2} \leq 135000$

$60x_{A3} + 50x_{B3} + 25x_{C3} \leq 30000$

$x_{ij} \geq 0$

19. min $z = y$

s.t. $20x_{1a} + 25x_{2a} + 20x_{3a} - y \geq 0$

$25x_{1b} + 20x_{2b} + 5x_{3b} - y \geq 0$

$x_{1a} + x_{1b} \leq 100$

$x_{2a} + x_{2b} \leq 150$

$x_{3a} + x_{3b} \leq 200$

$x_{ij}, y \geq 0, \ i = 1, 2, 3; \ j = 1, b$

20. The first constraint $6x_1 + 3x_2 + 2x_2$ in answer to Problem 16 is replaced by

$x_1 + 2x_2 \leq 700$

$x_2 + 3x_3 \leq 1400$

21. $\max z = \min \left\{ \sum_{i=1}^{m} a_{i1}x_{i1}, \frac{1}{2} \sum_{i=1}^{m} a_{i2}x_{i2}, \ldots, \frac{1}{n} \sum_{i=1}^{m} a_{in}x_{in} \right\}$

s.t. $x_{i1} + x_{i2} + \cdots + x_{in} \leq b_i$

$x_{ij} \geq 0, \ j = 1, 2, \ldots, n$

22. Let x_j be the number of waiters recruited on jth day, $j = 1, 2, 3, 4, 5$

min $z = x_1 + x_2 + x_3 + x_4 + x_5$

s.t. $x_1 + x_4 + x_5 \geq 25$

$x_1 + x_2 + x_5 \geq 35$

$x_1 + x_2 + x_3 \geq 40$

$x_2 + x_3 + x_4 \geq 30$

$x_3 + x_4 + x_5 \geq 20$

$0 \leq x_j \leq 30$ and are integers

23. x_{ij} = number of buses of type i allocated to city j; $i = 1, 2, 3$, $j = 1, 2, 3, 4$. s_j = number of passenger not served for the cities $j = 1, 2, 3, 4$

$$\min z = 3000x_{11} + 2200x_{12} + 2400x_{13} + 1500x_{14}$$
$$+ 3600x_{21} + 3000x_{22} + 3300x_{23} + 2400x_{24}$$
$$+ 3500x_{31} + 4500x_{32} + 3600x_{33} + 2000x_{34}$$
$$+ 40s_1 + 50s_2 + 45s_3 + 70s_4,$$

subject to

$$\sum_{j=1}^{4} x_{1j} \leq 5, \quad \sum_{j=1}^{4} x_{2j} \leq 8, \quad \sum_{j=1}^{4} x_{3j} \leq 10$$

$300x_{11} + 280x_{21} + 250x_{31} + s_1 = 1000$

$200x_{12} + 210x_{22} + 250x_{32} + s_2 = 2000$

$200x_{13} + 210x_{23} + 200x_{33} + s_3 = 900$

$100x_{14} + 140x_{24} + 100x_{34} + s_4 = 1200$

all $x_{ij} \geq 0$, and $s_j \geq 0$

24. x_{iA} = amount invested in year i under Scheme A; x_{iB} = amount invested in year i under Scheme B

max $z = 2.4x_{2B} + 1.6x_{3A}$

s.t. $x_{1A} + x_{1B} \leq 2,00,000$

$x_{2A} + x_{2B} \leq 1.6x_{1A}$

$x_{3A} \leq 2.4x_{1B} + 1.6x_{2A}$

$x_{iA}, x_{iB} \geq 0, \ i = 1, 2, 3$

25. Define $x_1 =$ number of units of P_1 made on regular time; $x_2 =$ number of units of P_2 made on regular time; $x_3 =$ number of units of P_1 made on overtime; $x_4 =$ number of units of P_2 made on overtime; $x_5 =$ number of units of P_1 made on regular time on M_1 and overtime on M_2; $x_6 =$ number of units of P_2 made on regular time on M_1 and overtime on M_2.

$$\max\ z = 8x_1 + 10x_2 + 4x_3 + 8x_4 + 6x_5 + 9x_6$$

s.t. $5x_1 + 4x_2 + 5x_5 + 4x_6 \leq 120$ (regular time of M_1)

$\qquad 5x_3 + 4x_4+ \leq 50$ (overtime of M_1)

$\qquad 3x_1 + 6x_2 \leq 150$ (regular time on M_2)

$\qquad 3x_3 + 6x_4 + 3x_5 + 6x_6 \leq 40$ (overtime on M_2)

$\qquad x_j \geq 0,\ j = 1, 2, \ldots 6$

26. $\max\ z = (x_1 + x_{11} + x_{12})p_1 + (y_1 + y_{11} + y_{12})p_2$

s.t. $24x_1 + 36y_1 - 0.75x_{11} - 0.75y_{11} - 0.33y_{12} \leq G_1$

$\qquad 8x_1 + 12y_1 - 0.5x_{11} - 0.5y_{11} \leq G_2$

$\qquad 100x_1 + 50y_1 - 0.9x_{11} - 0,9y_{11} - 0.6y_{12} \leq P_1$

$\qquad x_1, x_{11}, x_{12}, y_1, y_{11}, y_{12} \geq 0$

27. $x_1 =$ number of units of P_1, $x_2 =$ number of units of P_2

$\max\ z = 10x_1 + 30x_2$

s.t. $-0.6x_1 + 0.4x_2 \leq 0$

$\qquad x_1 \leq 100$

$\qquad x_1 + 2x_2 \leq 120$

$\qquad x_1, x_2 \geq 0$

28. $\max\ z = 4000(x_{1A} + x_{1B} + x_{1C}) + 3000(x_{2A} + x_{2B} + x_{2C}) + 4000(x_{3A} + x_{3B} + x_{3C})$

subject to

Availability of acreage for each crop

$x_{1A} + x_{1B} + x_{1C} \leq 700$

$x_{2A} + x_{2B} + x_{2C} \leq 800$

$x_{3A} + x_{3B} + x_{3C} \leq 300$

Availability of usable acreage in each farm

$$x_{1A} + x_{2A} + x_{3A} \leq 400$$

$$x_{1B} + x_{2B} + x_{3B} \leq 600$$

$$x_{1C} + x_{2C} + x_{3C} \leq 300$$

Water available (in acre feet) constraints

$$5x_{1A} + 4x_{2A} + 3x_{3A} \leq 1500$$

$$5x_{1B} + 4x_{2B} + x_{3C} \leq 2000$$

$$5x_{1C} + 4x_{2C} + x_{3C} \leq 900$$

To ensure that the percentage of usable acreage is same for each farm

$$\frac{x_{1A} + x_{1B} + x_{1C}}{400} = \frac{x_{2A} + x_{2B} + x_{2C}}{600} = \frac{x_{3A} + x_{3B} + x_{3C}}{300} \text{ or}$$

$$3(x_{1A} + x_{1B} + x_{1C}) = 2(x_{2A} + x_{2B} + x_{2C})$$

$$x_{2A} + x_{2B} + x_{2C} = 2(x_{3A} + x_{3B} + x_{3C})$$

$$x_{ij} \geq 0 \ i = 1, 2, 3; \ j = A, B, C$$

29. max $z = \sum_{j=1}^{8} u_j (1 - p_{1j})^{x_{ij}} p_{2j}^{x_{2j}} p_{3j}^{x_{3j}}$, subject to

$$\sum_{j=1}^{8} x_{ij} \leq a_i, \ i = 1, 2, 3$$

$$\sum_{i=1}^{3} x_{ij} \geq b_j, \ j = 1, 2, \dots, 8$$

$x_{ij}, p_{ij} \geq 0,$

where p_{ij} is the probability that target j will be undamaged by weapon i and x_{ij} is the number of weapons i assigned for target j.

Problem Set 2

1. (a) Not convex; (b) convex; (c) not convex; (d) convex; (e) not convex; (f) not convex

2. min $z = 2x_1 + x_2$,

s.t. $x_1 + x_2 \geq 1$

$\quad x_1 + 2x_2 \geq 2$

$\quad x_1, x_2 \geq 0$

7. The equivalent system with inequalities is

$$x + 1 + x + 2 \leq 1$$

$$2x_1 - 4x_3 \leq -5$$

$$-3x_1 - x_2 + 4x_3 \leq 4$$

10. (a) Triangle with vertices at $(0, -1)$, $(2, 0)$, $(1, 2)$; (b) $\mathbb{R}^2$; (c) Disc $\{(x_1, x_2) : x_1^2 + x_2^2 \leq 1\}$; (d) quadrilateral with vertices as these points.

12. (a) $(0, 2, 0, 0, 0)$; (b) $(-6, 14, 0, 0, 0)$; (c) for x_3 and x_4 as basic variables: infinite solutions; (d) if we take x_1 and x_5 as basic variables: nonexisting solution

13. $(1, 2, 0, 0, 0, 0)$, $(0, 5/3, 4/3, 0, 0, 0)$, $(3, 0, 0, 0, 0, 8)$

17. No

19. $\binom{n-p}{m-p}$

20. Problem 18 ensures the existence of three bases corresponding to the degenerate BFS $(0, 2, 0, 0)$. Two bases are given in Example 2, Section 2.3 and the third one is

$$\begin{bmatrix} 1 & -8 \\ -1 & -1 \end{bmatrix}$$

21. $(-11, 5, 2, 8)$

24. (a) $x_1 = 2$, $x_2 = 0$, $z = -2$; (b) Unbounded solution

25. The optimum will not exist until the objective function is constant.

29. min $z = -x_1 - x_2 + 14$,

s.t. $-3x_1 + x_2 \leq 3$

$x_2 \leq 7$

$x_1 - x_2 \leq 2$

$x_1, x_2 \geq 0$

Optimal solution: $x_1 = 9$, $x_2 = 7$, $z = -2$

30. $(1, 1, 0, 0)$, $(1, 0, 1, 0)$, $(1, 0, 0, 1)$, $(0, 7/2, 0, -3/2)$, $(0, 0, -7/2, -3/2)$

Problem Set 3

2. (a) $x_1 = 2/3$, $x_2 = 7/3$, $z = 16/3$; (b) $x_1 = 2$, $x_2 = 2$, $z = 14$;
(c) $x_1 = 0$, $x_2 = 1$, $z = 1$; (d) $x_1 = 0$, $x_2 = 30$, $z = 18$

3. max $z = -4x_1 + x_2$

s.t. $7x_1 - 3x_2 \leq 3$

$\quad -2x_1 + x_2 \leq 1$

$\quad x_1, x_2 \geq 0$

4. Basic solution $(4, 0, -6, 0)$, but not feasible

5. Unbounded solution

6. $x_1 = 0, x_2 = -2, z = 6$

8. (i) x_3, x_1 and s_3 are the basic variables in that order; (ii) The
LPP is

$\quad$ max $\quad z = 6x_1 + 2x_2 + 10x_3 + 2x_4,$

$\quad$ s.t. $x_2 + 2x_3 \leq 5$

$\qquad 3x_1 - x_2 + x_3 + x_4 \leq 10$

$\qquad x_1 + x_3 + x_4 \leq 8$

$\qquad x_1, x_2, x_3 \geq 0$

10. Unbounded solution

11. (a) $x_1 = 0$, $x_2 = 2$, $s_1 = 1$, $s_2 = 0$, $s_3 = 7$
(b) $x_1 = 1/8$, $x_2 = 9/4$, $s_1 = 11/8$, $s_2 = s_3 = 0$

12. (a) $(0, 2, 1, 0, 1)$ (b) $(1/8, 9/4, 11/8, 0, 0)$

13. $x_1 = x_2 = 0$, $x_3 = x_4 = 1$, $x_5 = 3$, $z = 5$

15. The solution space is unbounded in x_2 direction

16. $x_1 = 7$, $x_2 = -1$, $z = 22$

17. Optimal solution: $x_1 = 22/5$, $x_2 = 31/5$, $x_3 = 0$ $z = 8$

18. x_1 must be preferred if $\theta_1(z_1 - c_1) < \theta_2(z_2 - c_2)$ for min problem,
and the reverse inequality is considered for max problem

21. $x_1 = 1$, $x_2 = 0$, $x_3 = 1$, $x_4 = 0$, $z = -5/4$

Problem Set 4

1. $C = -b$ and A is skew-symmetric matrix

2. Yes, take (x_1, x_2) as the starting BFS

3. Dual:
$$\max \quad z = -y_1 - 2y_2 - 2y_3$$
$$\text{s.t.} \quad -y_1 + y_2 - y_3 \leq -2$$
$$-2y_1 + y_3 \leq -3$$
$$y_2 - 2y_3 \leq -1$$
$$y_1, y_2, y_3 \geq 0$$
Optimal solution of the primal $x_1 = 0$, $x_2 = 1/2$, $x_3 = 5/4$, $x_0 = -11/4$

4. Optimal solution of primal: $x_1 = 1/8$, $x_2 = 9/4$, $x_0 = 7/4$;
Optimal solution of the dual: $y_1 = 0$, $y_2 = 7/4$, $y_3 = 1/4$, $y_0 = 7/4$.

5. Dual:
$$\min \quad y_0 = 14y_1 + 17y_2 - 19y_3 + 100$$
$$\text{s.t.} \quad -3y_1 - 5y_2 + 2y_3 \leq -9$$
$$8y_1 - 2y_2 - 4y_2 \leq 6$$
$$-5y_1 + 6y_2 = -4$$
$$y_1, y_3 \geq 0, \ y_2 \text{ unrestricted}$$

6. Primal:
$$\min \quad z = -2x_1 - 3x_2$$
$$\text{s.t.} \quad x_1 - x_2 \geq 3$$
$$-x_1 + x_2 \geq 2$$
$$x_1, x_2 \geq 0$$
Dual:
$$\max \quad z' = 3y_1 + 2y_2$$
$$\text{s.t.} \quad y_1 - y_2 \leq -2$$
$$-y_1 + y_2 \leq -3$$
$$y_1, y_2 \geq 0$$

7. $x_1 = 0$, $x_2 = 5$, $x_0 = 15$; Dual:

min $y_0 = -3y_1 + 5y_2$

s.t. $2y_1 + 3y_2 \geq 2$

$-y_1 + y_2 \geq 3$

$y_1, y_2 \geq 0$

10. Converse is also true

11. min $x_0 = 2x_1 + 3x_2 + 4x_3$

s.t. $x_2 + x_3 \geq -2$

$-x_1 + 3x_3 \geq 3$

$-2x_1 - 3x_2 \geq 4$

$x_1, x_2, x_3 \geq 0$

13. yes, converse is also true

15. min $y_0 = 5y_1 + 10y_2 + 8y_3$

s.t. $3y_2 + y_3 \geq 6$

$y_1 - y_2 \geq 2$

$2y_1 + y_2 + y_3 \geq 10$

$y_2 + y_3 \geq 0$

$Y - 1, y_2, y_3 \geq 0$

Optimal solution: $y_1 = 4$, $y_2 = 2$, $y_3 = 0$, $y_0 = 40$

16. $x_1 = -4$, $x_2 = 4$, $x_0 = 0$

17. $y_1 = 3$, $y_2 = -1$, $y_0 = 5$

18. (b) $y_1 = 1$, $y_2 = -4/3$, $y_0 = 11/3$

19. Optimal solution is $x_1 = 1.286$, $x_2 = 0.476$, $x_3 = 0$, $z = 2.095$

20. No feasible solution

21. Optimal of primal: $x_1 = 0$, $x_2 = 0$, $x_3 = 3.57$, $x_4 = 1.43$, $x_5 = 0$, $x_6 = 0.86$, $x_0 = 2.71$, Optimal of dual: $y_1 = -0.36$, $y_2 = 0.07$, $y_3 = 0.93$, $y_0 = 2.71$

22. max $y_0 = b^T Y + \ell^T Y' - u^T Y''$, subject to $A^T Y + \mathbf{I}Y' - \mathbf{I}Y'' \leq C$, where $Y = (y_1, y_2, \ldots, y_m)$ unrestricted, $Y' = (y', y', \ldots y'_n) \geq 0$, $Y'' = (y''_1, y''_2, \ldots, y''_n) \geq 0$ and $\mathbf{I}$ is identity matrix of order n

23. max $y_0 = y_{m+1}$, subject to $A^T Y + e y_{m+1} \leq C$, where $Y = (y_1, y_2, \ldots, y_m)^T$ and y_{m+1} are unrestricted

Problem Set 5

1. $x_1 = 71/10$, $x_2 = 0$, $x_3 = 0$, $x_4 = 13/10$, $x_5 = 0$, $x_6 = 2/5$, $z = 71/10$

2. Infeasible solution (no solution)

3. $x_1 = 5/3$, $x_2 = 0$, $x_0 = 10/3$

4. $x_1 = 0$, $x_2 = 14$, $x_3 = 9$, $z = -9$

5. $x_1 = 4$, $x_2 = 15/4$, $x_3 = 0$, $z = 123/14$

6. $y_1 = 3$, $y_2 = 5$, $y_3 = 2$, $z = 34$

7. A-type belts $= 200$, B-type belts 800, maximum profit $= 2200$

8. max $z = 6x_1 + 7.50x_2 - 0.5x_3^-$

　　s.t. $10x_1 + 12x_2 + x_3^+ - x_3^- = 2500$

　　　　$150 \leq x_1 \leq 200$

　　　　$x_2 \leq 45$

　　　　all var ≥ 0

(a) Optimal solution $x_1 = 200$, $x_2 = 45$, $x_3^+ = 0$, $x_3^+ = 40$ and $z = \$1517.50$

(b) Optimal solution $x_1 = 196$, $x_2 = 45$, $x_3^+ = 0$, $x_3^+ = 0$ and $z = \$1513.50$, hence no overtime is recommended by the new solution.

9. $x_1 = 47/20$, $x_2 = 1/10$, $x_3 = 27/10$, $x_4 = 6/4$, $z = 479/10$

11. Taking $Q = 2^6 = 65$, $n = 6$ and $M > 0$ (large), the Karmarkar's standard form is

$$\min \ y_0 = 65y_1 + 65y_2 + 65y_3 - 195y_4 + 390y_5 + 260y_6 + My_9$$

$$\text{s.t. } y_1 + y_2 + 3y_4 - y_5 + 2y_6 - 6y_7 = 0$$

$$y_2 + y_3 - y_4 + 4y_5 + y_6 - 3y_7 - 3y_8 = 0$$

$$y_1 + y_3 - 2y_4 + y_5 + y_5 + 5y_6 - 5y_7 - y_8 = 0$$

$$y_1 + y_2 + y_3 + y_4 + y_5 + y_6 + y_7 - 64y_8 + 57y_9 = 0$$

$$y_1 + y_2 + y_3 + y_4 + y_5 + y_6 + y_7 + y_8 + y_9 = 1$$

$$\text{all var } \geq 0$$

12. With $\alpha = 1/\sqrt{6}$, $x^1 = (4/9, 5/18, 5/18)$, $\min = 5/9$. Continuing this iterative process, the Karmarkar's algorithm will stop at the optimal solution $x^1 = (1, 0, 0)$, $\min = 0$

Problem Set 6

1. For same optimal solution the variation in costs satisfies the inequality $\underline{\delta} \leq \delta \leq \bar{\delta}$, where

$$\underline{\delta} = \max_j \left\{ \max_k \left(\frac{z_j - c_j}{\alpha_k^j}, \ \alpha_k^j, \ j \in \overline{N} \right), -\infty \right\},$$

$$\bar{\delta} = \min_j \left\{ \min_k \left(\frac{z_j - c_j}{\alpha_k^j}, \ \alpha_k^j, \ j \in \overline{N} \right), +\infty \right\}$$

2. For same optimal basis the variation δ in b satisfies the inequality $\underline{\delta} \leq \delta \leq \bar{\delta}$, where $\underline{\delta} = \max_k \left\{ \max_i \left\{ \frac{X_{B_i}}{-\beta_{ik}}, \ \beta_{ik} > 0 \right\}, -\infty \right\}$,

$$\bar{\delta} = \min_k \left\{ \min_i \left\{ \frac{X_{B_i}}{-\beta_{ik}}, \ \beta_{ik} < 0 \right\}, +\infty \right\}$$

4. (a) $c_1 \in [1, 6]$ (b) $b_2 \in [5/3, 10]$

6. (a) $x_1 = 7/2$, $x_2 = 1/2$, $x_0 = 17/2$ (b) $x_1 = 12/5$, $x_2 = 6/5$, $x_0 = 42/5$ (c) $x_1 = 1$, $x_2 = 1$, $x_0 = 5$ or $x_1 = 5/2$, $x_2 = 0$, $x_0 = 5$

7. (a) $x_1 = 0$, $x_2 = 2$, $x_3 = 0$, $x_0 = -4$ (b) $x_1 = 0$, $x_2 = 6$, $x_3 = 4$, $x_0 = -8$ (c) Unbounded solution

9. $x_1 = x_2 = 0$, $x_3 = 8$, $z = -8$, Dual variables $y_1 = 1$, $y_2 = 0$, $y_0 = 8$; (a) optimal solution remains unchanged (b) optimal solution remains unchanged (c) $x_1 = x_2 = 0$, $x_3 = 4$, $x_0 = -4$

(d) Increase right side of the first constraint. Since $C_B^T B^{-1} = (-1, 0)$, the increase in right side of the first constraint will further reduce the objective function value, it is profitable to increase right side of the first constraint. Increase in right side of second constraint will cause no effect on the objective function value.

10. (i) Optimal solution $x_1 = 0$, $x_2 = 8$, $x_3 = 20/3$, $x_0 = 20/3$, $x_0 = 680/3$; (ii) $c_1 > 44/3$; (iii) $c_2 = 56/3$; (iv) $x_1 = 0$, $x_2 = 43/5$, $x_3 = 17/3$, $x_0 = 686/3$ when $b_1 = 103$, and $x_1 = 0$, $x_2 = 12$, $x_3 = 0$, $x_0 = 240$, when $b_1 = 200$

11. $x_1 = 7/5$, $x_2 = 2/5$, $z = 4$; $x_1 = x_2 = x_3 = 1$, $z' = 6$

12. (a) Deletion of first constraint renders the LPP with unbounded solution. Similar is the case when third constraint is deleted (b) The deletion of second constraint causes no effect on the optimal solution

13. $x_1 = 20$, $x_2 = 0$, $x_0 = 80$

Problem Set 7

1. $x_{12} = 5$, $x_{13} = 20$, $x_{21} = 20$, $x_{24} = 10$, $x_{32} = 15$, $x_{34} = 5$, $x_{43} = 10$, $x_{45} = 25$, $x_0 = 590$

3. $x_{13} = 50, x_{21} = 30, x_{23} = 120, x_{31} = 70, x_{32} = 130, x_{43} = 30$; $x_0 = 1460$

4. $x_{12} = 50, x_{22} = 50, x_{23} = 100, x_{31} = 100, x_{33} = 100, x_{42} = 30$; $x_0 = 1500$

5. $x_{12} = 15, x_{13} = 13, x_{21} = 5, x_{23} = 7, x_{24} = 6, x_{25} = 5, x_{34} = 19$; $x_0 = 148$

6. N-W rule: 1765, LCM: 1800, VAM: 1695, modified VAM: 1645. Clearly, modified VAM gives better initial basic feasible solution.

7. $\binom{m(n+1)}{m+n}$

8. (a) No (b) up to 4 (c) yes, $x_{12} = 12$, $x_{14} = 9$, $x_{1\text{dummy}} = 1$, $x_{23} = 10$, $x_{31} = 5$, $x_{3\text{dummy}} = 3$, $x_0 = 89$

9. 270

12. (a) Not an optimal schedule (b) 4 (c) no change

13. $x_{13} = 20, x_{21} = 30, x_{24} = 10, x_{32} = x_{33} = 25$; $x_0 = 495$

14. $W_1 \rightarrow M_4$, $W_2 \rightarrow M_3$, $W_3 \rightarrow M_2$, $W_4 \rightarrow M_1$, $W_5 \rightarrow M_5$, $W_5 \rightarrow M_6$

15. $M_1 \rightarrow J_1$, $M_2 \rightarrow J_2$, $M_3 \rightarrow J_3$, $M_4 \rightarrow J_4$; $x_0 = 19$

16. $1 \rightarrow 1$, $2 \rightarrow 2$, $3 \rightarrow 3$, $4 \rightarrow 4$, total cost: \$15,000

17. $M_1 \rightarrow A$, $M_2 \rightarrow C$, $M_3 \rightarrow B$, $M_4 \rightarrow D$, minimum cost of installation: 34

21. The formulation of the problem is

$$\min \quad x_0 = \sum_{i=1}^{l}\sum_{j=1}^{m}\sum_{k=1}^{n} c_{ijk}x_{ijk}$$

$$\text{s.t.} \quad \sum_{j=1}^{m} x_{ijk} \geq a_{ik}, \quad i = 1, 2, \ldots, l; \quad k = 1, 2, \ldots, n \quad \text{(demand)}$$

$$\sum_{i=1}^{l} x_{ijk} \geq b_{jk}, \quad j = 1, 2, \ldots, m; \quad k = 1, 2, \ldots, n \quad \text{(demand)}$$

$$\sum_{k=1}^{n} x_{ijk} \leq d_{ij}, \quad i = 1, 2, \ldots, l; \quad j = 1, 2, \ldots, m \quad \text{(supply)}$$

$$x_{ijk} \geq 0 \text{ for all } i, j, \text{ and } k.$$

Problem Set 8

1. (a) Connect nodes 1 and 2, 2 and 3, 3 and 4, 3 and 5, 3 and 6, 5 and 7, minimum length = 14 units (b) shortest path: $1 \rightarrow 2 \rightarrow 5 \rightarrow 7$, shortest distance = 10 units, alternate shortest path: $1 \rightarrow 3 \rightarrow 5 \rightarrow 7$

2. (a) $1 \rightarrow 3 \rightarrow 2 \rightarrow 5 \rightarrow 6 \rightarrow 4 \rightarrow 7$, distance 11 units (b) $7 \rightarrow 6 \rightarrow 5 \rightarrow 2 \rightarrow 3 \rightarrow 1$, distance 11 units (c) $2 \rightarrow 5 \rightarrow 6 \rightarrow 4 \rightarrow 7$, distance 7 units

3. max flow 25 units and optimal flow in arcs $1 \rightarrow 2$: 8 units, $1 \rightarrow 3$: 13, $1 \rightarrow 5$: 4, $2 \rightarrow 4$: 2, $2 \rightarrow 5$: 6, $3 \rightarrow 4$: 3, $3 \rightarrow 5$: 10, $4 \rightarrow 5$: 5

(a) surplus capacities: $1 \to 3$ 1 unit, $2 \to 4$ 2 units $3 \to 4$ 3 units;

(b) flow through node $2 = 8$ units, through node $3 = 13$ units, through node $4 = 5$ units;

(c) can be increased along $3 \to 5$ because there is surplus capacity at node 1, increase along $4 \to 5$ is also possible;

(d) alternative optimal solution: max flow 25 units and optimal flow in arcs $1 \to 2$: 7, $1 \to 3$: 14, $1 \to 5$: 4, $2 \to 4$: 3, $2 \to 5$: 6, $3 \to 2$: 4, $3 \to 4$: 3, $3 \to 5$: 10, $4 \to 5$: 5

4. maximum flow $= 110$ million bbl/day; (a) Refinery 1 $= 20$ million bbl/day, Refinery 2 $= 80$ million bbl/day, Refinery 10 million bbl/day; (b) Terminal 7 $= 60$ million bbl/day, Terminal 8 $= 50$million bbl/day; (c) Pump 4 $= 30$ million bbl/day, Pump 5 $= 50$ million bbl/day, Pump 6 $= 70$ million bbl/day

5. maximum flow $= 100$ million bbl/day; Pump 4 $= 30$ million bbl/day, Pump 5 $= 40$ million bbl/day, Pump 6 $= 60$ million bbl/day.

Problem Set 9

1. Denote (1,2)=A, (1,3)=D, (2,4)=B, (3,5)=E, (4,5)=C and join appropriate nodes in ascending order

2. Critical path: $BEGIJ$

3. Take $A = (1,2)$, $B = (1,3)$, $C = (1,4)$, $D = (2,3)$, $G = (3,4)$, $I = (3,5)$, $E = (2,6)$, $F = (2,7)$, $H = (4,6)$, dummy activity $= (6,7)$, $J = (6,9)$, $M = (7,8)$, $O = (8,9)$, dummy activity $(5,8)$, $L = (5,9)$, $K = (5,10)$, $P = (9,10)$ and join appropriate nodes in ascending order

4. A is initial node and H is terminal node, Also, introduce dummy activity from E to F. Critical path: $A \to B \to D \to E \to F \to H$, Normal duration 29 days

5. Critical path $ADFI$, 67 days

6. Critical Path: $1 \longrightarrow 2 \longrightarrow 5 \longrightarrow 6 \longrightarrow 7 \longrightarrow 8 \longrightarrow 10 \longrightarrow 12 \longrightarrow 13$. The normal duration is 320 hours

7. Critical path: $ABEG$ (dummy activity connects D and E), Normal duration = 26 days

8. The most economical schedule is for 21 days with total cost = 9535

Problem Set 10

1. $1 \rightarrow 3 \rightarrow 5 \rightarrow 4 \rightarrow 2$, Total elapsed time = 28 days, idle time for $M_1 = 2$ days, idle time for $M_2 = 6$ days

2. $1 \rightarrow 4 \rightarrow 3 \rightarrow 7 \rightarrow 2 \rightarrow 6 \rightarrow 5$, Total elapsed time = 485 hours, idle time for $M_1 = 20$ hours, idle time for $M_2 = 105$ hours

3. $A \rightarrow D \rightarrow C \rightarrow G \rightarrow B \rightarrow F \rightarrow E$, Total elapsed time = 85 hours, idle time for $M_1 = 12$ hours, idle time for $M_2 = 9$ hours

4. No method is applicable

5. (a) $3 \rightarrow 6 \rightarrow 5 \rightarrow 7 \rightarrow 2 \rightarrow 4 \rightarrow 1$, alternative sequence $3 \rightarrow 5 \rightarrow 7 \rightarrow 2 \rightarrow 6 \rightarrow 4 \rightarrow 1$, Total elapsed time = 54 hours (b) $1 \rightarrow 4 \rightarrow 3 \rightarrow 6 \rightarrow 2 \rightarrow 5 \rightarrow 7$, Total elapsed time = 88 hours

6. $D \rightarrow A \rightarrow C \rightarrow B$

8. Job 1 precedes job 2 on machine M_1, job 1 precedes job 2 on machine M_2, job 1 precedes job 2 on machine M_3, job 2 precedes on machine 1; Total elapsed time $22 + 10 + 1 = 33$ hours

9. $4 \rightarrow 1 \rightarrow 3 \rightarrow 6 \rightarrow 5 \rightarrow 2$, Total elapsed time = 159 minutes, idle time for $A = 17$ minutes, idle time for $B = 3$ minutes

Problem Set 11

1. $x_1 = 1$, $x_2 = 3$, min value = 4.

2. $x_1 = 15$, $x_2 = 2$, min value = 106

4. Optimal tour $1 \longrightarrow 2 \longrightarrow 4 \longrightarrow 3 \longrightarrow 1$, Cost of travel = 14

5. Optimal tour $1 \longrightarrow 4 \longrightarrow 2 \longrightarrow 3 \longrightarrow 1$, Travel cost = 13

6. Optimal tour $1 \longrightarrow 3 \longrightarrow 4 \longrightarrow 2 \longrightarrow 1$, Travel cost = 170

9. x_i = amount of the food item i to be loaded, $i = 1, 2, 3, 4, 5$.

max $40x_1 + 50x_2 + 60x_3 + 55x_4 + 60x_5$

s.t. $20x_1 + 30x_2 + 40x_3 + 55x_4 + 80x_5 \le 105$

$x_i = 0$ or 1

Problem Set 12

2. $x = 0, y = 2$, min value: 2

3. $y_1 = 18/3$, $y_2 = 12/13$ min value: 30/13

4. $y_1 = 1, y_2 = 1, y_3 = 18$ max value: 326, For alternate solutions any two variables at unity level and third variable at 18.

5. $y_1 = 4, y_2 = 5, y_3 = 1$, $y_0 = 38$

7. $x_1 = 7$, $x_2 = 11$, $z_{\max} = 106$

8. $x_1 = 2$, $x_2 = 6$, $z_{\max} = 36$

Problem Set 13

1. (a) Indefinite (b) positive semi-definite

2. (a) $\{(x_1, x_2) : x_1 > 1/3, \ x_2 > 3 - 7x_1\}$ (b) $\{(x_1, x_2) : x_1 > 0, \ x_1 x_2 > 1\}$

3. Not convex

4. Minimum points: $X_1 = (2, 2, 1)$, $X_2 = (2, -2, 1)$, $X_3 = (2.8, 0, 1.4)$ is not an extreme point

9. Optimal solution $x_1 = 2/3$, $x_2 = 14/9$, $z_{\max} = 22/9$.

10. Optimal solution $x_1 = 0$, $x_2 = 1$, $f_{\max} = 1$

11. Infeasible solution

12. Optimal solution $x_1 = 4/11$, $x_2 = 3/11$, $x_0 - 5/11$.

13. $x_A = 0$, $x_B = 85/2$; $f(X) = 80325/4$, x_i = number of units of product i, $i = A, B$

15. min $z = 3x_1^2 + 3x_2^2 + 8x_3$

    s.t. $x_1 - x_3 = 200$

        $x_2 + x_3 = 200$

        $x_1, x_2, x_3 \geq 0$

$x_1 = 748/3$, $x_2 = 752/3$, $x_3 = 148/3$, where $x_1 =$ number of units produced in January, $x_2 =$ number of units produced in February, $x_3 =$ number of units sold at the end of January.

Problem Set 14

4. (a) 34 (b) 20 (c) 16 (d) 16

5. (a) $X_{\min} = 1.49975$, $f_{\min} = 0.943$ (b) $X_{\min} = 1.845$, $f_{\min} = 0.25$
(c) $X_{\min} = 1.932$, $f_{\min} = 0.0.26$

6. For both methods with $n = 8$, $X_{\min} = 0.01$, $f_{\min} = 2.01$

10. Minimum point $X_3 = (-1, 3/2)^T$ and $f_{\min} = 0.25$

11. Minimum point $X_2 = (4/3, 0)^T$ and $f_{\min} = 248/27$, only one iteration is required, since $\nabla f(X_2) = (0, 0)^T$.

Problem Set 15

1. (a) $\sqrt{2}$ (b) 1

5. minimum cost $= 2c_4V$; each side $= [2c_4V/(c_1 + c_2 + c_3)]^{1/3}$ $L = [Vc_1^2/(\pi c_2^2 N)]^{1/3}$, $R = [Vc_2/(\pi c_1 N)]^{1/3}$, L and R are length and radius of the pipe, respectively

10. $100[1 - n(1 + 0.01k)]$

11. $x_0 = (2^{11/4}/5^{5/8})c_1^{5/8}c_2^{1/8}c_3^{1/4}$, $D = 5x_0/8c_1$, $Q = 4c_3/x_0$

12. $P_2 = (P_1^2 P_4)^{1/3}$, $P_3 = (P_1 P_4^2)^{1/3}$

13. $D = (5c_2/c_1)^{1/6}$

Problem Set 16

1. $x_1 = 25/3$, $x_2 = 0$, $x_3 = 5/3$, $s_2^+ = 25/3$. Since $s_1^- = s_1^+ = 0$ and $s_3^+ = 0$, the first and third goals are fully satisfied. However, the employment level exceeds by $25/3$, i.e., 833 employees.

4. (b) The formulation of the problem is

$$\min \ S = P_1(s_1^- + s_1^+) + P_2 s_2^+ + P_3 s_3^+ + P_4 s_4^- + P_5 s_5^+ + P_6 s_6^+$$
$$+ P_7 s_7^- + P_8 s_8^+$$

s.t.

$$2x_1 + 5x_2 + 4x_3 + d_1 = 600$$
$$3x_1 + 7x_2 + 5x_3 + d_2 = 500$$
$$2x_1 + 5x_2 + 6x_3 + s_1^- - s_1^+ = 450$$
$$4x_1 + 3x_2 + 3x_3 + s_2^- - s_2^+ = 500$$
$$2x_1 + 3x_2 + 4x_3 + s_3^- - s_3^+ = 600$$
$$3x_1 + 3x_2 + 5x_3 + s_4^- - s_4^+ = 500$$
$$3x_1 + 4x_2 + 7x_3 + s_5^- - s_5^+ = 900$$
$$5x_1 + 3x_2 + 3x_3 + s_6^- - s_6^+ = 800$$
$$3x_1 + 6x_2 + 4x_3 + s_7^- - s_7^+ = 400$$
$$3x_1 + 2x_2 + 2x_3 + s_8^- - s_8^+ = 500$$
$$d_1, d_2, x_j, s_i^-, s_i^+ \geq 0, j = 1, 2, 3; i = 1, 2, \ldots, 8.$$

The optimal solution is $x_1 = 91.66$, $x_2 = 0$, $x_3 = 44.44$, $d_1 = 238.88$, $d_2 = 2.78$, $s_3^- = 238.88$, $s_4^- = 2.78$, $s_5^- = 313.88$, $s_6^- = 208.33$, $s_7^+ = 52.77$, $s_8^- = 136.11$, $S = 2.78$. The algorithm requires 4 iterations to find the optimal solution of the problem whereas, the number of iterations required by the lexicographic minimization method using LINDO software is 8.

6. The optimal solution is $x_1 = 27/13$, $x_2 = 0, x_3 = 3/26, N_g = 4$. The goal constraints with the respective priorities being assigned

are:

$$x_1 - x_2 - x_3 \quad \leq 2$$
$$x_1 + x_2 + 2x_3 \quad \leq 3$$
$$4x_1 - x_2 + 6x_3 \quad = 9$$
$$3x_1 - 2x_2 - 2x_3 \geq 4$$
$$-2x_1 - x_2 + x_3 \geq 2 \quad P_5$$
$$x_1, x_2, x_3 \geq 0$$

Here the goal constraints (1), (2), (3) and (4) may be assigned any priorities from P_1 to P_4.

Problem Set 17

1. $x_1 = 1/2$, $x_2 = 1/2$, value of game $= 2.5$; $y_1 = 1/2$, $y_2 = 1/2$, $y_3 = 0$

3. $x_1 = 1/2$, $x_2 = 1/2$, value of game $= 0$; $y_1 = 2/3$, $y_2 = 1/3$, $y_3 = 0$

4. $x_1 = 2/3$, $x_2 = 1/3$, value of game $= 4/3$; $y_1 = 2/3$, $y_2 = 1/3$

5. $x_1 = 0$, $x_1 = 2/3$, value of game $= 7/3$; $y_1 = 8/9$, $y_2 = 1/9$, $y_3 = 0$

6. $p = (5, \infty)$, $q = (-\infty, 5)$

7. x_1, x_2 and x_3 are the investments on Deposits, Bonds and Stocks, respectively. (a) $x_1 = 1,000,000$, $x_2 = 0$, $x_3 = 0$, value of the game $= 8,000$; (b) $x_1 = 0$, $x_2 = 87,500$, $x_3 = 12,500$, value of the game $= 8,750$; (c) $x_1 = 0$, $x_2 = 1,000,000$, $x_3 = 0$; value of the game $= 1,000$.

8. Let x_1 and x_2 correspond to frequency the Australian and West Indies teams use lineup-i and lineup-j, respectively, $i = 1, 2, 3$ and $j = 1, 2, 3$. (a) $x_1 = 0$, $x_2 = 2/5$, $x_3 = 3/5$, $y_1 = 4/5$, $y_2 = 0$, $y_3 = 1/5$, value of the game $= 23/50$; (b) $x_1 = 23/49$, $x_2 = 9/49$, $x_3 = 17/49$, value of the game $= 22/49$

Problem Set 18

1. $x_{11} = 10$, $x_{12} = 0$, $x_{13} = 15$, $x_{14} = 0$, $x_{15} = 5$, $x_{21} = 0$, $x_{22} = 0$, $x_{23} = 10, x_{24} = 30$, $x_{25} = 0$, $x_{31} = 10$, $x_{32} = 20$, $x_{33} = 0$, $x_{34} = 0$, $x_{35} = 0$.

This solution yields values of 14, 20 and 21 for the first, second and third secondary objective functions, respectively. The solution also happens to be the optimal solutions for the second and third secondary objective functions.

Appendix: Answers to Objective Type Questions

1. (a) **2.** (a) **3.** (a) **4.** (c) **5.** (c) **6.** (b) **7.** (b) **8.** (a) **9.** (c) **10.** (c) **11.** X_1 **12.** (a) **13.** (d) **14.** X_2 **15.** (c) **16.** (c) **17.** (c) **18.** (b) **19.** (d) **20.** (b) **21.** (a) **22.** (b) **23.** (a) **24.** (b) **25.** (d) **26.** (c) **27.** (c) **28.** (d) **29.** (c) **30.** (a) **31.** (b) **32.** (a) **33.** (c) **34.** (b) **35.** (b) **36.** (d) **37.** (c) **38.** (c) and (b) depending whether the rule $\theta_j(z_j - c_j)$ is followed **39.** (c) **40** -5 **41.** (a) **42.** (b) **43.** (a) **44.** (d) **45.** (c) **46.** (b) **47.** (a) **48.** (c) **49.** (a) **50.** (b) **51.** $(0, -1)^T$ **52.** $y_1 = -16/5, y_2 = -11/10$ **53.** (b) **54.** (a) **55.** (b) **56.** (c) **57.** (c) **58.** (c) **59.** (b) **60.** (c) **61.** (a) **62.** (a) **63.** (a) **64.** (d) **65.** (a) **66.** 3 **67.** $8x_1 + 13x_2$ **68.** $(13, 8)^T$ **69.** $(3, 2)^T$ **70.** $x_1 = 2, x_2 = 1$ **71.** $x_1 = 4, x_2 = 0$ **72.** (d) **73.** (b) **74.** (b) **75.** (c) **76.** (a) **77.** (c) **78.** (a) true (b) False (c) False (d) True (e) True (f) True, if solution space is bounded; False, if solution space is unbounded (g) True (h) False (i) True (j) False (k) False (l) False (m) True **79.** (c) **80** (c) **81.** (d) **82.** (d) **83.** (d) **84.** (b) **85.** (a) **86.** (a) **87.** (d) **89.** (b) **90.** (d) **91.** (a) **92.** (b) **93.** (b) **94.** (d) **95.** (b **96.** (b) **97.** (a) True (b) False (c) False (d) False (e) True (f) True (g) false (h) true (i) False (j) True (k) True (l) False **98.** (c) **99.** (c) **100.** (a) **101.** (d) **102.** (b) **103.** (a) **104.** (d) **105.** (a) **106.** (c) **107.** (b) **109.** (b)

Index